Host-Pathogen Interactions

METHODS IN MOLECULAR BIOLOGY™

John M. Walker, SERIES EDITOR

489. Dynamic Brain Imaging: Methods and Protocols, edited by Fahmeed Hyder, 2008
484. Functional Proteomics: Methods and Protocols, edited by Julie D. Thompson, Christine Schaeffer-Reiss, and Marius Ueffing, 2008
483. Recombinant Proteins From Plants: Methods and Protocols, edited by Loïc Faye and Veronique Gomord, 2008
482. Stem Cells in Regenerative Medicine: Methods and Protocols, edited by Julie Audet and William L. Stanford, 2008
481. Hepatocyte Transplantation: Methods and Protocols, edited by Anil Dhawan and Robin D. Hughes, 2008
480. Macromolecular Drug Delivery: Methods and Protocols, edited by Mattias Belting, 2008
479. Plant Signal Transduction: Methods and Protocols, edited by Thomas Pfannschmidt, 2008
478. Transgenic Wheat, Barley and Oats: Production and Characterization Protocols, edited by Huw D. Jones and Peter R. Shewry, 2008
477. Advanced Protocols in Oxidative Stress I, edited by Donald Armstrong, 2008
476. Redox-Mediated Signal Transduction: Methods and Protocols, edited by John T. Hancock, 2008
475. Cell Fusion: Overviews and Methods, edited by Elizabeth H. Chen, 2008
474. Nanostructure Design: Methods and Protocols, edited by Ehud Gazit and Ruth Nussinov, 2008
473. Clinical Epidemiology: Practice and Methods, edited by Patrick Parfrey and Brendon Barrett, 2008
472. Cancer Epidemiology, Volume 2: Modifiable Factors, edited by Mukesh Verma, 2008
471. Cancer Epidemiology, Volume 1: Host Susceptibility Factors, edited by Mukesh Verma, 2008
470. Host-Pathogen Interactions: Methods and Protocols, edited by Steffen Rupp and Kai Sohn, 2008
469. Wnt Signaling, Volume 2: Pathway Models, edited by Elizabeth Vincan, 2008
468. Wnt Signaling, Volume 1: Pathway Methods and Mammalian Models, edited by Elizabeth Vincan, 2008
467. Angiogenesis Protocols: Second Edition, edited by Stewart Martin and Cliff Murray, 2008
466. Kidney Research: Experimental Protocols, edited by Tim D. Hewitson and Gavin J. Becker, 2008.
465. Mycobacteria, *Second Edition*, edited by Tanya Parish and Amanda Claire Brown, 2008
464. The Nucleus, Volume 2: Physical Properties and Imaging Methods, edited by Ronald Hancock, 2008
463. The Nucleus, Volume 1: Nuclei and Subnuclear Components, edited by Ronald Hancock, 2008
462. Lipid Signaling Protocols, edited by Banafshe Larijani, Rudiger Woscholski, and Colin A. Rosser, 2008
461. Molecular Embryology: Methods and Protocols, Second Edition, edited by Paul Sharpe and Ivor Mason, 2008
460. Essential Concepts in Toxicogenomics, edited by Donna L. Mendrick and William B. Mattes, 2008
459. Prion Protein Protocols, edited by Andrew F. Hill, 2008
458. Artificial Neural Networks: Methods and Applications, edited by David S. Livingstone, 2008
457. Membrane Trafficking, edited by Ales Vancura, 2008
456. Adipose Tissue Protocols, Second Edition, edited by Kaiping Yang, 2008
455. Osteoporosis, edited by Jennifer J. Westendorf, 2008
454. SARS- and Other Coronaviruses: Laboratory Protocols, edited by Dave Cavanagh, 2008
453. Bioinformatics, Volume 2: Structure, Function, and Applications, edited by Jonathan M. Keith, 2008
452. Bioinformatics, Volume 1: Data, Sequence Analysis, and Evolution, edited by Jonathan M. Keith, 2008
451. Plant Virology Protocols: From Viral Sequence to Protein Function, edited by Gary Foster, Elisabeth Johansen, Yiguo Hong, and Peter Nagy, 2008
450. Germline Stem Cells, edited by Steven X. Hou and Shree Ram Singh, 2008
449. Mesenchymal Stem Cells: Methods and Protocols, edited by Darwin J. Prockop, Douglas G. Phinney, and Bruce A. Brunnell, 2008
448. Pharmacogenomics in Drug Discovery and Development, edited by Qing Yan, 2008.
447. Alcohol: Methods and Protocols, edited by Laura E. Nagy, 2008
446. Post-translational Modifications of Proteins: Tools for Functional Proteomics, *Second Edition*, edited by Christoph Kannicht, 2008.
445. Autophagosome and Phagosome, edited by Vojo Deretic, 2008
444. Prenatal Diagnosis, edited by Sinhue Hahn and Laird G. Jackson, 2008.
443. Molecular Modeling of Proteins, edited by Andreas Kukol, 2008.
442. RNAi: Design and Application, edited by Sailen Barik, 2008.
441. Tissue Proteomics: Pathways, Biomarkers, and Drug Discovery, edited by Brian Liu, 2008
440. Exocytosis and Endocytosis, edited by Andrei I. Ivanov, 2008
439. Genomics Protocols, Second Edition, edited by Mike Starkey and Ramnanth Elaswarapu, 2008
438. Neural Stem Cells: Methods and Protocols, Second Edition, edited by Leslie P. Weiner, 2008
437. Drug Delivery Systems, edited by Kewal K. Jain, 2008
436. Avian Influenza Virus, edited by Erica Spackman, 2008
435. Chromosomal Mutagenesis, edited by Greg Davis and Kevin J. Kayser, 2008
434. Gene Therapy Protocols: Volume 2: Design and Characterization of Gene Transfer Vectors, edited by Joseph M. Le Doux, 2008
433. Gene Therapy Protocols: Volume 1: Production and In Vivo Applications of Gene Transfer Vectors, edited by Joseph M. Le Doux, 2008
432. Organelle Proteomics, edited by Delphine Pflieger and Jean Rossier, 2008
431. Bacterial Pathogenesis: Methods and Protocols, edited by Frank DeLeo and Michael Otto, 2008
430. Hematopoietic Stem Cell Protocols, edited by Kevin D. Bunting, 2008
429. Molecular Beacons: Signalling Nucleic Acid Probes, Methods and Protocols, edited by Andreas Marx and Oliver Seitz, 2008
428. Clinical Proteomics: Methods and Protocols, edited by Antonia Vlahou, 2008
427. Plant Embryogenesis, edited by Maria Fernanda Suarez and Peter Bozhkov, 2008
426. Structural Proteomics: High-Throughput Methods, edited by Bostjan Kobe, Mitchell Guss, and Thomas Huber, 2008
425. 2D PAGE: Sample Preparation and Fractionation, Volume 2, edited by Anton Posch, 2008

METHODS IN MOLECULAR BIOLOGY™

Host-Pathogen Interactions

Methods and Protocols

Edited by

Steffen Rupp

Fraunhofer Institute for Interfacial Engineering and Biotechnology
Stuttgart, Germany

and

Kai Sohn

Fraunhofer Institute for Interfacial Engineering and Biotechnology
Stuttgart, Germany

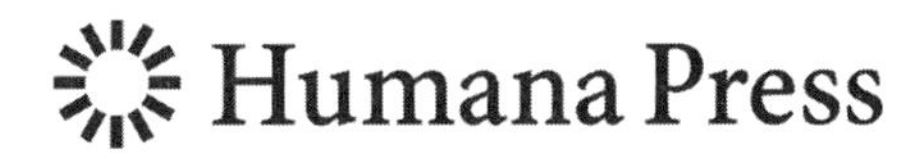

Editor
Steffen Rupp
Fraunhofer Institute for Interfacial
Engineering & Biotechnology
Nobelstrasse 12
70569 Stuttgart, Germany
steffen.rupp@igb.fraunhofer.de

Kai Sohn
Fraunhofer Institute for Interfacial
Engineering & Biotechnology
Nobelstrasse 12
70569 Stuttgart, Germany
kai.sohn@igb.fraunhofer.de

Series Editor
John M. Walker
University of Hertfordshire
Hatfield Herts
UK

ISBN: 978-1-58829-886-7
ISSN: 1064-3745
e-ISBN: 978-1-59745-204-5
e-ISSN: 1940-6029

Library of Congress Control Number: 2008937208

Printed on acid-free paper

9 8 7 6 5 4 3 2 1

springer.com

Preface

Infectious diseases have long been regarded as losing their threat to mankind. However, in recent decades, infectious diseases have been regaining ground and are back in the focus of research. Infectious diseases result from the interplay between pathogenic microorganisms and the hosts they infect, especially the host's defense system. The appearance and severity of disease resulting from any pathogen depends upon the ability of that pathogen to damage the host as well as the ability of the host to react to the pathogen. Therefore, the interaction between the host and the pathogen is crucial for the outcome of an infection with regard to the establishment of disease or asymptomatic, commensal colonization by the organisms.

In Host-Pathogen Interactions, we try to give an overview of the different methods used to study host-pathogen interaction on a molecular level in order to understand the mechanisms of infection and to identify the "Achilles' heel" of the pathogen of interest. We included examples of fungal and bacterial pathogens, as well as parasites, in order to show a broad spectrum of different methodologies that can be used to study different classes of pathogens. On the other hand, methods to study host reactions or to identify factors required for host defense are described. The methods presented in this book can be directly applied for the respective organisms as described. However, they also may serve to give an idea of the different possibilities for investigation of the factors that are responsible for the pathogenicity of the microorganism of interest, and may serve as an entry point to choose and adapt the most appropriate technology for your specific question.

Steffen Rupp
Kai Sohn

Contents

Contributors

FRANCINE AUMONT • *Department of Microbiology and Immunology, Faculty of Medicine, University of Montreal, Montreal, Quebec, Canada*
JIN WOO BOK • *Department of Medical Microbiology and Immunology, University of Wisconsin, Madison, Wisconsin, USA*
CHRISTELLE BOURGEOIS • *Department of Medical Biochemistry, Medical University Vienna, Max F. Perutz Laboratories, Vienna, Austria*
KLAUS BREHM • *Institute for Hygiene and Microbiology, Würzburg, Germany*
SHAOJI CHENG • *Julius-Maximilians-University Würzburg, Department of Medicine, Division of Infectious Disease, University of Florida, Gainesville, Florida, USA*
CORNELIUS J. CLANCY • *University of Florida College of Medicine and North Florida/South Georgia Veterans Health System, Gainesville, Florida, USA*
LIWANG CUI • *Department of Entomology, The Pennsylvania State University, University Park, Pennsylvania, USA*
DAVID ERMERT • *Department of Cellular Microbiology, Max Planck Institute for Infection Biology, Berlin, Germany*
SCOTT G. FILLER • *Division of Infectious Diseases, Los Angeles Biomedical Research Institute at Harbor-UCLA Medical Center, Torrance, California, USA; and The David Geffen School of Medicine at UCLA, Los Angeles, California, USA*
EDAN FOLEY • *Department of Medical Microbiology and Immunology, University of Alberta, Edmonton, Alberta, Canada*
INGRID FROHNER • *Department of Medical Biochemistry, Medical University Vienna, Max F. Perutz Laboratories, Vienna, Austria*
MATTHIAS FROSCH • *Institute for Hygiene and Microbiology, Würzburg, Germany*
ANDREA GAIS • *Julius-Maximilians-University Würzburg, Institute for Medical Microbiology, Georg-August-University, Göttingen, Germany*
CONCHA GIL • *Department of Microbiology II, Faculty of Pharmacy, Complutense University of Madrid, Madrid, Spain*
UWE GROSS • *Institute for Medical Microbiology, Georg-August-University, Göttingen, Germany*
SVEN HAMMERSCHMIDT • *Max von Pettenkofer-Institute, Faculty of Bacteriology, Ludwig Maximilians University Munich, Munich, Germany*
ZAHER HANNA • *Department of Medicine, Laboratory of Molecular Biology, Clinical Research Institute of Montreal Division of Experimental Medicine, McGill University, Montreal, Quebec, Canada*
CHRISTOF R. HAUCK • *Department of Cell Biology, University of Konstanz, Konstanz, Germany*
ROSA HERNANDEZ • *Fraunhofer Institute for Interfacial Engineering and Biotechnology, Stuttgart, Germany*
DIANA HIPPE • *Institute for Medical Microbiology, Georg-August-University, Göttingen, Germany*

BERNHARD HUBE • *Department of Microbial Pathogenicity Mechanisms, Leibniz Institute for Natural Product Research and Infection Biology – Hans Knoell Institute Jena (HKI), Friedrich Schiller University Jena, Jena, Germany*

ALEXANDER D. JOHNSON • *Department of Microbiology and Immunology, and Department of Biochemistry and Biophysics, University of California, San Francisco, San Francisco, California, USA*

PAUL JOLICOEUR • *Department of Microbiology and Immunology, Laboratory of Molecular Biology, Clinical Research Institute of Montreal, Division of Experimental Medicine, McGill University, Montreal, Quebec, Canada*

THIERRY JOUAULT • *Inserm U 799; Laboratoire de Mycologie Fondamentale et Appliquée Université de Lille II, Faculté de Médecine H. Warembourg, Pôle Recherche, Lille Cedex, France*

NANCY P. KELLER • *Department of Medical Microbiology and Immunology, Plant Pathology Department, University of Wisconsin, Madison, Wisconsin, USA*

KARL KUCHLER • *Max F. Perutz Laboratories, Department of Medical Biochemistry, Medical University Vienna, Vienna, Austria*

KATHARINA KUESPERT • *Department of Biology, University of Konstanz, Konstanz, Germany*

STUART M. LEVITZ • *Department of Medicine and Molecular Genetics and Microbiology, University of Massachusetts Medical School, Worcester, Massachusetts, USA*

DANIEL LEWANDOWSKI • *Department of Microbiology and Immunology, Faculty of Medicine, University of Montreal, Montreal, Quebec, Canada*

CARSTEN G. K. LÜDER • *Institute for Medical Microbiology, Georg-August-University, Göttingen, Germany*

OLIVIA MAJER • *Max F. Perutz Laboratories, Department of Medical Biochemistry, Medical University Vienna, Vienna, Austria*

MARIA MARTÍNEZ-ESPARZA • *Departamento de Bioquímica y Biología Molecular (B) e Inmunología, Facultad de Medicina, Universidad de Murcia, Murcia, Spain*

M. HONG NGUYEN • *College of Medicine, University of Florida, Gainesville, Florida, USA*

CÉSAR NOMBELA • *Department of Microbiology II, Faculty of Pharmacy, Complutense University of Madrid, Madrid, Spain*

PATRICK H. O'FARRELL • *Department of Biochemistry and Biophysics, University of California, San Francisco, San Francisco, California, USA*

QUYNH T. PHAN • *Division of Infectious Diseases, Los Angeles Biomedical Research Institute at Harbor-UCLA Medical Center, Torrance, California, USA*

AIDA PITARCH • *Department of Microbiology II, Faculty of Pharmacy, Complutense University of Madrid, Madrid, Spain*

DANIEL POULAIN • *Inserm U 799; Laboratoire de Mycologie Fondamentale et Appliquée, Université de Lille II, Faculté de Médecine H. Warembourg, Pôle Recherche, Lille Cedex, France*

LOUIS DE REPENTIGNY • *Department of Microbiology and Immunology, Sainte-Justine Hospital and University of Montreal, Montreal, Quebec, Canada*

STEFFEN RUPP • *Fraunhofer Institute for Interfacial Engineering and Biotechnology, Stuttgart, Germany*

AURORE SARAZIN • *Inserm U 799; Laboratoire de Mycologie Fondamentale et Appliquée, Université de Lille II, Faculté de Médecine H. Warembourg, Pôle Recherche, Lille Cedex, France*

JETSUMON SATTABONGKOT • *Department of Entomology, AFRIMS, Bangkok, Thailand*

MARTIN SCHALLER • *Department of Dermatology, Eberhard Karls University, Tübingen, Germany*
CHRISTOPH SCHOEN • *Institute for Hygiene and Microbiology, Würzburg, Germany*
ANJA SCHRAMM-GLÜCK • *Institute for Hygiene and Microbiology, University of Würzburg, Germany*
KLAUS SCHRÖPPEL • *Institute of Medical Microbiology and Hygiene, University of Tübingen, Tübingen, Germany*
ALEXANDRA SCHUBERT-UNKMEIR • *Institute for Hygiene and Microbiology, Julius-Maximilians-University Würzburg, Würzburg, Germany*
ANJA SCHWEIZER • *Comparative Genomics Centre, Cellular Immunology Group, James Cook University, Townsville, Australia*
OLAGA SHEVCHUK • *Institute for Microbiology, Technische Universität Braunschweig, Braunschweig, Germany*
KAI SOHN • *Fraunhofer Institute for Interfacial Engineering and Biotechnology, Stuttgart, Germany*
MARKUS SPILIOTIS • *Institute for Hygiene and Microbiology, Julius-Maximilians-University Würzburg, Würzburg, Germany*
MICHAEL STEINERT • *Institute for Microbiology, Technische Universität Braunschweig, Braunschweig, Germany*
SHANNON L. STROSCHEIN-STEVENSON • *Department of Microbiology and Immunology, University of California, San Francisco, San Francisco, California, USA*
NAMTIP TRONGNIPATT • *Department of Entomology, AFRIMS, Bangkok, Thailand*
DIMITRIOS I. TSITSIGIANNIS • *Laboratory of Plant Pathology, Department of Crop Science, Agricultural University of Athens, Athens, Greece*
RACHANEE UDOMSANGPETCH • *Department of Pathobiology, Mahidol University, Bangkok, Thailand*
CONSTANTIN URBAN • *Department of Cellular Microbiology, Max Planck Institute for Infection Biology, Berlin, Germany*
GÜNTHER WEINDL • *Department of Dermatology, Eberhard Karls University, Tübingen, Germany*
KAREN L. WOZNIAK • *Department of Medicine and Microbiology, University of Massachusetts Medical School, Worcester, Massachusetts, USA*
ARTURO ZYCHLINSKY • *Department of Cellular Microbiology, Max Planck Institute for Infection Biology, Berlin, Germany*

Part I
Bacterial Pathogens

Chapter 1

Introduction: Bacterial Pathogens

Christof R. Hauck

Though vertebrates are able to deal with a large and varied resident microbial flora (which outnumbers the cells of the human body), contact with pathogenic bacteria can cause life-threatening harm. The appreciation that infections by specialized bacteria are a major source of human disease and the development of vaccination and antibiotic treatment are the paramount medical achievements of the past century. Studies on the molecular basis of infectious diseases have illuminated common features that are shared by many bacterial pathogens such as the ability to adhere to host tissues, to evade innate and acquired immune defenses, to secrete toxins, to overcome tissue barriers, and to invade host cells. However, a close-up look reveals that each single pathogen species has developed its own variation of these broad themes and, speaking in molecular terms, has acquired specific virulence factors allowing these microbes to interfere with defined host processes or to recognize particular structures.

For example, adhesive proteins of bacteria (i.e., the large assortment of fimbrial and afimbrial adhesins) not only are structurally divergent but also specifically bind to a wide variety of distinct molecules, ranging from carbohydrates and glycolipids to all kinds of proteins on their host cells. Often, these adhesive interactions determine if a pathogen is able to cause disease in a particular organism. Therefore, the identification and characterization of bacterial adhesins is an intense field of current research and is discussed by Hammerschmidt (**Chapter 3**) and by Kuespert and Hauck (**Chapter 5**).

In many instances, virulence factor expression is tightly regulated, and bacterial pathogens respond with specific gene regulation events during their interaction with different host compartments.

Steffen Rupp, Kai Sohn (eds.), *Host-Pathogen Interactions*, DOI: 10.1007/978-1-59745-204-5_1,

Accordingly, methods to monitor bacterial gene expression during infection are critically needed to understand the adaptive response of microorganisms. To unravel novel and potentially useful sites of intervention as a sort of microbial "Achilles' heel" for future therapeutic exploitation, these studies should allow the analysis of the complete bacterial transcriptome. One such global approach of gene expression profiling using DNA-oligonucleotide microarrays is detailed by Schubert-Unkmeir, Schramm-Glück, Frosch, and Schoen (**Chapter 2**).

Many of the virulence-associated traits of bacteria can be studied during infection of isolated eukaryotic cells making *in vitro* cell culture systems easily standardizable and manipulable models. Even for bacteria that are found in the environment together with particular host organisms and accidentally infect humans, such as *Legionella pneumophila*, genetically accessible cell culture models exist and are discussed by Schevchuk and Steinert (**Chapter 4**).

Together, the methods presented in Part I of this volume should provide a useful selection of current approaches to identify and study virulence factors and adaptations of bacterial pathogens that might be applicable to other pathogenic organisms as well.

Chapter 2

Transcriptome Analyses in the Interaction of *Neisseria meningitidis* with Mammalian Host Cells

Alexandra Schubert-Unkmeir, Anja Schramm-Glück, Matthias Frosch, and Christoph Schoen

Abstract

As in many other areas of basic and applied biology, research in infectious diseases has been revolutionized by two recent developments in the field of genome biology: first, the sequencing of the human genome as well as that of many pathogen genomes; and second, the development of high-throughput technologies such as microarray technology, proteomics, and metabolomics. Microarray studies enable a deeper understanding of genetic evolution of pathogens and investigation of determinants of pathogenicity on a whole-genome scale. Host studies in turn permit an unprecedented holistic appreciation of the complexities of the host cell responses at the molecular level. In combination, host-pathogen studies allow global analysis of gene expression in the infecting bacterium as well as in the infected host cell during pathogenesis providing a comprehensive picture of the intricacies of pathogen-host interactions. This chapter briefly explains the principles underlying DNA microarrays including major points to consider when planning and analyzing microarray experiments and highlights in detail their practical application using the interaction of *Neisseria meningitidis* with endothelial cells as an example.

Key words: *Neisseria meningitidis*, infection biology, cellular microbiology, gene expression, transcriptomics, microarrays.

1. Introduction

The large amounts of data from genome sequencing projects have led to a wealth of information about many organisms. In particular, the completion of the human genome project *(1, 2)* together with emerging information on the genomes of around 500 bacteria including all major human pathogenic species offers new possibilities for exploring the molecular pathogenesis of infectious diseases *(3)*. In particular, the introduction of

Steffen Rupp, Kai Sohn (eds.), *Host-Pathogen Interactions*, DOI: 10.1007/978-1-59745-204-5_2,

the DNA microarray made it possible to study the expression of every single gene within a genome at once *(4)*. This genome-wide study of the complete mRNA expression profile (the transcriptome) has been termed accordingly *transcriptomics*. Compared with conventional gene expression technologies such as RT-PCR or Northern blotting, microarrays allow for the interaction between different molecular pathways in the cell to be studied—arguably one of the most important advantages of this novel technology. In addition, microarray-based approaches allow researchers to interrogate host and pathogen genomes without prior bias regarding which genes might be involved in a disease process.

This chapter only very briefly describes the basic principles of DNA microarray technology and highlights in particular the application of microarrays to the analysis of the changes in gene expression in *Neisseria meningitidis* upon adhesion to endothelial cells *(5, 6)*, as well as the transcriptional response in the endothelial cell upon infection with the bacterium *(7)*. For a more general review of host gene expression profiling in host-pathogen interactions, see Ref. *8*, and for the results of microarray experiments in the analysis of the infection biology of *N. meningitidis*, see Refs. *7* and *9*.

1.1. Introducing Microarray Technology

A microarray is a densely packed array of DNA probes that are fixed to a solid substrate in a predetermined matrix on a small surface area. The arrays are then amenable to hybridization to RNA or, more commonly, cDNA nucleotide samples isolated from experimental cultures. The nucleotides are labeled either fluorescently or radioactively before hybridization, and afterwards the fluorescent intensity or radioactivity from each probe spot can be measured. When compared with a control, the spot intensities can be taken as a measure of the relative level of the complementary nucleotide in the sample. Currently, there are two main types of microarrays: spotted DNA and *in situ* synthesized oligonucleotide arrays (Affymetrix gene chips).

Affymetrix (http://affymetrix.com) uses a method that directly synthesizes an oligonucleotide sequence on a silicon slide/chip using a technique known as light-directed combinatorial chemical synthesis *(10)*. Although still quite expensive, Affymetrix chips can thus achieve very high spot densities allowing the incorporation of, for example, up to 47,000 probes as in the latest human oligonucleotide gene chip. However, for the whole-genome analysis of prokaryotes, which have a considerably smaller genome, the more easily obtainable spotted arrays are sufficient for most projects in microbiology.

Spotted arrays most commonly use coated glass slides (microarray) or nylon membranes (macroarray) as the solid substrate. The coating chemistry of microarray slides can vary, but the common feature of the activated glass is that it allows cross-linking between the positively charged surface and the negatively charged nucleotide probes. The spotted array uses either PCR amplicons or presynthesized oligonucleotides of unique fragments from each of the target genes as probes. The oligonucleotide probes are in general melting-temperature-normalized and are commonly 50 to 70 nucleotides long. As they are specifically designed to locations in each open reading frame (ORF), oligonucleotides have an increased specificity and a reduced cross-reactivity with other probes when compared with PCR-based microarrays. In addition, predesigned sets of probes covering the entire genome of many bacterial pathogens are commercially available (e.g., Operon Inc.; http://www.operon.com). The PCR product or oligonucleotide probes are then spotted onto the array surface by an automated robot that uses specially designed pins to pick up the sample from a well of a microtiter plate and deposit it onto the array surface in a prearranged grid formation. This allows the analysis software to link each spot with the gene it represents.

In this chapter, we will focus on spotted oligonucleotide arrays as they are the most common platform for microbial gene expression analysis using microarrays. For more comprehensive details of microarray technology, the reader is referred to Ref. *11*.

1.2. Designing Microarray Experiments

Producing useful data from microarrays depends heavily on the manner in which the study was performed so that potential sources of error are reduced as far as possible. There are two fundamental sources of error that occur during a microarray experiment: random errors and systematic errors *(12)*. Random errors are minimized by obtaining an adequate number of technical as well as biological repeats to reduce extraneous influencing factors. Systematic error is reduced as far as possible by using an appropriate experimental design. With respect to the reference used, microarray experiments can in principle be divided into type 1 and type 2 experiments *(13)*. A type 1 experiment uses control and experimental cDNA labeled with two different fluorescent dyes to make a direct comparison between the two on one array (see later). The identification of genes upregulated or downregulated at a certain time point during infection compared with *in vitro* culture would be an example of a type 1 experiment. Alternatively, radioactively labeled samples are hybridized separately to two different slides and the results are compared with each other computationally. This design is ideal when directly comparing one or two different strains or experimental conditions. But if the candidates to be analyzed are greater in number, then this design becomes tedious. Therefore, in a type 2 experiment, an indirect comparison is made between two samples. In such a common reference design, one strain or condition is a common reference and all other strains or conditions are indirectly compared with this common reference, which also allows comparisons between a large number of various experimental strains/conditions.

As an empirical rule of thumb, one should include at least three biological replicates of each experiment. In addition, a first technical replicate is to have at least two spots for each probe on each microarray. For fluorescent labeling with two different dyes (see later), a second technical replicate is to include a dye swap in each experiment to compensate for different labeling efficiencies of the dyes. Therefore, even a simple type 1 experiment comparing just two different samples/conditions already results in $3 \times 2 \times 2 = 12$ data sets for each probe that have to be analyzed.

However, a detailed discussion of the biostatistical issues involved in any microarray experiment is beyond the scope of this chapter; only rough guidelines can be given here. For a detailed discussion of the bioinformatical and biostatistical methods required for the proper planning and analysis of DNA microarrays, the reader is referred to Refs. *14* and *15*.

For detailed information especially on all aspects of microarray study design, the reader is referred to Ref. *12*. In addition, it is of major importance to make contact in a timely manner with the local biostatistics or bioinformatics department for advice on statistically state-of-the-art design and analysis of the intended microarray-based experiment(s) *(16)*.

A graphical overview over the entire process of planning, performing, analyzing, and validating a microarray experiment is given in **Fig. 2.1**.

1.3. Using Microarrays

There are four laboratory steps in using a microarray to measure gene expression in a sample:

1. Sample preparation
2. Hybridization
3. Washing
4. Image acquisition

1.3.1. Sample Preparation and Labeling

There are a number of different ways in which a sample can be prepared and labeled for microarray experiments. In any case, rapid RNA extraction is always a crucial step if it aims to assess gene expression levels in a meaningful manner. Additionally, mRNA has a short half-life, and speed of isolation through to analysis is thus critical. A number of commercial kits have become available to be used for RNA extraction, but essentially the RNA is purified by precipitation and resuspended in RNase-free water. To ensure that the RNA prepared is free of contaminating genomic DNA, which would give a false-positive signal in any array experiment, the RNA is treated with DNase before it is used for further analysis.

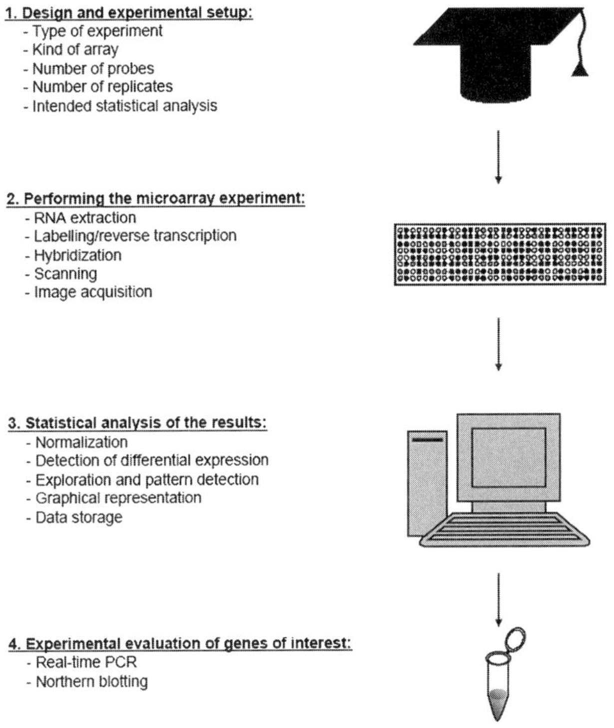

1. Design and experimental setup:
- Type of experiment
- Kind of array
- Number of probes
- Number of replicates
- Intended statistical analysis

2. Performing the microarray experiment:
- RNA extraction
- Labelling/reverse transcription
- Hybridization
- Scanning
- Image acquisition

3. Statistical analysis of the results:
- Normalization
- Detection of differential expression
- Exploration and pattern detection
- Graphical representation
- Data storage

4. Experimental evaluation of genes of interest:
- Real-time PCR
- Northern blotting

Fig. 2.1. Schematic scheme of the steps involved and points to be considered in planning and performing a microarray experiment. For details, see text.

Although it is possible to hybridize complementary RNA, it is much more common to hybridize complementary DNA to the arrays. Therefore, total RNA is reverse-transcribed to cDNA at which point it is labeled in a process known as *direct labeling* either by incorporation of radioactively labeled nucleotides such has, for example, [^{33}P]αATP or dCTP to which the fluorescent dyes Cy3 (excited by a green laser) or Cy5 (excited by a red laser) have been covalently attached.

1.3.2. Hybridization

Hybridization is the step in which the DNA probes on the glass and the labeled DNA target form heteroduplexes via Watson-Crick base-pairing. Before hybridization, the cDNA/DNA samples have to be denatured to single-stranded molecules with a brief incubation at 95°C and added to the hybridization mixture, which guarantees a stringent environment for the binding reaction. In addition, to provide nonspecific binding of nucleotides, the slides must also be prehybridized before use. During the hybridization reaction, the samples are hybridized to the array overnight at a desired temperature (45°C or 65°C) depending on the type of array used, the bacterial species, and the G/C content of the genome.

1.3.3. Washing

After hybridization, the slides are washed to remove excess hybridization solution and unbound labeled cDNA from the array.

1.3.4. Image Acquisition

The final step of the laboratory process is to produce an image of the surface of the hybridized array. For radioactively labeled samples, the hybridized and washed microarray is exposed with a low-energy phosphoimaging screen that is used in conjunction with a suited detection system such as, for example, the PhosphorImager, Molecular dynamics, Sunnyvale, CA instrument. Phosphoimaging screens retain energy from beta particles, x-rays, and gamma rays generated by the radioactive decay of the labeled sample. Upon laser-induced stimulation in the PhosphorImager instrument, for each spot light is emitted from the storage phosphor screen in proportion to the amount of radioactivity in the sample. The emitted light is then quantified and further analyzed by an image analysis software such as the commercial AtlasImage (BD Biosciences Clontech; http://www.bdbiosciences.com).

For fluorescently labeled samples, the levels of fluorescence from each spot for both Cy3 and Cy5 are determined using a confocal scanner. The scanner contains two (or more) lasers that are focused onto the array. After a prescan to check for the negative and positive controls and that the unspecific background is not too imposing, the array is scanned at a gain just below the threshold where any of the spots become saturated. The array is scanned at a wavelength of 635 nm for Cy5 and 532 nm for Cy3 resulting in two monochrome images captured as TIFF (Tagged Image File Format) graphics files that reflect the fluorescent intensities on a black-and-white gradient. Consecutively, the monochromatic TIFF images are exported to image analysis software such as the free Spotfinder by TIGR (http://www.tigr.org/software/tm4/spotfinder.html) or the commercial ImaGene (http://www.biodiscovery.com/ imagene.asp) to create the familiar red-green false-color images of microarrays and to quantify the fluorescent intensities.

1.4. Computational Analysis and Storage of Microarray Data

For fluorescently labeled samples, the quantified spot intensities have to be further processed computationally by applying a suitable form of data normalization method such as variance stabilization (VSN) *(17)* or locally weighted regression scatterplot smoothing (LOWESS) *(18)* to further reduce systematic errors. Useful software tools for microarray analysis are, among others, the freely available software package Bioconductor (http://www.bioconductor.org/) *(18)* or the TIGR software MIDAS (http://www.tm4.org/midas.html). After normalization, statistical testing for differentially regulated genes, clustering of genes with similar expression patterns, and detailed database analyses are necessary before a well-founded biological interpretation of the data is possible. For more information on the biostatistical analysis of microarray data, the reader is referred to Refs. *14* and *15*.

As microarrays provide a vast amount of complex data, the proper storage and interpretation of these data is a subject matter in its own right. To enable a reinterpretation of these data and to allow other researchers to use them for meta-analyses, a standardized storage format is pivotal. One attempt to facilitate the access and usability of microarray data has been made in the form of the MIAME (Minimum Information About a Microarray Experiment; http://www.mged.org/) microarray data file format *(19)*.

1.5. Experimental Validation of Microarray Data

Microarray data are often heavily influenced by a large number of different parameters such as growth media used for cultivation, method of RNA extraction, type of microarray platform used, and so forth. As a result, there is a lot of uncertainty about using microarray data as concrete evidence. Therefore, microarrays are often used as mainly exploratory tools that generate leads that can be followed up individually with simpler and robust low-throughput techniques such as real-time quantitative PCR, Northern blots, or after suitable phenotypic validation of the interesting candidates identified.

1.6. Application of Microarrays in the Study of N. meningitidis Host Cell Interaction

In the following sections, first the analysis of gene expression in *N. meningitidis* serogroup B strain MC58 upon adhesion to human brain microvascular endothelial cells is presented using glass slides spotted with oligonucleotides specific to the open reading frames of strain MC58 provided by Operon (see previous) *(6)*. Second, the analysis of the transcriptome of the host cell in response to infection with *N. meningitidis* MC58 will be described using radioactively labeled cDNA probes and plastic membranes containing more than 11,000 human oligonucleotides spotted in duplicate (BD Atlas Plastic Microarrays; BD Biosciences Clontech; http://www.clontech.com/) *(7)*.

2. Materials

2.1. Bacterial Strains and Growth Conditions

1. *N. meningitidis* serogroup B strain MC58 (serotyped as: B:15:P1.7,16) is a clinical isolate of the ST-32 complex, which was isolated in 1983 in the United Kingdom (R. Borrow, personal communication) and was kindly provided by E.R. Moxon. The strain is stored in a 20% (v/v) glycerol solutio at −70°C. GC agar is the medium of choice for cultivation of meningococci *(20)*.
2. Meningococcal medium: PPM (proteose peptone medium): to 1000 mL dH_2O add 15 g Proteose peptone, 5 g NaCl, 0.5 g starch, and 20 mL stock solution (200 g KH_2PO_4, 50 g K_2HPO_4 to 1000 mL dH_2O).

2.2. Cell Culture

1. Human brain microvascular endothelial cells (HBMECs): HBMECs have been described by Stins and colleagues *(21, 22)*. The cells are positive for factor VIII-Rag, carbonic anhydrase IV, Ulex Europeus Agglutinin I, take up fluorescently labeled acetylated low-density lipoprotein, and express gamma glutamyl-transpeptidase, demonstrating their brain endothelial cell characteristics.
2. RPMI 1640 medium (Gibco Life Technologies, Karlsruhe, Germany) supplemented with fetal calf serum (10%) (FCS; Gibco Life Technologies); heat-inactivate FCS at 56°C for 30 min, then store at −20°C.
3. Nu Serum IV (10%) (Becton Dickinson, Bedford, MA, USA), heat-inactivate at 56°C for 30 min, then store at −20°C ; wear gloves during work with Nu Serum.
4. Vitamins (1%), nonessential amino acids (1%), sodium pyruvate (1 mM), L-glutamine (2 mM), heparin (5 U/mL) (all reagents from Biochrom, Berlin, Germany).
5. Endothelial cell growth supplement (ECGS; Cell Systems Clonetics, St. Katharinen, Germany).
6. Solution of trypsin and ethylenediamine tetraacetic acid (EDTA) from Biochrom.

2.3. Infection Assays

1. T75 cell culture flasks (Sarstedt).
2. Human serum, heat-inactivated (30 min at 56°C) from a serum pool (voluntary staff) (*see* **Note 3**).
3. Phosphate-buffered saline (PBS; Biochrom): It is useful to make a 10x stock solution during frequent use and for infection assays prepare a 1x PBS solution.

2.4. RNA Isolation

2.4.1. Isolation of Mammalian RNA

1. Saturated phenol: for 160 mL use 100 g Phenol (e.g., Sigma; #P1037); in a fume hood heat phenol in 170°C water bath for 30 min until it is completely melted. Add 95 mL phenol directly to the saturation Buffer and mix well.
2. Chloroform (Sigma; #C2432).
3. Isopropanol (Sigma; #I9516).
4. Dry ice and refrigerated centrifuge capable of 15,000 × *g*.

2.4.2. Isolation of Bacterial RNA

1. RNeasy Mini Kit (Qiagen).
2. Lysing tubes containing Lysing Matrix B (Q-BIOgene), Fast Prep machine (Q-BIOgene).

2.4.3. Quality Analysis of the RNA

1. 10x MOPS buffer: 0.2 M MOPS, pH 7.0, 0.1 M NaOAc, pH 7.0, 10 mM EDTA, pH 7.0, dissolve components in RNase-free water, store at room temperature and protect

from light. Working solution of 1x MOPS is prepared in RNase-free water.

2. Ethidium bromide (10 mg/mL).
3. 2x RNA-sample solution: 50% formamide, 2.2 M formaldehyde, 1x MOPS-buffer, 4% Ficoll 400, 0.001% bromophenol blue in RNase-free water. Store in aliquots at –20°C.
4. 1% RNA-MOPS-agarose-gel: Prepare a 1% agarose solution in 1x MOPS by bringing it to boiling. Cool down to 50°C to 60°C and add 40 μL 1 M guanidine thiocyanate (*see* **Note 4**) and 7.5 μL 30% ethidium bromide. Mix the gel and pour onto the gel support.

2.5. DNase Treatment

2.5.1. Bacterial RNA

1. RNase-free DNase I (10 U/μL, Roche).
2. DEPC-treated 0.05 M $MgSO_4$.
3. DEPC-treated 1 M sodium acetate (pH 5.0 acetic acid).
4. PCR reaction: NEB-Taq-Polymerase (5 U/μL, New England Biolabs; including 10x Thermopol buffer), 4 mM dNTP. Oligonucleotide primers:
   a. ORF NMB0829 (*hsdM*):
      NMB0829_F: 5′-AATCCGCCTTATTCCATCAAC-3′
      NMB0829_R: 5′-TTCGCGTGTGTCTTCGGCT-3′
   b. ORF NMB1972 (*groEL*):
      NMB1972_F: 5′-GTCGGTGCCGCGACCGAAGT-3′
      NMB1972_R: 5′-CATCATGCCGCCCATACCAC-3′

   Positive control: genomic DNA from *N. meningitidis* serogroup B strain MC58.

2.5.2. Mammalian RNA

1. RNase-free DNase I (10 U/μL, Roche).
2. 10x DNase I buffer: 400 mM Tris-HCl, pH 7.5, 100 mM NaCl, 60 mM $MgCl_2$.
3. Saturated Phenol: chloroform (Sigma).
4. 95% ethanol.
5. 2 M NaOAc (pH 4.5).

2.6. Reverse Transcription Reaction with Direct Cy-Labeling of Meningococcal RNA

1. CyDye-dNTP: Cy3-dCTP and Cy5-dCTP (GE Healthcare).
2. 1 mg/mL luciferase control RNA (Promega): Prepare a 10 ng/μL solution in RNase-free water. Store in aliquots of 20 μL at −80°C.
3. 100 mM dNTP Set (Invitrogen), containing the four deoxynucleotides, each 100 mM: Prepare a 20 mM dATP, a 20 mM dTTP, a 20 mM dGTP, and a 10 mM dCTP stock in RNase-free water and store at −20°C.

4. SuperScript II Reverse Transcriptase (200U/μL; Invitrogen): 0.1 M dithiothreitol (DTT) and 5x First-Strand buffer.
5. RNaseOut Recombinant Ribonuclease Inhibitor (40 U/μL; Invitrogen).
6. NONA-Random-Oligo (5′-NNNNNNNNN-3′; Operon) is resuspended to a final concentration of 5 μg/μL in RNase-free water.
7. RNase, DNase-free (500 μg/mL; Roche Applied Science).
8. AutoSeq G-50 (GE Healthcare).

2.7. cDNA Probe Synthesis of Mammalian RNA

1. [^{33}P] αATP (10 μCi/μL), >2500 Ci/mmol (Amersham; BF1001).
2. Reagents included with the Atlas Microarray: 10x dNTP, 10x Random Primer Mix, PowerScript Reverse Transcriptase, 5x PowerScript Reaction Buffer, DTT (100 mM).

2.8. Hybridization, Washing, and Analysis

2.8.1. N. meningitidis Whole-Genome Slides

1. 1% SDS.
2. 20x SSC: 3 M NaCl, 0.3 M sodium citrate.
3. Microarray hybridization chamber (Corning).
4. Washing solution I: 2x SSC, 0.1% SDS.
5. Washing solution II: 2x SSC.
6. Washing solution III: 0.2x SSC.
7. Dyesaver (Genisphere).

2.8.2. BD Atlas Plastic Microarrays

1. BD Atlas Plastic Array Hybridization Box.
2. 10x denaturing solution: 1 M NaOH, 10 mM EDTA.
3. 2x neutralization solution: 1 M $Na_2H_2PO_4$.
4. 20x SSC: 175.3 g NaCl, 88.2 g Na_3Citrate-H_2O; add 900 mL H_2O; adjust pH to 7.0 with 1 M HCl; add H_2O to 1 L. Store at room temperature.
5. 20% SDS.
6. High Salt Wash solution: 2x SSC, 0.1% SDS.
7. Low Salt Wash solution 1: 0.1x SSC, 0.1% SDS.
8. Low Salt Wash solution 2: 0.1x SSC.

2.9. Stripping

1. Stripping Solution: 0.1 M Na_2CO_3.

2.10. Equipment

1. Magnetic particle separator.
2. Polypropylene centrifuge tubes (1.5, 2, 15, and 50 mL; Greiner).

2.11. Microarray Platforms Used

1. *Neisseria meningitidis* whole-genome slides: Because of space limitations, only a very brief description of the fabrication of spotted microarrays can be given. For more information, see Ref. *11*. As a convenient alternative, custom-made microarrays can be used instead as these will be offered for most microorganisms by different vendors in the near future. *Neisseria meningitidis* whole-genome slides were obtained by robotically spotting (Omnigrid 100, GenMachines) 70mer oligonucleotides (70mer-Oligos including an aminolinker; Operon) in triplicate including luciferase-specific oligonucleotides as specificity and stringency controls onto pretreated epoxy-coated glass slides (Schott Nexterion) as described by the manufacturer (Omnigrid 100; GenMachines). The printed slides were stored protected from dust in the dark under dry storing conditions and at room temperature.
2. BD Atlas Plastic Microarrays for profiling expression in HBMECs: The methods described for host-cell transcriptome analysis are based on the use of the BD Atlas Plastic Microarrays kit (BD Biosciences Clontech; http://www.clontech.com). The plastic microarray consists of more than 11,000 long oligonucleotides immobilized by UV radiation on a rigid, translucent plastic support. In detail, they contain 384 repeating blocks of 64 spots. Furthermore, there are 3 blocks (blocks are labeled with "C") with identical sets of control spots plus experimental genes. The control spots comprise negative controls (various λ phage–specific oligonucleotides) and housekeeping genes.

3. Methods

3.1. Bacterial Strains and Growth Conditions

1. Streak out the *N. meningitidis* strain MC58 frozen in glycerol solution at –70°C onto fresh GC agar plates at 37°C for 18 h in an atmosphere of 5% (v/v) CO_2.
2. Resuspend the bacterial culture using cotton swabs in 10 mL PPM+ (proteose peptone + 1% PolyviteX (Biomerieux). Incubate at 37°C with shaking for 1.5 to 2 h.
3. After 1.5 to 2 h, prepare a 1:10 dilution of the culture and measure the absorbance at 600 nm. An absorbance at 600 nm of 0.1 is equivalent to 1×10^8 meningococci. Thus the inoculum is prepared using the formula:

$$\text{Total inoculum required (mL)} \times \frac{0.001}{A_{600\text{nm}} \text{ of 1:10 dilution}}$$

$$= \text{mL of neat culture required}$$

This volume will give 1 × 10^8 CFU (colony forming units) in the final inoculum, which corresponds with a multiplicity of infection (MOI) of 10 (the number of HBMECs in a T75 tissue flask grown to confluence is around 1 × 10^7; *see* **Section 3.2**).

4. Resuspend bacteria in 10 mL human serum (HS) supplemented RPMI cell culture medium (10% HS) or fetal calf serum (FCS) supplemented RPMI cell culture medium (10% FCS) (*see* **Note 3**).

3.2. Prepare Cells for Infection

HBMECs should be cultured in RPMI 1640 medium supplemented with 10% FCS, 10% Nu Serum, 1% vitamins, 1% nonessential amino acids, 1 mM sodium pyruvate, 2 mM L-glutamine, 5 U/mL heparin, and 30 μg/mL endothelial cell growth supplement (*see* **Section 2.2**). Cultures should be incubated in a humid atmosphere at 37°C with 5% CO_2 (*see* **Note 1**)

1. HBMECs are passaged when approaching confluence with trypsin/EDTA to provide new maintenance cultures on T75 tissue flasks. Two (for HBMEC transcriptome analysis) to six (for *N. meningitidis* transcriptome analysis) are required for each experimental data point. Split target cells into a T75 tissue culture flask at 0.5 × 10^7 mL and incubate for 48 h at 37°C at 5% CO_2. This will provide a confluent monolayer.
2. Aspirate the media from the plate and carefully wash cells with prewarmed 1x PBS.
3. Repeat this washing step twice.
4. Add bacteria in 10 mL HS supplemented RPMI medium (10% HS) to the monolayer at a MOI of 10. For analyzing gene expression profile in *N. meningitidis*, upon adhesion to HBMECs supplementation with 10% FCS is recommended (*see* **Note 3**).
5. Incubate the cells with bacteria at 37°C with 5% CO_2 for an 8-h period.
6. Incubate control cells (uninfected HBMECs) with supplemented RPMI (+ 10% HS) alone according to the protocol.
7. For isolation of control RNA from nonadherent bacteria, incubate bacteria (1 × 10^8) in 10 mL RPMI medium supplemented with 10% FCS without HBMECs in a fresh T75 tissue flask according to the protocol.
8. Stop infection of HBMECs with *N. meningitidis* at different time points: use a 4-h and 8-h incubation period for HBMEC transcriptome analysis and a 6-h period for *N. meningitidis* transcriptome analysis *(6, 7)*.
9. After 4 h, 6 h, and 8 h of infection, aspirate the infection media from the monolayer.

10. Carefully wash cells with 1x PBS.
11. Repeat step two times.
12. Trypsinize bacteria from cells: add 10 mL trypsin/EDTA for 10 min with easy shaking. This method has been described before to separate adherent bacteria from cells *(6)* (*see* **Note 5**).
13. Transfer cells to a 15-mL centrifuge tube and centrifuge at 1600 × *g* for 10 min at 4°C.
14. Transfer the supernatant containing bacteria to a fresh 15-mL centrifuge tube and centrifuge at 4000 rpm for 10 min at 4°C. Resuspend bacterial pellet in 1.0 mL RLT-βME buffer.
15. Resuspend the pellet (**step 13**) containing HBMEC cells in ice-cold PBS, centrifuge at 1000 × *g* for 10 min at 4°C, and shock freeze the pellets at −70°C for RNA isolation.
16. Continue with **step 3.4.**1.

3.3. RNA Isolation from N. meningitidis, Reverse Transcription, Hybridization, and Data Analysis

Figure 2.2 shows a flow sheet of the steps that have to be carried out to analyze gene expression in *N. meningitidis* upon adhesion to HBMECs and the host-cell's transcriptome in response to infection with *N. meningitidis.*

3.3.1. RNA Preparation of Adherent and Nonadherent Bacteria

1. Transfer the complete bacteria-RLT-β-ME mix of one condition into a lysing tube including Lysing Matrix B (*see* **Note 2**). To lyse the cells, "centrifuge" for 30 s at a speed of 6.5 in the Fast Prep machine. Incubate the lysing tubes for 30 s on ice and then repeat the step for another 20 s (*see* **Note 6**).
2. The lysing tubes are centrifuged for 2 min at a maximum speed (≥15,000 × *g*) in an Eppendorf Centrifuge (Biofuge pico, Hereus, Germany) depositing the Lysing Matrix B.
3. Transfer the supernatant to a RNase-free tube and add 550 μL absolute EtOH. Mix it by shaking (*see* **Note 7**).
4. Perform the further purification steps according to the manufacturer's cleanup protocol (RNeasy Mini Kit; Qiagen) (*see* **Note 8**). Transfer the purified RNA in a RNase-free cup.
5. Check the RNA using a 1% RNA-MOPS-agarose-gel (*see* **Note 9**): Add 7.5 μL RNA to 7.5 μL 2x RNA-sample solution in a RNase-free cup and incubate for 3 min at 65°C, chill on ice, and load onto the gel. Run gel at 5 V/cm distance from anode to cathode in running buffer (1x MOPS). The 16 and 23S rRNA bands should be clearly visible, whereas the mRNA will form a faint background smear.
6. Incubate the RNA in a final volume of 150 μL with 15 μL 0.05 M DEPC-treated $MgSO_4$, 15 μL 1 M DEPC-treated sodium acetate, pH 5.0, and 3 μL 10 U/μL DNase I for 1 h at 37°C to digest contaminating DNA.

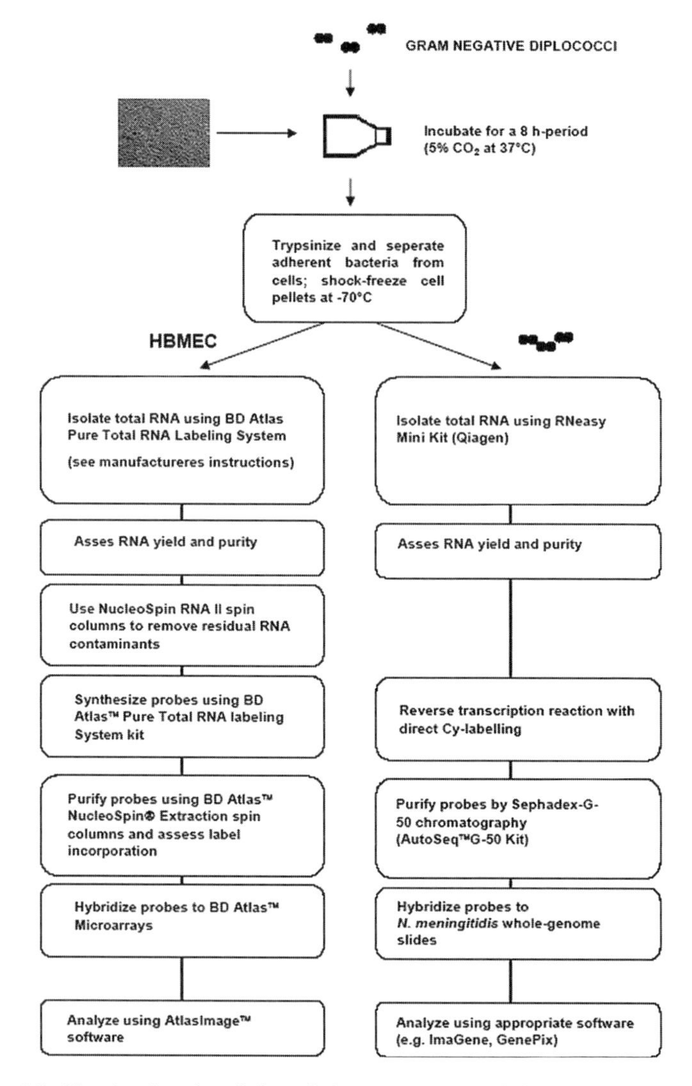

Fig. 2.2. Step-by-step description of the processes carried out to analyze gene expression in *N. meningitidis* upon adhesion to human brain microvascular endothelial cells (HBMECs) and vice versa host-cell transcriptome in response to infection with *N. meningitidis.*

7. Pipette the solution to 525 μL RLT-β-ME and add 375 μL absolute EtOH in a RNase-free cup. Purify the RNA again with RNeasy Midi Kit (Qiagen) according to the manufacturer's protocol. Transfer the RNA in a RNase-free cup and store at −70°C.
8. Recheck the RNA for integrity using a 1% RNA-MOPS-agarose-gel.
9. To test for traces of contaminating genomic DNA, perform a PCR reaction as follows:

For a 50-μL reaction volume, combine the following reagents:

5	μL	10x Thermopol buffer
2.5	μL	4 mM dNTP
0.25	μL	100 μM Primer 1 (*see* **Section 2.5.1** Bacterial RNA)
0.25	μL	100 μM Primer 2 (*see* **Section 2.5.1** Bacterial RNA)
0.25	μL	Taq DNA Polymerase (5 U/μl)
1	μL	Template (RNA, pos. and neg. control) (*see* **Note 8**)
Add 50	μL	ddH_2O

Place the reaction tube in a thermal cycler and subject to PCR with the following conditions: an initial denaturation at 95°C for 10 min, followed by 32 cycles each at 95°C for 60 s, 58°C for 45 s, and 72°C for 45 s, and additional 72°C for 5 min; store at 4°C.

10. Load 10 μL of the PCR-product onto a 1% TBE-agarose gel (an example result is shown in **Fig. 2.3**). Make sure that no contaminating genomic DNA is present in the preparation.
11. The concentration of the RNA is determined by a spectrophotometer at an absorbance of 260 nm. An absorbance of 1 at 260 nm corresponds with 40 μg RNA/mL if the measurement is done in pure water (*see* **Note 10**).

3.3.2. Reverse Transcription Reaction with Direct Cy-Labeling and Hybridization of a Microarray

1. Accurate pipetting during all working steps is important to get reproducible results.
2. From **step 3** to **step 10**, RNase-free working conditions have to be ensured.
3. For the labeling reaction, 20 μg total RNA for each condition (i.e., adherent versus nonadherent) in 14.1 μL RNase-free water (*see* **Note 11**).
4. Add 1 μL of 10 ng/μL luciferase-RNA (Promega) to each labeling reaction as spike-in control.
5. Add 0.9 μL of 5 μg/μL NONA random primer and heat the probes to 70°C for 5 min to dissolve secondary and tertiary structures of the RNA and primer dimers. Chill on ice.
6. Prepare the master mix by mixing 20 μL 5x First-Strand buffer, with 1 μL 0.1 M DTT, 2 μL 10 mM dCTP, 2.5 μL 20 mM dATP, 2.5 μL 20 mM dTTP, 2.5 μL 20 mM dGTP, 14.5 μL DEPC-H_2O, 2.5 μL 40 U/μL RNase Out and 2.5 μL 200 U/μL reverse transcriptase.
7. Add 20 μL of the master mix to both reactions and 2 μL Cy3-dCTP to one of the both labeling reactions and 2 μL Cy5-dCTP to the other (*see* **Note 12**).

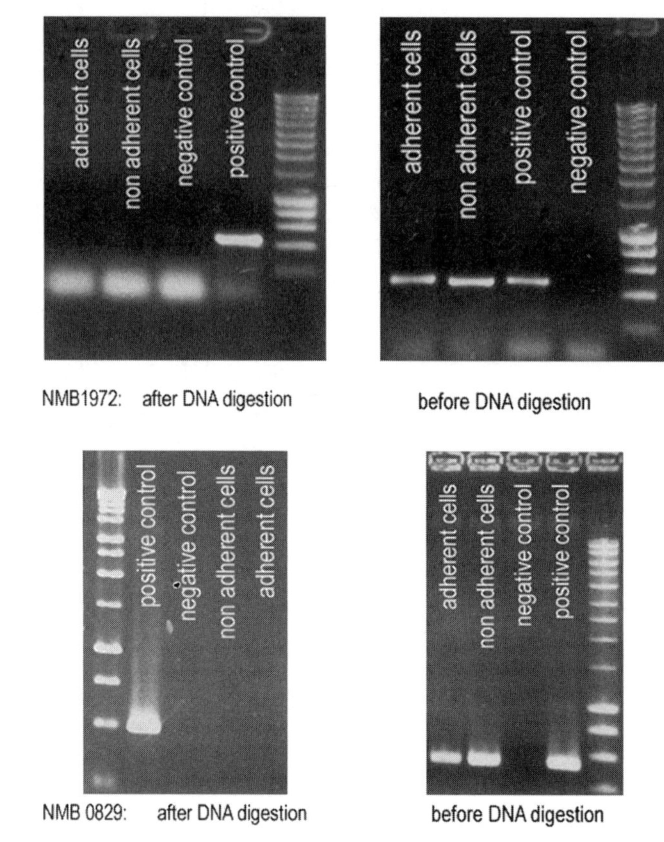

Fig. 2.3. Detection of contaminating genomic DNA in the RNA sample preparation from adherent as well as nonadherent bacteria using PCR specific for the genes NMB1972 (upper half) and NMB0829 (lower half) as described in **Section 3** with the primer pairs NMB1972_F /NMB1972_R and NMB0829_F /NMB0829_R, respectively, before (right) and after (left) treatment of the RNA samples with RNAse.

8. The probes are incubated at room temperature for 10 min to anneal the primer and 2 more hours at 42°C to proceed the reverse transcription to completion.
9. Inactivate the reaction, especially the RNase inhibitor, by heating up to 70°C for 15 min. Then add 2 μL of 500 μg/mL RNase, DNase-free to the samples to degrade the RNA. The RNA-digestion is performed at 37°C for 45 min.
10. The Cy3 respectively Cy5 labeled cDNA is purified by Sephadex-G-50 chromatography (AutoSeq G-50 Kit): Centrifuge the columns for 1 min at 2000 × *g* and discard the flow-through. Place the columns in new collection tubes and pipette the complete labeling reaction directly onto the Sephadex matrix. Centrifuge for 1 min at 2000 × *g*.
11. Pool both purified fluorescent labeled samples and concentrate the resulting mixture to a final volume of 30 μL in a Speed Vac. Add 6 μL of 20 x SSC and 4 μL 1% SDS.

12. Denature the mixture of the labeled samples by heating at 95°C for 3 min, perform a quick spin in a microcentrifuge, and chill on ice.
13. Pipette the mixture without the inclusion of air bubbles onto a preprocessed microarray slide (*see* **Note 13**). Place a coverslip onto the solution and place the microarray into a hybridization chamber and ensure a humid atmosphere. Hybridize at 50°C for 16 h.
14. Wash the slides according to the manufacturer's protocol. Dry the slides by centrifugation at 390 × *g* for 3 min.
15. Dipping the slides at this point for 10 s in the Dyesaver solution to prevent degradation of the fluorescent Cy dyes is recommended (*see* **Note 14**).

3.3.3. Scanning of the Slide and Image Acquisition

Scan the slide with a microarray scanner (e.g., from Axon, Tecan, PerkinElmer) at 543 nm for Cy3 and 633 nm for Cy5 to yield two TIFF-formatted images. To achieve two comparable images (low background, similar spot intensity of most spots [~80%] in both channels, a signal maximum of 80% of the brightest spots), adjust the photo multiplier tube gain and the laser settings.

Analyze the images using appropriate analysis software (e.g., ImaGene [Biodiscovery], GenePix [Axon]). The spots have to be examined manually first and if necessary flagged (high background, stray fluorescence, etc.). Then the images are quantified by measuring the mean or median signal intensity of each spot and the local background of the spots. The actual signal intensity is calculated by subtracting the local background intensity from the signal intensity of each spot. Negative signal values (signal background > signal spot) were excluded from further analysis. After global normalization of the data using LOWESS (see previous), the fold difference in mRMA levels in adherent versus nonadherent bacteria are calculated for each gene. Finally, for each gene, this ratio is transformed by its logarithm to the base 2 to center the data symmetrically around zero. Therefore, transcriptionally upregulated genes have a positive and downregulated genes a negative log ratio. Results are saved in a tab-delimited format and further analyses are performed using Microsoft Excel.

3.4. RNA Isolation from HBMECs, RNA Enrichment, Probe Synthesis, and Data Analysis

3.4.1. RNA Isolation of Infected and Noninfected HBMECs

1. Isolate total RNA using the BD Atlas Total RNA Labeling System (according to the manufacturer's instructions). The key components are streptavidin-coated magnetic beads and biotinylated oligo(dT), which allow one to carry out both poly A+ RNA enrichment and probe synthesis in a single procedure.
2. To 1×10^7 to 3×10^7 cells, add 3 mL denaturation solution (Box 1). Pipette up and down vigorously and vortex well until cell pellet is completely resuspended.

3. Incubate on ice for 5 to 10 min.
4. Centrifuge homogenate at 15,000 × *g* for 5 min at 4°C. This step removes cellular debris.
5. Transfer the entire supernatant to new tubes.
6. Add 6 mL saturated phenol.
7. Vortex for 1 min and incubate on ice for 5 min.
8. Add 1.8 mL chloroform.
9. Shake sample and vortex vigorously for 2 min. Again, incubate on ice for 5 min.
10. Centrifuge the homogenate at 15,000 × *g* for 10 min at 4°C.
11. Transfer the upper aqueous phase containing the RNA to a new tube. Do not pipette any material from the white interface or lower organic phase.
12. Perform a second round of phenol:chloroform extraction.
13. Add 1.8 mL isopropanol.
14. Mix the solution well and incubate on ice for 10 min.
15. Centrifuge the samples at 15,000 × *g* for 5 min at 4°C.
16. Remove the supernatant without disturbing the RNA pellet.
17. Add 3 mL 80% ethanol.
18. Centrifuge at 15,000 × *g* for 5 min at 4°C. Quickly and carefully discard the supernatant.
19. Air dry the pellet and resuspend the pellet in appropriate RNase-free H_2O to ensure an RNA concentration of 1 to 2 μg/mL.
20. Store RNA samples at −70°C until ready to proceed with DNase treatment.

3.4.2. DNase Treatment of Total HBMEC-RNA (Optional)

1. Treat the isolated RNA with DNase I using the following protocol:

500	μL	total RNA
100	μL	10x DNase I buffer
50	μL	DNase I (1 U/μL)
350	μL	deionized H_2O
1	mL	total volume

   Use Ambion's ANTI-RNase (#2692) at a concentration of 1 U/μL with this protocol if you still see degradation.
2. Follow the manufacturer's instructions.

3.4.3. Assessing Yield and Purity of Total RNA

The yield of total RNA will vary depending on the cells from which it is obtained: 1×10^7 to 3×10^7 HBMECs will yield around 50 μg total RNA. Determine the quality and quantity of

RNA as described in **Section 3.3.1**. On a RNA-agarose gel, total RNA from mammalian cells should appear as two bands at approximately 4.5 and 1.9 kb (28S and 18S ribosomal RNA). The ratio of intensities of 28S and 18S rRNA should be 1.5 to 2.5:1. Lower ratios indicate degradation.

3.4.4. Streptavidin Magnetic Bead Preparation and Poly A^+ RNA Enrichment

Carry out poly A^+ RNA enrichment using streptavidin-coated magnetic beads and biotinylated oligo(dT) according to the manufacturer's instructions (see Atlas Pure Total RNA Labeling System User Manual).

3.4.5. cDNA Probe Synthesis

1. Prepare a Master Mix for all labeling reactions plus one extra reaction

5x PowerScript Reaction Buffer	4 μL
10x dNTP Mix (for dATP label)	2 μL
[^{33}P] αATP (10 μCi/μL); >2500 Ci/mmol	5 μL
DTT (100 mM)	0.5 μL
Total volume	11.5 μL

2. Continue with the manufacturer's protocol.

3.4.6. Column Chromatography

This step separates labeled cDNA from unincorporated ^{33}P-labeled nucleotides and small (<0.1 kb) cDNA fragments with Nucleospin extraction spin columns. All reagents required are provided with the BD Atlas NucleoSpin Extraction Kit.

1. Add 180 μL Buffer NT2 to dilute the probe synthesis reaction to 200 μL total volume; mix well by pipetting.
2. Place a Nucleospin Extraction Spin Column into a 2-mL collection tube and pipette the sample into the column. Centrifuge again at maximum speed for 1 min. Reserve the flow-through for RT reaction efficiency estimation. Discard collection tube in an appropriate container for radioactive waste.
3. Insert the NucleoSpin column into a fresh 2-mL collection tube. Add 400 mL Buffer NT3 (make sure that 95% ethanol was added before use). Centrifuge again at maximum speed for 1 min. Discard collection tube and flow-through.
4. Repeat **step 3** twice.
5. Transfer the NucleoSpin column to a clean 1.5-mL microcentrifuge tube. Add 100 μL Buffer NE and allow column to soak for 2 min.
6. Centrifuge again at maximum speed for 1 min to elute purified probe.

7. Check the radioactivity incorporation in the probe by scintillation counting:
   i. Add 5 μL of each purified probe to 5 mL of scintillation fluid in separate scintillation-counter vials.
   ii. Add 5 μL of the flow-through from each purified probe to 5 mL of scintillation fluid in separate scintillation-counter vials.
   iii. Count samples on the ^{33}P channel. Multiply counts by a dilution factor of 20.

   Probes should yield a minimum of 5×10^6 to 25×10^6 cpm.
8. Discard flow-through fractions, columns, and elution tubes in the appropriate container for radioactive waste.

3.4.7. Hybridization Procedure

1. The microarrays should be washed twice with 2x SSC, 0.1% SDS and twice with 0.1x SSC, 0.1% SDS for 5 min at 60°C in all cases, and then twice with 0.1x SSC at 30°C for 5 min each time (*see* **Note 15**).
2. During the hybridization procedure, ensure that the printed surface of the microarray is facing up. Furthermore, ensure that the hybridization box is well sealed and continuously shaking.
3. Allow the microarrays to air-dry completely (about 15 min).
4. Expose the printed surface of the microarray to a phosphoimaging screen suitable for ^{33}P detection.
5. We recommend exposure times of 12 h and 72 h and if necessary of 1 week.
6. Scan the phosphoimaging screen at a resolution of 50 μm.

3.4.8. Array Analysis

For cDNA microarray analysis, the BD AtlasImage 2.7 software was used according to the manufacturer's guidelines. The arrays are aligned with BD AtlasImage Grid Template (BD Sciences Clontech) automatically and then fine-tuned for each gene using manual adjustment options. The background is calculated based on the median intensity of the "blank spaces" between different panels of the array (default method of calculation), and the raw signal intensity of each spot was measured. A raw intensity (before normalization) of twofold over background was taken as an indication that a gene was expressed at significant level.

For comparing the expression patterns of infected cells and uninfected cells, the signal intensities are normalized by the global normalization-sum method, which is best suited for the comparison of two similar tissue samples. Signal values in arrays hybridized with meningococci-infected cDNA should be normalized with respect to those from arrays reprobed with cDNA of uninfected cells. The adjusted signal intensities of each of the individual cDNA spots should be compared for infected and uninfected cells. The ratio is calculated as adjusted intensity of array 2 (infected HBMECs):adjusted intensity of array 1 (uninfected cells) according to the manufacturer's guidelines. Differences are estimated when a gene signal either in array 1 or array 2 is at background level.

Instead of numerical values, we indicated these genes as upregulated (↑) or downregulated (↓) in infected HBMECs *(7)*. Results were saved in a tab-delimited format and further analyzed using Microsoft Excel.

4. Notes

1. The splitting of cells, changing of media, preparation of infection conditions, and infection of cells must be done using sterile techniques in a biological safety cabinet to prevent contamination.
2. Standard procedures to minimize RNase contamination should be used for all solutions and glassware *(23)*. Whenever possible, disposable sterile plasticware should be used for any steps prior to the RNA being converted to single-stranded cDNA. Wear disposable gloves during preparation of materials and solutions that will be used for isolation and analysis of RNA and during manipulations involving RNA. Change gloves frequently.
3. The recommendation for using human serum supplemented RPMI cell culture medium is based on the observation that meningococcal entry is supported by binding of the outer membrane protein Opc via fibronectin to integrins on HBMECs *(24)*. Human serum should be pooled and heat-inactivated at 56°C for 30 min. For analyzing gene expression profile in *N. meningitidis* upon adhesion to HBMECs, supplementation with 10% FCS is recommended, as there were no detectable differences in adherence efficacy using HS and FCS.
4. Guanidine thiocyanate denatures proteins and inhibits RNases in the presence of reducing agents. It is used as a substitute for toxic formaldehyde.
5. Cells should not be incubated longer than 10 min with trypsin/EDTA, because of cell-surface protein release and toxic effects.
6. The cells are disrupted and simultaneously homogenized by a combination of turbulence and mechanical shearing. After the lysis of the cells, there should be no foam on the sample.
7. It is important to take care that no lysing matrix is transferred.

8. To elute RNA from the column, pipette 100 μL RNase-free water directly onto the membrane of the column and let it stand for at least 10 min, then centrifuge for 1 min at ≥8000 × *g*.
9. The application of a Bioanalyzer (Agilent) to check the integrity and purity of the RNA preparation is more accurate than is the analysis using conventional agarose gel electrophoresis.
10. The quality of the RNA is the most important factor influencing the sensitivity and reproducibility of the experiments. Contamination by genomic DNA is a critical point, therefore all RNA samples must be treated with RNase-free DNase I. We recommend consideration of the guidelines for working with RNA *(23)*. At an absorbance of 230 nm, organic solvents, salt, and protein are determined, and at 280 nm, protein is determined. The ratios of the absorbance values at 260 nm and 280 nm, and at 260 nm and 230 nm, respectively, indicate the purity of the RNA. The results of both ratios should be more than 1.8. For exact determination of the ratios, the measurement should be done in 10 mM Tris-HCl pH 7.5 or in buffered deionized water as the pH has a strong effect on these values at 280 nm.
11. If the volume of 20 μg RNA is larger than 14.1 μL, concentrate the sample in a Speed Vac.
12. It should be taken into consideration that both dyes are photosensitive. Therefore, exposition to light should be reduced to a minimum during all working steps involving these dyes.
13. The preprocessing of the slide was made according to the manufacturer's instructions for epoxy-coated slides (Schott Nexterion).
14. The fluorescent dye Cy5 bleaches rapidly in the presence of high ozone concentration such as during the summertime.
15. BD Atlas Plastic Array Hybridization Box is recommended by the manufacturer. We also strongly recommend using this hybridization box because plastic microarrays precisely fit in them and are kept flat during hybridization and washing. Roller bottles curl the plastic film and lead to poor images after postprocessing of the arrays.

References

1. Lander, E. S., Linton, L. M., Birren, B., et al. (2001) Initial sequencing and analysis of the human genome. *Nature* **409,** 860–921.
2. Venter, J. C., Adams, M. D., Myers, E. W., et al. (2001) The sequence of the human genome. *Science* **291,** 1304–51.

3. Hacker, J. and Dobrindt, U., eds. (2006) *Pathogenomics*. Weinheim: Wiley-VCH.
4. Schena, M., Shalon, D., Davis, R. W. and Brown, P. O. (1995) Quantitative monitoring of gene expression patterns with a complementary DNA microarray. *Science* **270,** 467–70.
5. Grifantini, R., Bartolini, E., Muzzi, A., et al. (2002) Previously unrecognized vaccine candidates against group B meningococcus identified by DNA microarrays. *Nat Biotechnol* **20,** 914–21.
6. Dietrich, G., Kurz, S., Hubner, C., et al. (2003) Transcriptome analysis of *Neisseria meningitidis* during infection. *J Bacteriol* **185,** 155–64.
7. Schubert-Unkmeir, A., Sokolova, O., Panzner, U., Eigenthaler, M. and Frosch, M. (2007) Gene expression pattern in human brain endothelial cells in response to *Neisseria meningitidis*. *Infect Immun* **75,** 899–914.
8. Hossain, H., Tchatalbachev, S. and Chakraborty, T. (2006) Host gene expression profiling in pathogen-host interactions. *Curr Opin Immunol* **18,** 422–29.
9. Claus, H., Vogel, U., Swiderek, H., Frosch, M. and Schoen, C. (2007) Microarray analyses of meningococcal genome composition and gene regulation: a review of the recent literature. *FEMS Microbiol Rev* **31,** 43–51.
10. Lipshutz, R. J., Fodor, S. P., Gingeras, T. R. and Lockhart, D. J. (1999) High density synthetic oligonucleotide arrays. *Nat Genet* **21,** 20–24.
11. Bowtell, D. and Sambrook, J., eds. (2002) *DNA Microarrays*. Cold Spring Harbor: Cold Spring Harbor Laboratory Press.
12. Churchill, G. A. (2002) Fundamentals of experimental design for cDNA microarrays. *Nat Genet* **32,** 490–95.
13. Yang, Y. H. and Speed, T. (2002) Design issues for cDNA microarray experiments. *Nat Rev Genet* **3,** 579–88.
14. Stekel, D. (2003) *Microarray Bioinformatics*. Cambridge: Cambridge University Press.
15. Parmigiani, G., Garret, E. S., Irizarry, R. A. and Zeger, S. L., eds. (2003) *The Analysis of Gene Expression Data*. Berlin: Springer.
16. Tilstone, C. (2003) DNA microarrays: vital statistics. *Nature* **424,** 610–2.
17. Huber, W., von Heydebreck, A., Sultmann, H., Poustka, A. and Vingron, M. (2002) Variance stabilization applied to microarray data calibration and to the quantification of differential expression. *Bioinformatics* **18,** 96–104.
18. Cleveland, W. S. (1979) Robust locally weighted regression and smoothing scatterplots. *J Am Statist Assoc* **74,** 829–36.
19. Brazma, A., Hingamp, P., Quackenbush, J., et al. (2001) Minimum information about a microarray experiment (MIAME)-toward standards for microarray data. *Nat Genet* **29,** 365–71.
20. Tinsley, C. R. and Heckels, J. E. (1986) Variation in the expression of pili and outer membrane protein by *Neisseria meningitidis* during the course of meningococcal infection. *J Gen Microbiol* **132,** 2483–90.
21. Stins, M. F., Gilles, F. and Kim, K. S. (1997) Selective expression of adhesion molecules on human brain microvascular endothelial cells. *J Neuroimmunol* **76,** 81–90.
22. Stins, M. F., Badger, J. and Sik Kim, K. (2001) Bacterial invasion and transcytosis in transfected human brain microvascular endothelial cells. *Microb Pathog* **30,** 19–28.
23. Sambrook, J., Fritsch, E. F., and Maniatis, T., eds. (1989) *Molecular Cloning: A Laboratory Handbook*, 2nd ed. Cold Spring Harbour: Cold Spring Harbor Laboratory Press.
24. Unkmeir, A., Latsch, K., Dietrich, G., et al. (2002) Fibronectin mediates Opc-dependent internalization of *Neisseria meningitidis* in human brain microvascular endothelial cells. *Mol Microbiol* **46,** 933–46.

Chapter 3

Surface-Exposed Adherence Molecules of *Streptococcus pneumoniae*

Sven Hammerschmidt

Abstract

Surface-exposed proteins of pathogenic bacteria are considered as potential virulence factors through their direct contribution to host-pathogen interactions. The specific interaction of bacterial proteins with host proteins often subverts the physiologic function of host-derived proteins, and therefore the bacterial proteins are considered as key players in the infectious process. The direct binding of host proteins is exploited by the pathogens for colonization, host tissue invasion, or immune evasion. Strikingly, surface proteins such as ABC transporters are also implicated in bacterial pathogenesis through their role in maintenance of bacterial fitness. Here, we are interested in surface-exposed proteins of *Streptococcus pneumoniae*, which interact with host proteins including proteins of the extracellular matrix, serum proteins, or ectodomains of cellular host receptors. These bacterial proteins are termed collectively adhesins or MSCRAMMs (microbial surface components recognizing adhesive matrix molecules). We have shown that choline-binding proteins and proteins that lack classic features of surface proteins such as a signal peptide that is required for protein secretion or a membrane anchor motif represent a major class of adhesins produced by *S. pneumoniae*.

Key words: adhesins, binding assays, adherence, invasion, flow cytometry, surface plasmon resonance.

1. Introduction

Streptococcus pneumoniae (the pneumococcus) are versatile Gram-positive bacteria that colonize as commensals the upper and lower respiratory tract of humans without causing clinical symptoms. On the other hand, pneumococci convert under appropriate conditions into highly pathogenic microorganisms and cause serious and life-threatening infections including pneumonia, septicemia, and meningitis *(1)*.

Pneumococci are encased by a capsular polysaccharide (CPS), which has been recognized as a *sine qua non* of virulence.

Steffen Rupp, Kai Sohn (eds.), *Host-Pathogen Interactions*, DOI: 10.1007/978-1-59745-204-5_3,

The CPS has an important function in immune evasion *(2, 3)*. However, the high amount of CPS masks protein-based adherence molecules and is despite its requirement for colonization reduced upon contact of the pneumococci with epithelial cells *(3, 4)*. The surface of pneumococci is decorated with different clusters of proteins. The bioinformatics analysis of the genome of *S. pneumoniae* R6 (the rough derivative of serotype 2 strain D39) and TIGR4 (a Norwegian clinical isolate of serotype 4) predicted 153 (R6) and 256 (TIGR4) proteins, respectively, with a leader peptide. The leader peptide, which is essential for secretion of the proteins, was identified in lipoproteins (42 in R6 and 47 in TIGR4), in choline-binding proteins (10 in R6 and 15 in TIGR4), and proteins with an LPXTG motif (13 in R6 and 19 in TIGR4) *(5, 6)*. In addition to these predicted surface proteins, the cell envelope of pneumococci is decorated with another cluster of proteins that lack a classic leader peptide and membrane-anchoring motifs. These proteins are collectively termed nonclassic surface proteins. Adhesive functions have in particular been attributed to members of the choline-binding protein (CBP) family and to the nonclassic surface proteins *(7, 8)*. In contrast, the typical Gram-positive surface proteins that contain the peptidoglycan binding LPXTG motif and anchored covalently in a sortase-dependent reaction to the cell wall possess mainly enzymatic activities. These activities have been shown to unmask host cellular receptors and facilitate the interaction with host tissue cells *(8)*. CBPs are noncovalently anchored to the phosphorylcholine (*P*Cho), which is an unusual and physiologically important component of the pneumococcal cell wall. Interestingly, the *P*Cho functions like some of the CBPs also as an adhesin by recognizing the platelet-activating factor receptor (PAFr) of host cells *(9)*.

In order to identify potential pneumococcal adhesins, binding of proteins to pneumococci was assayed with radiolabeled host proteins or by flow cytometry. In another approach, binding of fluorochrome-labeled pneumococci to immobilized host proteins is analyzed. Once the interaction with a host protein has been indicated, the identification of binding motifs within the bacterial protein and host protein are elucidated in order to identify functional active sites and host specificity as has been done for the major pneumococcal adhesin PspC *(10–12)*. A prerequisite for these assays is often the generation of isogenic pneumococcal mutants and the purification of recombinant and tagged adhesins. Finally, the impact of the individual adhesin on adherence and invasion or degradation of the extracellular matrix can be explored in cell culture infection experiments or degradation experiments. Confocal immunofluorescence laser scanning microscopy (CLSM) and electron microscopy (EM) allow discrimination between extracellular and intracellular pneumococci.

2. Materials

2.1. Radiolabeling of Proteins

1. Iodine-125 (1 mCi; Amersham, Germany).
2. Silicone-coated 2 mL glass tubes. The glass tubes are treated with SIGMACOTE (Sigma) under agitation and dried under the extractor hood.
3. Chloramin T (1 mg/mL; Sigma) and sodium metabisulfide (1 mg/mL; Sigma) are dissolved immediately before use in 0.05 M phosphate buffer pH 7.5 (PB: stock solutions of A: 0.2 M monobasic sodium phosphate (28.8 g in 1000 mL) and B: 0.2 M dibasic sodium phosphate (53.65 g of $Na_2HPO_4 \times 7H_2O$ or 71.7 g of $Na_2HPO_4 \times 12\ H_2O$ in 1000 mL). For pH 7.5, mix approximately 16 mL of stock solution A with 84 mL of stock solution B and dilute to a total of 500 mL. Autoclave PB before storage at room temperature (RT) and control pH before use.
4. PBS-Tween: 37 mM NaCl, 2.7 mM KCl, 80 mM Na_2HPO_4, 1.8 mM KH_2PO_4, adjusted to pH 7.5 with HCl if necessary, supplemented with 0.05% Tween 20 (Applichem, Darmstadt, Germany).
5. Trichloroacetic acid: 10% TCA (v/v).
6. PD10 column (Amersham) (*see* **Note 1**).

2.2. Flow Cytometry

1. Fixation solution: 1.0% paraformaldehyde (PFA) and 1.0% fetal calf serum (FBS; Invitrogen) in PBS, pH 7.5. Prepare a 3.7% (w/v) solution in PBS and store at −20°C. Thaw the PFA solution at high temperature (80°C) before use and dilute in PBS.
2. The secondary antibodies that are fluorochrome-labeled (FITC-conjugated or Alexa Fluor 488-conjugate) can be purchased from MoBiTec Göttingen, Germany, Dianova Hamburg, Germany, or Invitrogen Karlsruhe, Germany. The dilution of the antibodies has to be checked out separately for each antibody.
3. Fluorescence is measured, for example, with a FACSCalibur (Becton Dickinson) using an excitation wavelength of 488 nm. The FACScan was used in standard configuration with a 530-nm bandpass filter as described *(13)* (*see* **Note 2**).

2.3. Binding of Bacteria to Immobilized Proteins

1. Labeling of the bacteria with fluorescein isothiocyanate (FITC) is performed under alkaline conditions using a 0.1 M sodium carbonate buffer pH 9.2 (use the stock solution of 1 M Na_2CO_3 and 1 M $NaHCO_3$ to prepare the sodium carbonate buffer of pH 9.2) with 1 mg/mL of FITC (*see* **Note 3**).
2. Maxisorb F96 microtiter plates (Nunc, Wiesbaden, Germany) are used to immobilize the host proteins on a polystyrene surface (*see* **Note 4**).

3. Bound fluorescence per 96-well is measured in a fluorophotometer at 485 nm/538 nm (excitation/emission) using, for example, a Fluoroskan Ascent (Thermo Labsystems).

2.4. Cell Culture

1. Dulbecco's modified Eagle's medium (DMEM; PAA Laboratories) supplemented with 10% fetal bovine serum (FBS; PAA Laboratories), 2 mM glutamine, penicillin G (100 IU/mL), and streptomycin (100 μg/mL) (all Invitrogen).
2. Supplements for Calu-3 lung epithelial cells are 1 mM sodium pyruvate and 0.1 mM nonessential amino acids (Invitrogen).
3. RPMI 1640 (PAA Laboratories) supplemented with 10% FBS (PAA Laboratories), 2 mM glutamine, and 1 mM sodium pyruvate.
4. RPMI 1640 is supplemented with 10% FBS, 10% Nu-Serum IV (Becton Dickinson), 1% nonessential amino acids, 1% MEM vitamins (PAA Laboratories), 1 mM sodium pyruvate, 2 mM glutamine, penicillin (100 units/mL), and streptomycin (0.1 mg/mL) when human brain-derived microvascular endothelial cells (HBMECs) *(14)* are cultured (*see* **Note 5**).
5. 24-well tissue culture plates (Greiner).
6. Cell culture flasks (25 cm^2 or 75 cm^2) (Greiner) for cell culturing.
7. DMEM-HEPES (PAA Laboratories).

2.5. Immunofluorescence and Antibiotic Protection Assay

1. Microscope coverslips (diameter, 12 mm) for fluorescence microscopy.
2. Polyclonal anti-pneumococci antiserum. This antiserum was generated in rabbits by a standard subcutaneous immunization with heat-killed pneumococci. A booster injection was performed twice. Antibodies (polyclonal or monoclonal antibodies) recognizing surface-exposed structures on bacteria can also be used for immunofluorescence.
3. Blocking solution: 10% FBS (v/v) in PBS.
4. Permeabilization solution: 0.1% Triton-X-100 (v/v) in PBS.
5. Secondary antibodies: Alexa Fluor 488–labeled goat anti-rabbit Ig and Alexa Fluor 568–labeled goat anti-rabbit Ig (MoBiTec).
6. Moviol (Hoechst) is used as medium for long-time storage at 4°C or −20°C and nail polish for fixation of the microscope cover glasses.

2.6. Surface Plasmon Resonance Experiments

1. Biosensor activation chemicals: 0.05 M *N*-hydroxysuccinimide (NHS) and 0.2 M *N*-ethyl-*N'*-(diethylaminopropyl) carbodiimide (EDC) *(15)*.

2. Coupling buffer: 20 mM sodium acetate buffer. The ligand (110 to 200 μg/mL) that is coupled on the CM5 biosensor surface is solved in sodium acetate or dialyzed against this buffer. The pH of the coupling buffer is one pH-unit below the isoelectric point of the ligand. The activated ester on the surface reacts with the ligand, and an amide bound is formed.
3. Blocking buffer: 1 M ethanolamine hydrochloride pH 8.5. This treatment deactivates reactive ester groups of the activated biosensor surface (*see* **Note 6**).
4. HBS-EP-BIAcore running buffer: 10 mM HEPES, 150 mM sodium chloride, 1.4 mM ethylenediamine tetraacetic acid (EDTA), 0.05% Tween 20, pH 7.4 at 20°C using a flow rate of 10 μL/min. The biosensor is highly susceptible against changes in puffer properties, therefore the analytes are dialyzed against the running buffer prior to their use.
5. Chemicals for regeneration of the biosensor surface (i.e., for elimination of bound analyte. The CM5 biosensor chip can be regenerated using low pH (10 mM glycine-HCl, pH 3.0 to pH 1.5), $MgCl_2$ (1 to 4 M), salt solutions, detergents, high pH (1 to 100 mM NaOH), or ethylene glycol (50%, 75%, and 100%).

2.7. Screening of Synthetic Peptide Arrays Membranes

1. Tris base saline (TBS) buffer: 50 mM Tris Base (6.1 g per liter), 8 g sodium chloride (NaCl), 0.2 g potassium chloride (KCl), add to 1 L, adjust to pH 7.0 with HCl. Autoclave TBS and store at 4°C.
2. T-TBS: TBS supplemented with 0.05% Tween 20 (Applichem).
3. Buffer A: 8 M urea, 1.0% SDS, 0.5% β-mercaptoethanol (add directly before use).
4. Buffer B: 10% acetic acid, 50% ethanol, 40% H_2O (v/v) (*see* **Note 7**). If possible, stripping should be avoided and identical membranes should be used for testing the background of the applied antibodies. This will show unspecific binding of the antibodies to the immobilized peptides and is important to discriminate for specific binding of the analyte.
5. Blocking solution: 2.0 mL Genosys blocking buffer (10x; Sigma), 8 mL T-TBS pH 8.0, 0.5 g saccharose (*see* **Note 8**).
6. The blocking buffer is stored in aliquots at −20°C during the course of the experiment and later at −70°C to −80°C. The stripping is very important, and the background reactions with the antibodies will show you unspecific binding.

3. Methods

3.1. Binding of Radiolabeled Host Proteins to Pneumococci

1. Host proteins are radiolabeled with iodine-125 by a standard chloramin T method *(16)* or by using commercially available reagents such as the IODO-BEADS Iodination Reagent according to the protocol of the manufacturer (PerbioScience, Erembodegem, Belgium).
2. For the chloramin T method, 100 μg of the host protein (1 mg/mL) in 0.05 M phosphate buffer pH 7.5 (PB) are mixed in silicone-coated 2 mL glass tubes (SIGMACOTE; Sigma) with approximately 0.5 mCi iodine-125. The oxidation reaction is started by adding 20 μL of chloramin T, which is freshly solved before use in PB (1 mg/mL).
3. The reaction is stopped after 5 min by adding freshly solved 20 μL sodium metabisulfide (1 mg/mL in PB).
4. One microliter is taken out and measured in a gamma counter. Thereafter, the protein is precipitated with 25 μL FBS (fetal bovine serum) and 1 mL TCA. The radioactivity of the sedimented protein is measured in the gamma-counter. The first value is the total activity in 1 μL and the second value shows the labeling efficiency.
5. The total volume of the reaction is adjusted to 2.5 mL with PBST (PBS with 0.05% Tween 20) and loaded on a PD 10 column. The loaded column is washed with 3.5 mL PBST and the flow-through, which contains the labeled protein, is collected.
6. Measure the cpm (counts per minute) of 10 μL to 25 μL of the labeled protein. Then, the amount of labeled protein can be calculated for 100,000 cpm.
7. Pneumococci are grown in Todd-Hewitt broth supplemented with 0.5% yeast extract (THY) to mid-log phase (OD_{600} = 0.40 to 0.50) and resuspended in PBS containing 0.1% Tween 20.
8. The binding reaction is conducted by incubation of 1×10^9 bacteria with 100,000 cpm (approximately 20 nCi) of iodinated protein in a total volume of 300 μL for 30 min at room temperature.
9. Pneumococci are sedimented and the supernatant aspirated by gentle suction. Pellet-bound radioactivity representing bound protein can be measured in a gamma counter. Protein binding can be expressed as a percentage of total radioactivity added and bound to fetal calf serum. Binding of the radiolabeled pellet-bound protein can further be illustrated by subjecting the total cell lysate to SDS-PAGE. Binding is detected by exposure of the gel to X-ray film.

3.2. Binding of Host Proteins and Extracellular Matrix Proteins to Bacteria Quantified by Flow Cytometry

1. Pneumococci are cultured in THY to mid-exponential phase (OD_{600} = 0.40 to 0.50) at 37°C under 5% CO_2 and washed with PBS.
2. A pneumococcal suspension of 1 × 10^7 in 100 μL PBS is incubated with different amounts of FITC-labeled host protein or unlabeled host protein for 30 min at 37°C.
3. Pneumococci were washed with PBS and resuspended in PBS (viable pneumococci) or alternatively in 1% PFA in PBS (fixed pneumococci) using a total volume of 200 μL.
4. Binding of unlabeled host protein is followed by incubation with anti-host proteins antibodies in PBS for 30 min at 37°C.
5. For detection, the PBS washed bacterial suspension is incubated with a fluorochrome-conjugated secondary antibody (FITC-conjugate or Alexa 488).
6. Bacteria are washed with PBS and resuspended in PBS (viable pneumococci) or alternatively in 1% PFA in PBS (fixed pneumococci) using a total volume of 200 μL.
7. The fluorescence intensity of the bacteria (as a read out for the amount of host protein bound to the bacterial cell surface) is analyzed by flow cytometry using, for example, a FACSCalibur (Becton Dickinson) using an excitation wavelength of 488 nm. The FACScan was used in standard configuration with a 530-nm bandpass filter as described *(13)*.
8. The bacteria are detected using log-forward and log-side scatter dot-plots, and a gating region is set to exclude debris and larger bacterial aggregates (*see* **Note 2**).

3.3. Binding of Fluorochrome-Labeled Pneumococci to Immobilized Host Proteins

1. Pneumococci were grown in THY to mid-log phase (OD_{600} = 0.40 to 0.50), and after washing with 0.1 M sodium carbonate buffer (pH 9.2), 1 × 10^9 bacteria were labeled with fluorescein isothiocyanate (FITC) in 500 μL of a sodium carbonate buffer FITC solution (1 mg/mL) for 1 h in the darkness under agitation.
2. Various amounts of host proteins (0.5 μg up to 2.0 μg per well) were coated overnight onto the surface of a 96-well microtiter plate (polystyrene surface) at 4°C. Bovine serum albumin (BSA) is used as a negative control.
3. The surface of the wells is blocked with 1% BSA for at least 3 h at room temperature.
4. Extensively washed FITC-labeled pneumococci (2 × 10^8 bacteria in 200 μL and serial dilutions) are incubated in darkness with the immobilized host proteins for 1 h at 37°C to allow binding of the bacteria to the host proteins *(17)*.
5. Immediately after incubation and prior to the first washing step with PBS, the fluorescence is measured in a fluorophotometer

at 485 nm/538 nm (excitation/emission) using, for example, a Fluoroskan Ascent (Thermo Labsystems). The values measured at this stage represent the total amount of applied FITC-labeled pneumococci (100%) and is defined as 100% (here, unbound bacteria are not eliminated) *(17)*.

6. Subsequently, unbound bacteria are eliminated by washing (1 to 3 times) the wells with PBS.
7. Fluorescence is measured at 485 nm/538 nm (excitation/emission), and binding activity can be expressed as the percentage of the total applied pneumococci. When counting the number of applied FITC-labeled pneumococci after plating sample aliquots on blood agar plates, the number of pneumococci bound to the immobilized protein can also be calculated using the percentage of binding *(18)*.

3.4. Cell Culture Infection Experiments with *Streptococcus pneumoniae*

1. Pneumococci are cultured in THY to an optical density of OD_{600} = 0.35 to 0.40 at 37°C under 5% CO_2.
2. In pneumococcal infection experiments, human nasopharyngeal epithelial cells Detroit 562 (ATCC CCL 138), lung alveolar carcinoma epithelial cell line A549 (type II pneumocyte; ATCC CCL-185), HEp-2 larynx carcinoma cell line (ATCC CCL-23), and Calu-3 cells (human lung epithelium; ATCC HTB-55) are most commonly used as epithelial cell lines. HBMECs (*14*), representing a model for the blood-brain barrier, are used as endothelial cell line. These cells lines are all cultured 37°C under 5% CO_2.
3. A549 and HEp-2 are grown in DMEM supplemented with 10% FBS, 2 mM glutamine, penicillin G (100 IU/mL), and streptomycin (100 μg/mL). The medium for Calu-3 cells is further supplemented with 1 mM sodium pyruvate and 0.1 mM nonessential amino acids. Detroit 562 are grown in RPMI 1640 supplemented with 10% fetal bovine serum, 2 mM glutamine, and 1 mM sodium pyruvate *(18, 19)*.
4. HBMECs are cultured in RPMI 1640–based medium supplemented with 10% FBS, 10% Nu-Serum IV (Becton Dickinson), 1% nonessential amino acids, 1% MEM vitamins (Invitrogen), 1 mM sodium pyruvate, 2 mM glutamine, penicillin (100 units/mL) and streptomycin (0.1 mg/mL) *(18)*.
5. The host cells are seeded in antibiotic-free medium on 24-well tissue culture plates (Greiner) or on glass coverslips (diameter, 12 mm) at a cell density of approximately 5×10^4 cells per well *(18)*.
6. Confluent cell layers with approximately 2×10^5 cells are washed prior to infections with the infection medium, and the cells are infected with pneumococci in 500 μL DMEM with HEPES (1x; PAA Laboratories) and 1% FBS per well

with an MOI of 10 to 100 pneumococci per cell at 37°C under 5% CO_2.

7. After the infection, the cells are rinsed several times with DMEM-HEPES or phosphate-buffered saline (PBS) to remove unbound bacteria.

3.5. Double Immunofluorescence Staining of Pneumococci and Microscopy

1. Pneumococci-infected host tissue cells are fixed with 1.0% paraformaldehyde on the microscopic coverslip after rinsing the cells several times with DMEM-HEPES or PBS.
2. For differentiation between extracellular and intracellular pneumococci, the fixed samples are blocked with 10% FBS in PBS for 1 h at room temperature.
3. The extracellular bacteria attaching to cells are incubated for 45 min with rabbit anti-pneumococcal antiserum, followed by 30 min staining with an Alexa Fluor 488–labeled goat anti-rabbit Ig (MoBiTec) *(10, 18)*.
4. The host cells are permeabilized with 0.1% Triton-X-100 for 5 min at room temperature.
5. Both the extracellular and intracellular pneumococci are incubated with anti-pneumococcal antibody, followed by staining with Alexa Fluor 568–labeled goat anti-rabbit Ig (MoBiTec) to stain intracellular and extracellular bacteria.
6. After final washing steps with PBS, the coverslips are embedded "upside down" in Moviol, sealed with nail polish, and stored at 4°C.
7. The number of extracellular (red/green = yellow) and intracellular pneumococci (red) can be counted using a fluorescence microscope. The schematic model and flowchart of the double immunofluorescence technique is shown in **Fig. 3.1**.

3.6. Quantification of Pneumococcal Invasion

1. The number of viable intracellular pneumococci can be determined by applying the gentamicin protection assay. After removal of unbound pneumococci, the extracellular and host cell attached bacteria are killed by treatment of the cells with gentamicin (200 μg/mL) and penicillin G (100 units) for 1 h at 37°C under 5% CO_2.
2. The intracellular and therefore invasive pneumococci can be recovered after incubating the cells in 300 μL DMEM-HEPES/1% saponin (1% w/v) for 10 min at 37°C. Saponin treatment results in partial permeabilization of the host cells. For pneumococci, Triton-X-100 is not recommended, because viability of pneumococci is affected by 0.1% Triton-X-100 *(19)*.
3. The number of intracellular recovered and viable pneumococci is quantified by plating serial dilutions on blood-agar plates.

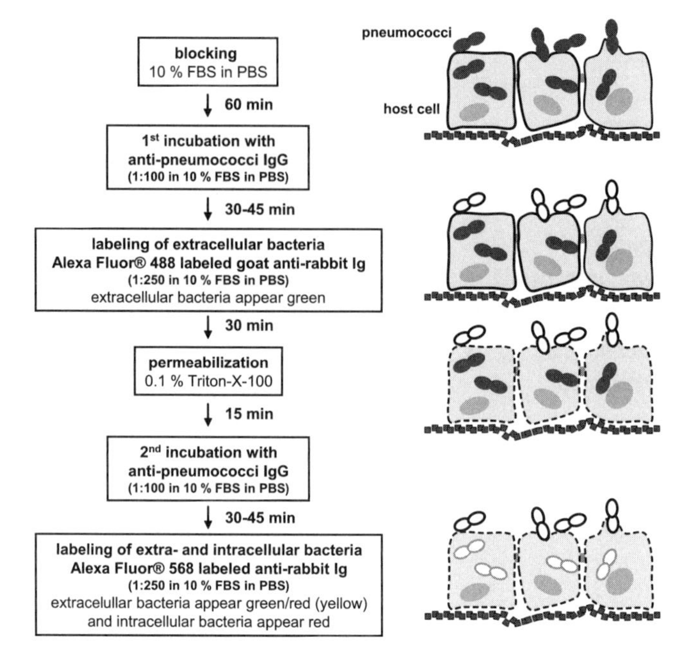

Fig. 3.1. Schematic model and flowchart of double immunofluorescence microscopy technique of host cell–attached pneumococci and invasive pneumococci. BM, basal membrane.

3.7. Protein-Protein Interactions Analyzed by Surface Plasmon Resonance

In protein-protein interactions, the specificity and the affinity constants are of particular interest. For the characterization of protein-protein interactions, binding assays with radiolabeled proteins or fluorescence resonance energy transfer (FRET) can be used. FRET is employed to determine the distances of two molecules and hence has a broad application in biochemistry and cell biology. Here, we have used surface plasmon resonance (SPR) to determine the rates of association (k_a), dissociation (k_d), and the equilibrium constant (K_D). SPR is a powerful technique in which the nonlabeled ligand is immobilized and the nonlabeled analyte is free in solution and passed over the ligand. This technique allows us to measure the interaction between two molecules in real-time, and only low amounts of proteins or molecules are required for measurements. In addition to protein-protein interactions, other interactions such as DNA-DNA or DNA-protein interactions can also be measured. The knowledge of binding epitopes and critical amino acids in protein-protein interactions is an invaluable advantage if not a prerequisite when evaluating the interaction. In addition, the purity of the used molecules should be of highest grade to prevent unspecific binding. SPR machines are distributed by different companies. Here, we have used the Biacore system *(10, 17)*. This system is equipped with a continuous flow system in which four channels are coupled in series. As an example, we have analyzed the interaction of the pneumococcal enolase with its host receptor protein plasminogen. In this interaction, the enolase was used as ligand and the glu-plasminogen as analyte.

1. Covalent immobilization of the ligand enolase proteins was performed essentially as described *(15)*.
2. The CM5 biosensor chip is activated with 70 μL *N*-hydroxysuccinimide (NHS; 0.05 M)/*N*-ethyl-*N'*-(diethylaminopropyl)carbodiimide (EDC; 0.2 M) using a flow rate of 10 μL/min.
3. The ligand-coated CM5 biosensor chip is blocked by 1 M ethanolamine hydrochloride using a flow rate 10 μL/min for 7 min.
4. The N-terminally His-tagged pneumococcal enolase (0.1 mg/mL in 20 mM sodium acetate pH 4.0) is coupled by covalent amine coupling at 10 μL min^{-1} onto the NHS/EDC-activated sensor chips. The amount of coupled ligand was monitored in real-time. Depending on the molecular weight of the ligand and analyte, a maximum of 100 response units (RU) should be reached after coupling the ligand for kinetic experiments.
5. One of the flow cells is used as a reference surface. This surface is used in its unmodified form or is activated and deactivated without ligand coupling. This provides selectivity and specificity of analyte binding and eliminates the buffer effects.
6. The availability of different sensor chips allows diverse coupling techniques. The direct coupling methods include amine coupling, ligand thiol coupling, surface thiol coupling, and aldehyde coupling. In addition, indirect coupling methods are also applicable. These include the coupling of capture antibodies or streptavidin.
7. Binding of the analyte (e.g., plasminogen) is performed in HBS BIAcore running buffer (10 mM HEPES, 150 mM NaCl, 1.4 mM EDTA, 0.05% Tween 20, pH 7.4) at 20°C using a flow rate of 10 μL/min.
8. The sample injection is followed by regeneration of the flow cells (i.e., elimination of the bound analyte). In case of an ideal regeneration, the analyte response is consistent after repeated injections. According to the manufacturer, a gentle regeneration is recommended. The CM5 biosensor chip can be regenerated using low pH (10 mM glycine-HCl, pH 3 to pH 1.5), $MgCl_2$ (1 to 4 M), salt solutions, detergents, high pH (1 to 100 mM NaOH), or ethylene glycol (50%, 75%, and 100%).
9. The analysis of BIAcore sensorgram data is performed by using the BIAevaluation software provided by the manufacturer. The interaction kinetics is analyzed from raw data of the Biacore sensorgrams using kinetics models included in the

software. For every evaluation, a minimum of six data sets corresponding with the binding reactions at concentrations between 0.1 nM and 500 nM are required (*see* **Note 9**). However, the Biacore experiments do not allow per se a prediction of the number of binding sites or whether conformational changes occur during the interaction. A detailed description of the kinetic models is provided in the software manual and cannot be discussed here.

10. An example is shown in **Fig. 3.2**. Here, binding of human plasminogen is assayed to immobilized wild-type enolase of *S. pneumoniae* and enolase^{del}, in which the C-terminal lysine residues are deleted.

3.8. Mapping and Analysis of Linear Protein-Protein Binding Sites by Using Synthetic Peptide Arrays

In protein-protein interactions, the binding regions and critical amino acids in the binding regions are of particular interest. The identification of epitopes is important for the elucidation of binding specificity and can help to explain species specificity. Although the construction of truncated protein derivatives and random site-directed mutagenesis of protein encoding sequences is a standardized method to map binding epitope, the use of peptide arrays is an elegant and easy to use method to map binding sites in protein-protein interactions. Finally, this method can also be applied to

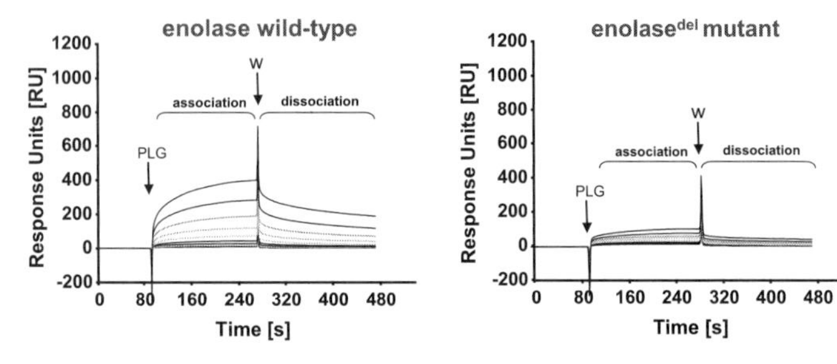

Fig. 3.2. Surface plasmon resonance (SPR) measurements of the pneumococcal enolase and plasminogen interactions. Recombinant wild-type enolase and recombinant and C-terminally truncated enolase, named enolase^{del}, are coated on a BIAcore CM5 sensor chip, and plasminogen is used as an analyte. Changes in plasmon resonance are shown as relative response units (RU). Sensorgrams show the binding kinetics and concentration dependence of rates of binding of plasminogen to wild-type enolase and modified enolase^{del} proteins. In enolase^{del}, the C-terminal lysyl residues (position 433 and 434), which were thought to be involved in plasminogen binding, were deleted by mutagenesis. However, the results of the SPR measurements demonstrated binding of plasminogen to wild-type enolase and enolase^{del}. This suggested that another region in enolase is important for the enolase-PLG interaction *(17)*. Plasminogen was used in concentrations of 250 nM, 125 nM, 62.5 nM, 31.25 nM, 15.625 nM, 7.8 nM, and 1.9 nM. The blank run was subtracted from each sensorgram. PLG, start of injection of plasminogen; w, stop of injection and start of its dissociation. (From Bergmann, S., Wild, D., Diekmann, O., Frank, R., Bracht, D., Chhatwal, G. S., and Hammerschmidt, S. 2003. Identification of a novel plasmin(ogen)-binding motif in surface displayed alpha-enolase of *Streptococcus pneumoniae. Mol. Microbiol.* **49,** 411–423. Reproduced with permission from Blackwell Publishing.)

identify critical amino acids in the binding epitope(s). An example is shown in **Fig. 3.3**. Here, a novel plasminogen-binding motif is identified in the pneumococcal enolase by using the complete sequence of the enolase (**Fig. 3.3A**). In addition, the minimal binding region is identified by using a peptide array with varying length of the immobilized peptides (**Fig. 3.3C**). Of course, the functional activity should finally be confirmed by site-directed

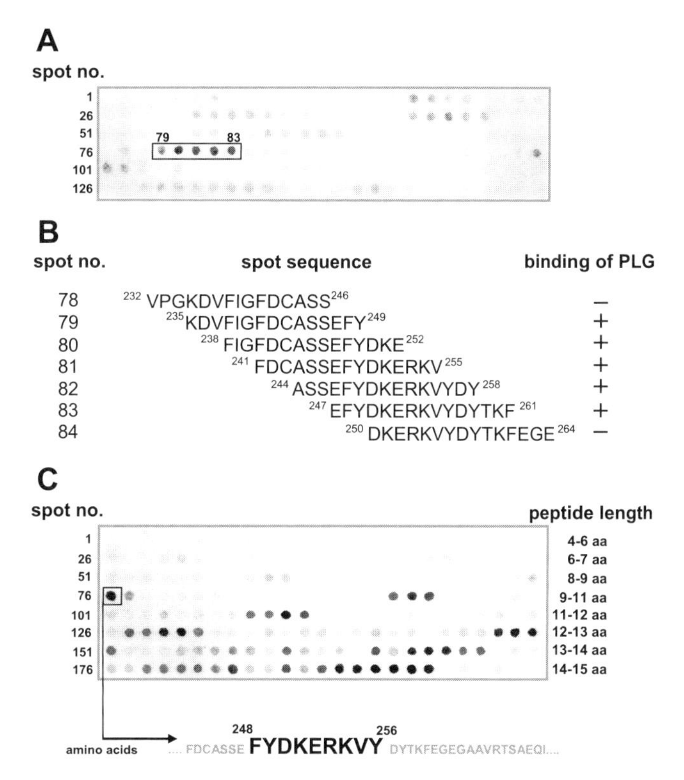

Fig. 3.3. Determination of the minimal plasminogen-binding motif of the pneumococcal enolase. **(A)** Spot membrane of the 434-amino-acid sequence of eno divided into 141 overlapping peptides of 15 amino acids each. The peptides overlap with 12 amino acids, and binding of human plasminogen is analyzed. Twenty-five peptides were immobilized in one line. Binding of human plasminogen is detected to spot number 79, 80, 81, 82, and 83. Reactivity of other spots is due to unspecific binding of anti-plasminogen antibody and secondary antibody used. **(B)** Sequences of spots (78→84) and binding reactivity with plasminogen. **(C)** Spot membrane with 198 overlapping peptides, with peptide length from 4-amino-acid up to 15-amino-acid residues per spot analyzed with plasminogen. The amino acid sequence used to construct the spot-membrane corresponds with amino acids 232 to 267 (VPGKDVFIGFDCASSEFYDKERKVYDYTKFEGEGAA) of pneumococcal enolase. The peptide FYDKERKVY (spot no. 76) located between amino acid 248 and amino acid 256 is identified as the minimal internal binding motif for plasminogen. The length of the peptides spotted on the membrane is indicated by numbers and arrows in the figure. (From Bergmann, S., Wild, D., Diekmann, O., Frank, R., Bracht, D., Chhatwal, G. S., and Hammerschmidt, S. 2003. Identification of a novel plasmin(ogen)-binding motif in surface displayed alpha-enolase of *Streptococcus pneumoniae*. *Mol. Microbiol.* **49,** 411–423. Reproduced with permission from Blackwell Publishing.)

mutagenesis and functional analysis. However, this application has also some limitations, in particular when the protein or the antibody does not recognize a linear epitope in the receptor protein. Peptides (with length varying from 4 amino acids up to 15 amino acids per spot) are synthesized on cellulose paper by a method named SPOT-synthesis *(20)*. The peptide array contains the peptide sequence of interest divided into several peptides, each peptide spot consists of 13 to 15 amino acids, with an overlap of 10 to 12 amino acids. The application of this method allows a rapid and highly parallel *in situ* screening of potential binding sites in the receptor protein when the ligand is used as soluble analyte *(21)*. In case the analyte is detected by an antibody, the background of the antibody reactions has to be investigated prior to binding assays with the analyte.

1. The wet membrane (activated with EtOH) is equilibrated with TBS buffer and washed 3 times with TBS buffer for 10 min.
2. Blocking of the peptide array is performed overnight in 8 mL blocking solution.
3. The peptide array is washed with T-TBS and incubated with the analyte (the host protein or bacterial protein when the host protein is divided into SPOT-synthesized peptides) for at least 5 h in blocking solution followed by three washes with T-TBS at room temperature.
4. For detection of the bound analyte, the membrane is washed 3 times with T-TBS for 10 min followed by incubation with an analyte-specific antibody in blocking solution.
5. After washing in T-TBS, the membrane is incubated with the secondary antibody (e.g., with a horseradish peroxidase–conjugated immunoglobulin) in blocking solution for 1.5 h.
6. The substrate solution was applied after washing the membrane once in T-TBS and twice in buffer recommended for your detection method. Binding can be detected using a substrate solution containing 1 mg/mL 4-chloro-1-naphthol and 0.1% H_2O_2 in PBS or by enhanced chemiluminescence detection.
7. Scanning of the membrane or the x-ray allows densitometric analysis of the peptide spots, and the intensity of the peptide spots can be calculated.

4. Notes

1. The radiolabeled proteins can be stored in sample aliquots at 4°C or −20°C. This strongly depends on the properties of the protein and, hence, a general recommendation cannot be

provided here. In case the binding experiments show for all reaction samples high binding values, this is most likely due to unspecific binding of the radiolabeled protein to the bacteria. Therefore, if possible, a positive control (i.e., a strain that has been shown to be negative for binding of the protein) and a positive control are included in each binding experiment.

2. The evaluation of the data measured by flow cytometry occurred generally by using Cell Quest Software (Becton Dickenson, San Jose, CA) or the PC-compatible software Windows Multiple Document Interface software (WinMDI). The WinMDI software is designed by J. Trotter and can be downloaded at http://facs.scripps.edu as freeware, available for Windows 95/98/NT4 and Windows 3.1/NT 3.x.
3. The FITC solution is prepared freshly for each experiment and used immediately for FITC-labeling of the bacteria. Take care that the reaction is conducted in the dark.
4. Binding of the proteins to the polystyrene surface of the 96-well plates is controlled using specific anti-host protein antibodies by enzyme-linked immunosorbent assay. These antibodies are distributed from several companies and allow one to control the dose-dependent immobilization of the host proteins.
5. The tissue host cell lines are repeatedly seeded from one initial culture. It is important to notice the number of passages, and in particular passages of more than 18 times should be avoided for the HBMEC cell line. In order to compare the results of independently performed experiments, the use of host cells in similar passages is recommended.
6. The NHS/EDC activation is of the CM5 biosensor chip activation is not very stable and, hence, the coupling of the ligand on the activated surface is performed separately for each of the four channels of the biosensor chip. Immediately after coupling the ligand and receiving the desired response units, the channel is blocked with ethanolamine hydrochloride.
7. If possible, stripping should be avoided and identical membranes should be used for testing the background of the applied antibodies. This will show unspecific binding of the antibodies to the immobilized peptides and is important to discriminate for specific binding of the analyte. However, stripping off analytes and/or antibodies bound to the peptide array is often required because only one peptide array is available. In addition, the binding of the protein has to be verified by at least one repetition of the experiment. The stripping protocol is as follows: The membrane is washed 2 times with H_2O and once with dimethylformamide (DMF). For this

procedure, glassware is used (no plastic material), and the DMF wash is accompanied by sonication at room temperature. When binding of the protein was detected by a color reaction, the color diminishes. This step is repeated and conducted for at least 15 min. Disappearance of the color is not equivalent with removal of the bound host protein or antibody. This step is followed by three washes with H_2O. Buffer A is used 3 times for 5 min at room temperature followed by incubation for 5 min at 40°C with sonication. Finally, buffer B is used 3 times and the membrane is washed 3 times with ethanol. Thereafter, the dried membrane is stored at −20°C, and the next binding experiment is performed.

8. The original Genosys blocking buffer is not further distributed by Sigma Chemicals. Yet, other blocking solutions have to be used although their potential may differ from that of the Genosys blocking buffer.
9. The concentrations of the analyte employed in SPR experiments cannot per se be predicted and strongly depend on the equilibrium constant. In case of a lower affinity, a higher K_D is expected and, hence, the concentration of the analyte has to be higher as mentioned in **Section 3**.

Acknowledgments

This work is supported by grants of the Deutsche Forschungsgemeinschaft (Sonderforschungsbereich 479, Teilprojekt A7) and the Bundesministerium für Bildung und Forschung (CAPNETZ project C8).

References

1. Cartwright, K. (2002) Pneumococcal disease in western Europe: burden of disease, antibiotic resistance and management. *Eur. J. Pediatr.* **161,** 188–195.
2. Paterson, G. K. and Mitchell, T. J. (2006) Innate immunity and the pneumococcus. *Microbiology* **152,** 285–293.
3. Magee, A. D., and Yothe, J. (2001) Requirement for capsule in colonization by *Streptococcus pneumoniae. Infect. Immun.* **69,** 3755–3761.
4. Hammerschmidt, S., Wolff, S., Hocke, A., Rosseau, S., Muller, E., and Rohde, M. (2005) Illustration of pneumococcal polysaccharide capsule during adherence and invasion of epithelial cells. *Infect. Immun.* **73,** 4653–4466.
5. Hoskins, J., Alborn, W. E., Jr., Arnold, J., Blaszczak, L. C., Burgett, S., DeHoff, B. S., Estrem, S. T., Fritz, L., Fu, D. J., Fuller, W., Geringer, C., Gilmour, R., Glass, J. S., Khoja, H., Kraft, A. R., Lagace, R. E., LeBlanc, D. J., Lee, L. N., Lefkowitz, E. J., Lu, J., Matsushima, P., McAhren, S. M., McHenney, M., McLeaster, K., Mundy, C. W., Nicas, T. I., Norris, F. H., O'Gara, M., Peery, R. B., Robertson, G. T., Rockey, P., Sun, P. M., Winkler, M. E., Yang, Y., Young-Bellido, M., Zhao, G., Zook, C. A., Baltz, R. H., Jaskunas, S. R., Rosteck, P. R., Jr., Skatrud, P. L., and Glass, J. I. (2001) Genome of the bacterium Streptococcus pneumoniae strain R6. *J. Bacteriol.* **183,** 5709–5717.

6. Tettelin, H., Nelson, K. E., Paulsen, I. T., Eisen, J. A., Read, T. D., Peterson, S., Heidelberg, J., DeBoy, R. T., Haft, D. H., Dodson, R. J., Durkin, A. S., Gwinn, M., Kolonay, J. F., Nelson, W. C., Peterson, J. D., Umayam, L. A., White, O., Salzberg, S. L., Lewis, M. R., Radune, D., Holtzapple, E., Khouri, H., Wolf, A. M., Utterback, T. R., Hansen, C. L., McDonald, L. A., Feldblyum, T. V., Angiuoli, S., Dickinson, T., Hickey, E. K., Holt, I. E., Loftus, B. J., Yang, F., Smith, H. O., Venter, J. C., Dougherty, B. A., Morrison, D. A., Hollingshead, S. K., and Fraser, C. M. (2001) Complete genome sequence of a virulent isolate of *Streptococcus pneumoniae*. *Science* **293,** 498–506.
7. Hammerschmidt, S. (2006) Adherence molecules of pathogenic pneumococci. *Curr. Opin. Microbiol.* **9,** 12–20.
8. Bergmann, S., and Hammerschmidt, S. (2006) Versatility of pneumococcal surface proteins. *Microbiology* **152,** 295–303.
9. Cundell, D. R., Gerard, N. P., Gerard, C., Idanpaan-Heikkila, I., and Tuomanen, E. I. (1995) Streptococcus pneumoniae anchor to activated human cells by the receptor for platelet-activating factor. *Nature* **377,** 435–438.
10. Elm, C., Braathen, R., Bergmann, S., Frank, R., Vaerman, J. P., Kaetzel, C. S., Chhatwal, G. S., Johansen, F. E., and Hammerschmidt, S. (2004) Ectodomains 3 and 4 of human polymeric immunoglobulin receptor (hpIgR) mediate invasion of *Streptococcus pneumoniae* into the epithelium. *J. Biol. Chem.* **279,** 6296–6304.
11. Hammerschmidt, S., Tillig, M. P., Wolff, S., Vaerman, J. P., and Chhatwal, G. S. (2000) Species-specific binding of human secretory component to SpsA protein of *Streptococcus pneumoniae* via a hexapeptide motif. *Mol. Microbiol.* **36,** 726–736.
12. Lu, L., Lamm, M. E., Li, H., Corthesy, B., and Zhang, J. R. (2003) The human polymeric immunoglobulin receptor binds to *Streptococcus pneumoniae* via domains 3 and 4. *J. Biol. Chem.* **278,** 48178–48187.
13. Kolberg, J., Aase, A., Bergmann, S., Herstad, T. K., Rodal, G., Frank, R., Rohde, M., and Hammerschmidt, S. (2006) *Streptococcus pneumoniae* enolase is important for plasminogen binding despite low abundance of enolase protein on the bacterial cell surface. *Microbiology* **152,** 1307–1317.
14. Stins, M. F., Gilles, F., and Kim, K. S. (1997) Selective expression of adhesion molecules on human brain microvascular endothelial cells. *J. Neuroimmunol.* **76,** 81–90.
15. Nice, E. C., Mcinerney, T. L., and Jackson, D. C. (1996) Analysis of the interaction between a synthetic peptide of influenza virus hemagglutinin and monoclonal antibody using an optical biosensor. *Mol. Immunol.* **33,** 801–809.
16. Hunter, W.M., and Greenwood, F.C. (1964) A radio-immunoelectrophoretic assay for human growth hormone. *Biochem. J.* **91,** 43–56.
17. Bergmann, S., Wild, D., Diekmann, O., Frank, R., Bracht, D., Chhatwal, G. S., and Hammerschmidt, S. (2003) Identification of a novel plasmin(ogen)-binding motif in surface displayed alpha-enolase of *Streptococcus pneumoniae. Mol. Microbiol.* **49,** 411–423.
18. Pracht, D., Elm, C., Gerber, J., Bergmann, S., Rohde, M., Seiler, M., Kim, K. S., Jenkinson, H. F., Nau, R., and Hammerschmidt, S. (2005) PavA of *Streptococcus pneumoniae m*odulates adherence, invasion, and meningeal inflammation. *Infect. Immun.* **73,** 2680–2689.
19. Hermans, P. W., Adrian, P. V., Albert, C., Estevao, S., Hoogenboezem, T., Luijendijk, I. H., Kamphausen, T., and Hammerschmidt, S. (2006) The streptococcal lipoprotein rotamase A (SlrA) is a functional peptidyl-prolyl isomerase involved in pneumococcal colonization. *J. Biol. Chem.* **281,** 968–976.
20. Frank, R. (1992) Spot-synthesis: an easy technique for positionally addressable, parallel chemical synthesis on a membrane support. *Tetrahedron* **48,** 9217–9232.
21. Frank, R. (2002) The SPOT-synthesis technique. Synthetic peptide arrays on membrane supports—principles and applications. *J. Immunol. Methods* **267,** 13–26.

Chapter 4

Screening of Virulence Traits in *Legionella pneumophila* and Analysis of the Host Susceptibility to Infection by Using the *Dictyostelium* Host Model System

Olaga Shevchuk and Michael Steinert

Abstract

The social soil amoeba *Dictyostelium discoideum* has been established as a host model for several human pathogens including *Legionella pneumophila*. The complete genome sequence, the genetic tractability, and the phagocytic characteristics of *Dictyostelium* generate many opportunities for the study of host-pathogen interactions. Important applications of this haploid model organism are (i) the use of *Dictyostelium* cells as a screening system for bacterial virulence, (ii) the use of *Dictyostelium* mutant cells to identify genetic host determinants of susceptibility and resistance to infection, and (iii) experiments that allow the dissection of the complex cross-talk with infectious agents. Accordingly, this chapter describes a plaque assay to identify attenuated pathogens, an infection assay for the analysis of host cell mutants and pathogens, and a screening method for the isolation of *Legionella* mutants that are defective in the reprogramming of the phagolysosomal maturation of the host.

Key words: *Dictyostelium*, *Legionella*, plaque assay, infection assay, phagosome maturation, virulence.

1. Introduction

The social amoeba *Dictyostelium discoideum* has been used as a model for human-disease analysis *(1)*. Especially in infection biology, this organism helps us to understand mammalian macrophage functions. In its natural forest soil habitat, *D. discoideum* feeds on bacteria and grows by mitotic division of single cells. Exhaustion of food resources triggers a differentiation process that results

Steffen Rupp, Kai Sohn (eds.), *Host-Pathogen Interactions*, DOI: 10.1007/978-1-59745-204-5_4,

in the aggregation of about 100,000 *Dictyostelium* cells and the development of a multicellular fruiting body. Many basic aspects of *Dictyostelium* differentiation, signal transduction, cell motility, and phagocytosis are similar to the respective processes in mammalian professional phagocytes. Because of its haploid nature, a sophisticated cellular and genetic tool box, a comprehensive highly curated data base (www.dictybase.org) that contains the complete genome sequence, GenBank records, expressed sequence tags, PubMed references, and other resources, *Dictyostelium* offers unique possibilities to study host-pathogen interactions *(2–4)*.

Pathogens for which the *Dictyostelium* infection model has been established comprise *Pseudomonas aeruginosa, Vibrio cholerae, Mycobacterium* spp., and *Cryptococcus neoformans* *(5–9)*. The most advanced application of *Dictyostelium* as a host model occurred with *Legionella pneumophila*. The inhalation of aerosolized legionellae by humans and the replication of the bacteria in the alveolar macrophages can result in a severe atypical pneumonia called Legionnaires' disease. The comparison of infected macrophages and *Dictyostelium* cells demonstrated that in both host systems, virulent *L. pneumophila* grow intracellularly within organelle studded vacuoles that are associated with rough endoplasmic reticulum (ER). During early infection, the *Legionella*-containing phagosomes exclude endocytic and lysosomal markers, which is recognized as a key feature of *Legionella* pathogenicity *(2, 10)*. The analysis of growth rates of various well-defined *Legionella* mutants in *Dictyostelium* helped to characterize a number of bacterial virulence factors including the Dot/Icm type IV secretion system *(11)*. Similarly, custom-tailored *Dictyostelium* mutant cells have proved useful to identify genetic host determinants of susceptibility and resistance *(12–14)*. Especially interesting is the observation that the *Dictyostelium* homologue of human natural resistance–associated membrane protein 1 (Nramp1) influences the replication of *Legionella (15)*. Polymorphic variants of Nramp1 in humans have been associated with susceptibility to tuberculosis and leprosy. Disruption mutants of this phagosomal metal cation transporter in *Dictyostelium* were more permissive hosts than were wild-type cells, and Nramp1 overexpression protected *Dictyostelium* cells from *Legionella* infection. Whether this effect is due to changes of intraphagosomal iron concentrations is currently under investigation.

In the following sections, we describe (i) the use of *Dictyostelium* cells as a screening system (plaque assay) for attenuated or mutagenized *Legionella* strains. In addition, we present (ii) an infection assay that can be used to characterize host cell mutants. Finally, we provide (iii) a method for a genetic screen that facilitates the isolation of *Legionella* mutants defective in arresting the maturation of their phagosomes.

2. Materials

2.1. Bacterial and Cell Culture

1. Buffered charcoal-yeast (BCYE) agar for *Legionella* spp.: 5 g *N*-(2-acetamido)-2-amino-ethanesulfonic acid (ACES; Gerbu, Gaiberg, Germany) and 10 g yeast extract (Oxoid, Wesel, Germany) are dissolved in 900 mL double-distilled water (ddH_2O), and the pH is adjusted to 6.9 with 10 N KOH. Then, 2 g activated charcoal (Fluka, Seelze, Germany) and 15 g Agar (BD Difco, Hiedelberg, Germany) are added, and the volume is completed to 980 mL by adding ddH_2O (*see* **Note 1**). After the agar has been autoclaved and cooled down to 50°C, it is supplemented with 0.4 g L-cysteine in 10 mL ddH_2O and 0.25 g $Fe(NO_3)_3 \cdot 9H_2O$ in 10 mL ddH_2O (*see* **Note 2**).
2. YEB Liquid medium for *Legionella*: 10 g yeast extract (Oxoid) is dissolved in 1000 mL ddH_2O and autoclaved. Then, a vial of the *Legionella* BCYE Growth Supplement SR0110A (Oxoid) is added for every 100 mL of YEB medium.
3. Luria-Bertani (LB) medium for *Klebsiella mobilis*: 10 g tryptone, 5 g yeast extract (Oxoid), and 5 g sodium chloride (NaCl) are dissolved in 1000 mL ddH_2O and autoclaved.
4. HL5-medium for *Dictyostelium*: 7.15 g yeast extract (Oxoid), 14.3 g Bacto proteose peptone Nr.2 (BD Difco), 1.28 g Na_2HPO_4, and 0.49 g KH_2PO_4 are dissolved in 900 mL ddH_2O, and the pH is adjusted to 7.5 with 10 N KOH. After autoclaving and cooling, 15.4 g glucose monohydrate dissolved in 100 mL ddH_2O is added (*see* **Note 3**).
5. 50X stock Soerensen buffer: 99.86 g Na_2HPO_4 and 17.8 g KH_2PO_4 are dissolved in 1 L ddH_2O, and the pH is adjusted to 6.0 with 10 N KOH. After autoclaving, the stock solution can be kept at room temperature for extended time periods.
6. SM-agar: 10 g Bacto proteose peptone Nr.2 (BD Difco), 1 g yeast extract (Oxoid), 1 g $MgSO_4$, 2.2 g KH_2PO_4, and 1.3 g K_2HPO_4 are dissolved in 950 mL ddH_2O, and the pH is adjusted to 6.4 with 10 N KOH. After autoclaving and cooling, 10 g glucose monohydrate dissolved in 100 mL ddH_2O is added (*see* **Note 3**).
7. HB (Homogenization buffer): 10 mL 50 mM EGTA and 100 mL 200 mM Hepes pH 7.2 are dissolved in 750 mL ddH_2O. After autoclaving and cooling, 85.5 g sucrose dissolved in 250 mL ddH_2O is added (*see* **Note 3**).

2.2. Plaque Assay

1. 1X Soerensen buffer (dilute one part 50X Soerensen buffer (*see* **Section 2.1, item 5**) with 49 parts ddH_2O and autoclave).
2. YEB-medium (*see* **Section 2.1, item 2**).
3. LB-medium (*see* **Section 2.1, item 3**).

4. HL5-medium (*see* **Section 2.1, item 4**).
5. SM agar plates (*see* **Section 2.1, item 6**).

2.3. Infection Assay

1. Infection medium: This is a 1:1 mixture of HL5-medium and 1X Soerensen buffer (*see* **Section 2.2, item 1**) (*see* **Note 4**).
2. Autoclaved ddH_2O.
3. BCYE agar plates (*see* **Section 2.1, item 1**).

2.4. Screening for Legionella Mutants Defective in Arrest of Phagolysosomal Maturation

2.4.1. Generation of Legionella Transposon Library

1. LB-medium supplemented with 25 μg/mL kanamycin (*see* **Section 2.1, item 3**).
2. YEB Liquid *Legionella* medium supplemented with 25 μg/mL kanamycin and 10% glycerol for storage of *Legionella* transposon library kanamycin (*see* **Section 2.1, item 2**).
3. BCYE agar plates supplemented with 5% sucrose and 25 μg/mL kanamycin (*see* **Section 2.1, item 1**).

2.4.2. Preparation of Iron Particles

1. 1.2 M $FeCl_2$: 4.76 g $FeCl_2 \cdot 4H_2O$ is dissolved in 20 mL ddH_2O.
2. 1.8 M $FeCl_3$: 9.72 g $FeCl_3 \cdot 6H_2O$ is dissolved in 20 mL ddH_2O.
3. 25% NH_4OH: 5% NH_4 OH.
4. Dextran (64 to 76 kDa, Sigma, Schnelldorf, Germany).
5. 0.3 M HCl.
6. 0.22-μm sterile filter (Millipore, Billerica, MA, USA).

2.4.3. Screening for Legionella Mutants Defective in the Arrest of Phagosome Maturation

1. Infection medium: This is a 1:1 mixture of HL5-medium and 1X Soerensen buffer (*see* **Section 2.2, item 1**) (*see* **Note 4**).
2. Autoclaved ddH_2O.
3. Iron particles (*see* **Section 3.4.2.**).
4. 1X Soerensen buffer (*see* **Section 2.2, item 1**).
5. HB-buffer with EDTA-free protease inhibitor cocktail (Roche, Penzberg, Germany) (*see* **Section 2.1, item 7**).
6. 0.4% Triton X-100 in HB
7. BCYE agar plates (*see* **Section 2.1, item 1**) supplemented with 20 μg/mL kanamycin.
8. Dura Grind stainless-steel homogenizer.
9. Miltenyi Biotec MiniMACS columns, OctoMACS, Bergisch Gladbach, Germany.

3. Methods

Because *Legionella* species are infectious to humans, experimental work with this pathogen has to be performed under S2 conditions (*see* **Note 5**).

3.1. Cultivation of Legionella spp., Klebsiella mobilis, and D. discoideum

1. *L. pneumophila* is grown on BCYE agar at 37°C in 5% CO_2 atmosphere for 3 days (*see* **Note 6**).
2. *K. mobilis* is grown on LB agar at 37°C overnight.
3. The *D. discoideum* wild-type strain (AX2) was grown in 30 mL HL5-medium at 24.5°C either as shaking culture in a 100-mL flask or in 75-cm^2 cell culture flasks (*see* **Notes 7** and **8**).

3.2. Plaque Assay

The plaque assay reveals weather or not the pathogen displays virulence either by evading amoeboid killing or actively killing *Dictyostelium*. Bacterial predation by *Dictyostelium* is scored by plating amoebae on nutrient agar plates seeded with the respective bacterial strains. Successful predation by the amoebae is visualized by the appearance of clear plaques (e.g., food bacteria like *K. mobilis* or avirulent *Legionella* strains). The absence of plaques reveals resistance to *Dictyostelium* predation and may indicate a virulent phenotype (**Fig. 4.1**).

1. *D. discoideum* cells are collected by centrifugation (200 × *g*, 7 min, at room temperature), washed once with 1X Soerensen buffer, and suspended in infection medium. The final cell density is 1×10^4 cells/mL.
2. The overnight bacterial cultures of *Legionella* spp. and *K. mobilis* are pelleted by centrifugation (3000 × *g*, 5 min, at room temperature) and suspended in 2 mL sterile ddH_2O. The bacterial suspension is diluted to 10^9 cells/mL (*see* **Note 9**).

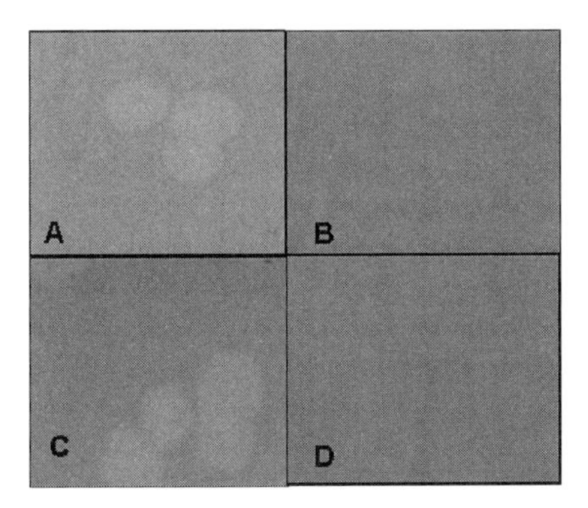

Fig. 4.1. *Dictyostelium* plaque assay to screen for bacterial virulence potential. *D. discoideum* cells were plated on SM agar plates seeded either with **(A)** *K. mobilis* alone or mixed with **(B)** *L. pneumophila* Corby, **(C)** *L. hackeliae*, and **(D)** *L. micdadei*, respectively. Successful predation by *D. discoideum* cells is scored by the appearance of clear plaques in the bacterial lawn.

3. 100 μL of the *Klebsiella* suspension and 100 μL of the *D. discoideum* suspension are mixed with 100 μL of the different *Legionella* suspensions respectively and plated onto SM-agar plates. After 3 to 5 days, the plates are examined for plaques formed by *Dictyostelium* amoebae.

3.3. Infection Assay

To analyze host cell mutants for increased resistance or susceptibility, a parental host strain and an isogenic *Dictyostelium* mutant should be compared with respect to the intracellular growth rate of well-characterized bacteria. Similarly, this assay can be used to compare *Legionella* mutants with their isogenic wild-type strains by using a well-defined host cell strain.

1. *D. discoideum* cells of a 3-day-old culture are harvested ($200 \times g$, 7 min, at room temperature) and resuspended in the same volume of infection medium. If mutants are grown in the presence of antibiotics, a washing step ($200 \times g$, 7 min, at room temperature) with infection medium is required.
2. 25×10^5 cells are seeded into 25-cm^2 cell culture flasks and the volume is adjusted to 5 mL with freshly mixed infection medium. The final cell density is 5×10^5 cells/mL. Before bacterial inoculation, the cells should have settled down for 30 min at 25.5°C (*see* **Note 8**).
3. A 3-day-old plate culture of *L. pneumophila* is suspended in 1 mL ddH$_2$O. The bacterial suspension is diluted to 10^6 cells/mL, and 50 μL of this dilution is added to every cell culture flask (multiplicity of infection [MOI] 0.02) (*see* **Note 10**).
4. After an invasion period of 2 h, the remaining extracellular bacteria are killed by a gentamicin treatment (100 μg/mL). Fifty microliters of a gentamicin stock solution are added to every cell culture flask. After 1 h incubation at 25.5°C, the *Dictyostelium* cells are washed 3 times with 5 mL Soerensen buffer (*see* **Note 11**).
5. 100 μL from each cell culture flask is transferred into a 1.5-mL reaction tube containing 900 μL ddH$_2$O. In order to determine the CFU/mL of the inoculum. 100 μL of this suspension and further dilutions are plated on BCYE agar.
6. After different time intervals (24, 48, 72, and 96 h), serial dilutions of cell lysates are plated on BCYE agar (*see* **Note 12**), and the CFU/mL is determined.

3.4. Screening for Legionella Mutants Defective in Arrest of Phagolysosomal Maturation

We have developed a screen that facilitates the isolation of *Legionella* mutants defective in arresting the maturation of their phagosomes (**Fig. 4.2**). The method should be applicable to other intracellular pathogens that block the progression of the phagosome maturation, survive transient exposure to lysosomal contents, and for which a mutational approach is feasible.

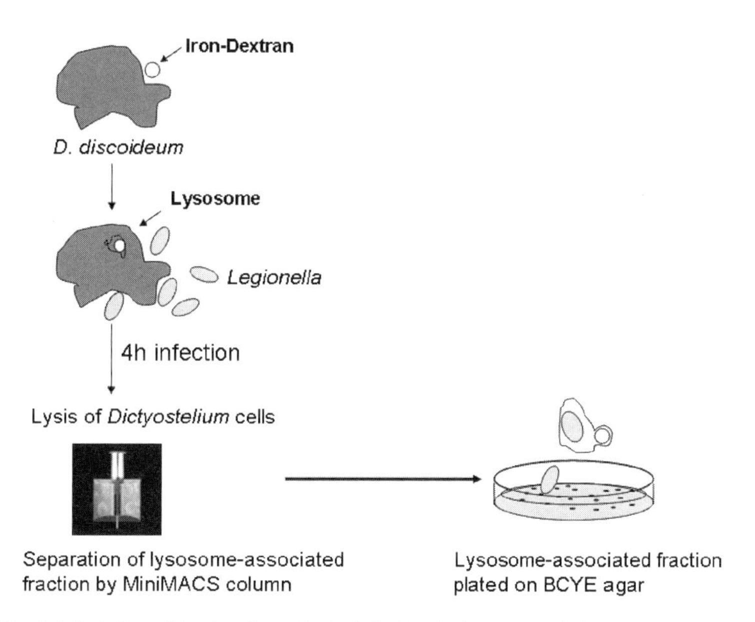

Fig. 4.2. Isolation of *Legionella* mutants defective in the arrest of phagosome maturation. *Dictyostelium* cells are incubated with iron-dextran that chases into lysosomes. After infection with *Legionella* from a library of transposon-mutagenized bacteria and host cell lysis, iron-dextran–loaded lysosomes are separated by MiniMACS columns. The lysosome-associated bacterial fraction is cultivated on BCYE agar. After four rounds of enrichment, the majority of mutants should reveal the desired phenotype.

3.4.1. Generation of the Legionella Transposon Library

1. The *L. pneumophila* Corby (serogroup 1) transposon mutant library is constructed as described previously *(16)*.
2. The random nature of transposon-mutants library is confirmed by Southern blot hybridization of 25 independent clones by using a kanamycin probe.
3. 5,700 individual mutant colonies are picked from multiple plates and, after culture in YEB for 5 days at 37°C in 5% CO_2, are frozen in 96-well plates with 10% glycerol.

3.4.2. Preparation of Iron Particles

1. 10 mL of 1.2 M $FeCl_2$ is mixed with 10 mL of 1.8 M $FeCl_3$ and agitated extensively while 10 mL of 25% NH_3 is added.
2. The suspension is placed on a magnet until the precipitate has gathered on the bottom of the tube. The precipitate is washed once with 5% NH_3 and twice with sterile ddH_2O.
3. The supernatant is decanted and the precipitate is suspended in 80 mL of 0.3 M HCl and stirred with a magnetic stirrer for 30 min. Four grams of dextran (64 to 76 kDa; Sigma) is added and stirred for 30 min.
4. The sample is dialyzed extensively against cold water for 2 days and filtered through the 0.22-μm sterile filter (*see* **Note 13**).

3.4.3. Screening for Legionella Mutants in D. discoideum Cells

1. *D. discoideum* cells of a 3-day-old culture are harvested (200 × *g*, 7 min, at room temperature) and resuspended in the same volume of infection medium (*see* **Section 2.3, item 1**).
2. 25 × 10^7 cells are seeded into 25-cm^2 cell culture flasks, and the volume is adjusted to 25 mL with freshly mixed infection medium. The final cell density is 1 × 10^6 cells/mL. Before bacterial inoculation, the cells should have settled down for 30 min at 25.5°C (*see* **Note 8**).
3. The *L. pneumophila* Corby transposon mutant library (5700 colonies) (*see* **Section 3.4.1**) is divided into 6 pools. A 5-day-old plate culture of each pool is suspended in 2 mL of sterile ddH_2O, and the cell density is adjusted to 10^9 cells/mL (*see* **Note 9**).
4. 250 μL of the prepared bacterial suspensions are added to each cell culture flask (MOI 10).
5. After an invasion period of 3.5 h, 2.7 mL of colloidal iron particles are added (final concentration 1 mg/mL).
6. After 4 h incubation at 25.5°C, the *Dictyostelium* cells are washed 3 times with 25 mL Soerensen buffer and once in HB supplemented with a protease inhibitor cocktail (*see* **Section 2.3.3, item 5**).
7. Cells are resuspended in 2 mL of HB and then broken by 12 strokes in a Dura Grind stainless-steel homogenizer. The lysate is subjected to low-speed centrifugation (400 × *g*, 5 min at 4°C) to remove nuclei and unbroken cells (*see* **Note 14**).
8. The supernatant is applied to a Miltenyi Biotec MiniMACS column. Then the column is washed with HB, and bound material is eluted with 1 mL 0, 4% Triton X-100 HB. A 100-μL aliquot of the eluate is immediately plated onto BCYE agar plates supplemented with kanamycin (20 μg/mL). The rest of the eluate should be frozen at −20°C.
9. After 5 days of culture, the bacteria are harvested from the agar plates, and selection is repeated. In order to enrich the amount of *Legionella* mutants defective in arrest of phagolysosomal fusion, the selection procedure should be performed 4 times.

4. Notes

1. Activated charcoal tends to aggregate. Hence, prepared agar should be mixed thoroughly before and after autoclaving.
2. $Fe(NO_3)_3{\cdot}9H_2O$ and L-cysteine precipitate if autoclaved. Therefore, both solutions should be freshly prepared. Store sterile filtered $Fe(NO_3)_3{\cdot}9H_2O$ solution (25 g/L) at 4°C.

3. The glucose and sucrose solutions should be autoclaved separately from other buffer components, as these sugars tend to caramelize in the presence of divalent cations.
4. The infection medium has to fulfil two criteria: (i) it should not promote the growth of *Dictyostelium* or extracellular bacteria; (ii) it should contain a residual amount of nutrients, so that differentiation and morphogenesis of *Dictyostelium* is prevented. An alternative infection medium is LoFlo (www.dictybase.org).
5. The inhalation of aerosolized legionellae can result in a severe pneumonia. Therefore, aerosol formation must be avoided. Furthermore, contaminated material has to be autoclaved and disinfected carefully.
6. The passaging of the bacteria should not be performed more than 4 times because *Legionella* tend to loose their virulence during this procedure. The stock culture is prepared by suspending a 2- to 3-day-old plate culture in 1 mL ddH_2O and adding 1 mL sterile 86% glycerine.
7. *Dictyostelium* preculture is prepared in 25-cm^2 cell culture flasks with 10 mL of HL5-medium inoculated with spores or cells from liquid nitrogen. It takes 2 to 3 days until the majority of *Dictyostelium* spores germinate. Cell cultures of *D. discoideum* are prepared by inoculating 30 to 50 mL HL5-medium with 1 to 2 mL of a fresh preculture.
8. Human infections occur at 37°C, but *Dictyostelium* does not survive temperatures above 27°C. The highest possible infection temperature in the *Dictyostelium* model is 25.5°C.
9. A 1:5 dilution of bacterial suspension is transferred into a photometer cuvette, and the absorption is measured at 550 nm. An absorption of 1.31 corresponds with 10^9 cells/mL.
10. The low MOI of 0.02 allows one to perform the assay without gentamicin treatment.
11. Washing in culture flasks requires careful removal of the supernatant and careful adding of Soerensen buffer. Avoid pipetting the liquid directly on the cell lawn.
12. *Legionella* is not able to replicate extracellularly in the infection medium. For other bacterial species, extracellular replication has to be considered.
13. The colloidal iron solution is stable for 3 months at 4°C.
14. Make every effort to avoid the development of air bubbles and foam in the preparation because this can lead to membrane and protein denaturation.

References

1. Saxe, C. L. (1999) Learning from the slime mold: *Dictyostelium discoideum* and human disease. *Am. J. Hum. Genet.* **65**, 25–30.
2. Farbrother, P., Wagner, C., Na, J., Tunggal, B., Morio, T., Urushihara, H., Tanaka, Y., Schleicher, M., Steinert, M., and Eichinger, L. (2006) *Dictyostelium* transcriptional host cell response upon infection with *Legionella*. *Cell. Microbiol.* **8**, 438–456.
3. Solomon, J. M., and Isberg, R. R. (2000) Growth of *Legionella pneumophila* in *Dictyostelium discoideum*: a novel system for genetic analysis of host-pathogen interactions. *Trends Microbiol.* **10**, 478–480.
4. Steinert, M., and Heuner, K. (2005) *Dictyostelium* as host for pathogenesis. *Cell. Microbiol.* 7, 307–314.
5. Cosson, P., Zulianello, L., Join-Lambert, O., Faurisson, F., Gebbie, L., Benghezal, M., van Delden, C., Kocjancic Curty, L. K., and Köhler, T. (2002) *Pseudomonas aeruginosa* virulence analysed in a *Dictyostelium discoideum* host system. *J. Bacteriol.* **184**, 3027–3033.
6. Pukatzki, S., Ma, A. T., Sturtevant, D., Krastins, B., Saracino, D., Nelson, W. C., Heidelberg, J. F., and Mekalanos, J. (2006) Identification of a conserved bacterial protein secretion system in *Vibrio cholerae* using the *Dictyostelium* host model system. *Proc. Natl. Acad. Sci.* **103**, 1528–1533.
7. Skriwan, C., Fajardo, M., Hägele, S., Horn, M., Wagner, M., Michel, R., Krohne, G., Schleicher, M., Hacker, J., and Steinert, M. (2002) Various bacterial pathogens and symbionts infect the amoeba *Dictyostelium discoideum*. *Int. J. Med. Micobiol.* **291**, 615–624.
8. Solomon, J. M., Leung, G. S., and Isberg, R. R. (2003) Intracellular replication of *Mycobacterium marinum* within *Dictyostelium discoideum*: efficient replication in the absence of host coronin. *Infect. Immun.* **71**, 3578–3586.
9. Steenbergen, J. N., Nosanchuk, J. D., Malliaris, S. D., and Casadevall, A. (2003) *Cryptococcus neoformans* virulence is enhanced after growth in the genetically malleable host *Dictyostelium discoideum*. *Infect. Immun.* **71**, 4862–4872.
10. Lu, H., and Clarke, M. (2005) Dynamic properties of *Legionella*-containing phagosomes in *Dictyostelium* amoebae. *Cell. Microbiol.* 7, 995–1007.
11. Hilbi, H., Segal, G., and Shuman, H. (2001) Icm/Dot-dependent upregulation of phagocytosis by *Legionella pneumophila*. *Mol. Microbiol.* **42**, 603–617.
12. Fajardo, M., Schleicher, M., Noegel, A., Bozzaro, S., Killinger, S., Heuner, K., Hacker, J., and Steinert, M. (2004) Calnexin, calreticulin and cytoskeleton associated proteins modulate uptake and growth of *Legionella pneumophila* in *Dictyostelium discoideum*. *Microbiology* **150**, 2825–2835.
13. Hägele, S., Köhler, R., Merkert, H., Schleicher, M., Hacker, J., and Steinert, M. (2000) *Dictyostelium discoideum*: a new host model system for intracellular pathogens of the genus *Legionella*. *Cell. Microbiol.* **2**, 165–171.
14. Li, Z., Solomon, J. M., and Isberg, R. R. (2005) *Dictyostelium discoideum* strains lacking the RtoA protein are detective for maturation of the *Legionella pneumophila* replication vacuole. *Cell. Microbiol.* 7, 431–442.
15. Peracino, B., Wagner, C., Balest, A., Balbo, A., Pergolizzi, B., Noegel, A. A., Steinert, M., and Bozzaro, S. (2006) Function and mechanism of action of *Dictyostelium* Nramp1 (Slc11a1) in bacterial infection. *Traffic* 7, 22–38.
16. Pope, C. D., Dhand, L., and Cianciotto, N. P. (1994) Random mutagenesis of *Legionella pneumophila* with mini-Tn10. *FEMS Microbiol. Lett.* **124**, 107–111.

Chapter 5

Characterizing Host Receptor Recognition by Individual Bacterial Pathogens

Katharina Kuespert and Christof R. Hauck

Abstract

A critical determinant of host range and specificity relies on the ability of pathogenic bacteria to recognize eukaryotic cell surface molecules via specialized adhesins. The specific adhesin-receptor interaction allows pathogens to tightly bind to their target cells, thereby facilitating the colonization of host tissues. Therefore, the identification and characterization of bacterial adhesins is a major topic in infection biology. This chapter focuses on a rapid and simple method for the analysis of adhesin-receptor interactions that permits the characterization of receptor binding properties at the level of single bacteria. Accordingly, this methodological approach is ideally suited for the analysis of adhesins expressed in a phase-variable manner and for the study of heterogeneous bacterial populations. Besides focusing on the receptor-binding assay, this chapter describes the production of fluorescence-tagged soluble host receptor domains required for conducting this assay.

Key words: Bacterial adhesin, CEACAM, flow cytometry, *Neisseria*, Opa protein, pull-down assay, receptor recognition.

1. Introduction

Pathogenic bacteria generally recognize eukaryotic cell surface receptors by specific proteins, so-called adhesins, to attach to host tissues *(1)*. This intimate contact on the molecular level often determines the ability of a bacterial pathogen to efficiently colonize and infect a certain range of species or a single host organism *(2)*. Therefore, characterization of the receptor-binding properties of a particular pathogen is a key question in infection biology.

Interestingly, almost each pathogenic microorganism analyzed to date uses a specific set of adhesive molecules and exploits distinct surface determinants of the eukaryotic cell ranging from

Steffen Rupp, Kai Sohn (eds.), *Host-Pathogen Interactions*, DOI: 10.1007/978-1-59745-204-5_5,

simple or complex carbohydrate structures to different membrane proteins including G-protein–coupled receptors, cadherins, integrins, or immunoglobulin-related cell adhesion molecules (IgCAMs) *(3)*. Prominent examples such as *Escherichia coli* FimH (binding to α-mannose residues), *E. coli* SfaS (binding to α-sialyl–2–3-β-galactose residues), *Helicobacter pylori* SabA (binding to sialyl-Le^X-residues), *Mycobacterium leprae* ML-LBP21 (binding to laminin2/α-dystroglycan), *Streptococcus pneumoniae* phosphorylcholine (binding to the PAF receptor), *Listeria monocytogenes* InlA (binding to E-cadherin), *Yersinia pseudotuberculosis* Inv (binding to integrin β_1), or *Neisseria gonorrhoeae* Opa_{CEA} protein (binding to carcinoembryonic antigen-related cell adhesion molecules (CEACAMs) highlight the broad range of unrelated bacterial adhesive factors recognizing a variety of host receptors. It is interesting to note that many pathogens express multiple adhesins that can act sequentially or that determine characteristic patterns of tissue tropism *(4)*. Furthermore, many bacterial adhesins are expressed in a phase-variable manner; that is, the expression is turned on or off in individual organisms *(5)*. Such a variation can be the result of stochastic events that affect gene transcription or translation of specific proteins and often results in antigenically heterogeneous bacterial populations *in vivo (6)*. Therefore, methods to analyze the receptor-binding capacity on the level of individual bacteria are most helpful in order to decipher the complete spectrum of host interactions potentially used by bacterial pathogens.

A number of strategies has been employed to identify and characterize host receptors involved in bacterial binding. In particular, affinity chromatography with purified bacterial adhesins, cell adhesion assays with wild-type and mutant bacteria, as well as infection of eukaryotic cell lines engineered to express the receptor in question are regularly conducted *(7–11)*. Though such approaches are ideal for the initial identification of unknown host receptors, they are laborious and mostly semiquantitative. Especially with regard to heterogeneous or phase-variable bacterial populations, additional methods have to be applied. Here, we describe a receptor-binding assay that uses soluble, fluorescence-tagged host receptor domains to analyze the binding profile of individual bacteria. The interaction is measured either in a pull-down format or by rapid quantitative determination using a flow cytometer (**Fig. 5.1**). As an example to illustrate the practicability of this method, the well-characterized interaction between Opa_{CEA} protein adhesins of *N. gonorrhoeae* and the human immunoglobulin–related glycoprotein CEACAM1 is used *(12)*. Opa_{CEA} proteins belong to a family of adhesins found in the outer membrane of the human-specific Gram-negative pathogens *N. gonorrhoeae* and *N. meningitidis*. Each bacterial strain encodes several Opa protein isoforms that are expressed in a phase-variable manner *(13)*. CEACAM1 is the corresponding

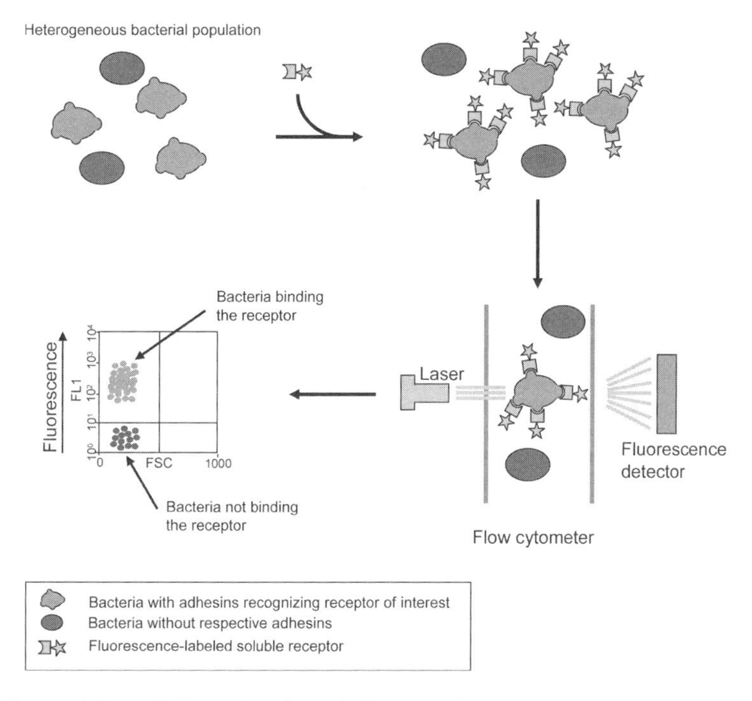

Fig. 5.1. Schematic illustration of the binding assay for rapid and quantitative analysis of receptor recognition by bacterial pathogens. Bacteria expressing or not expressing a receptor recognizing adhesin are incubated with a fluorescence-tagged soluble receptor construct. The receptor decorates adhesin-expressing bacteria with a fluorescence label, which can be quantitatively analyzed by flow cytometry to measure receptor-derived fluorescence associated with bacteria.

eukaryotic receptor and the extracellular, amino-terminal domain of this glycoprotein is the target of multiple Opa_{CEA} protein isoforms, whereas a closely related receptor, CEACAM8, is not recognized by Opa_{CEA} proteins *(14)*. Accordingly, soluble green fluorescent protein (GFP)-tagged CEACAM1 variants comprising the amino-terminal domain of CEACAM1, followed by either none (CEA1-N-GFP), two (CEA1-NA1B-GFP), or three (CEA1-NA1BA2-GFP) extracellular domains or the soluble GFP-tagged amino-terminal domain of CEACAM8 (CEA8-N-GFP; negative control), were expressed in human cells. Bacteria were incubated with cell-free culture supernatants containing soluble, GFP-tagged receptor domains, and after several washing steps, the bacterial population was analyzed by flow cytometry (**Fig. 5.2**). Whereas nonopaque *N. gonorrhoeae* (Ngo Opa^{-}) do not associate with any receptor construct, Opa_{CEA}-expressing gonococci (Ngo Opa_{CEA}) are labeled by all receptors harboring the amino-terminal domain of CEACAM1, but not by the CEACAM8-derived amino-terminal domain (**Fig. 5.2**). This straightforward approach yields fast results and can be easily quantified. Therefore, it might also be worthwhile to adapt this strategy to other pathogen-host interactions, where soluble receptor domains can be produced.

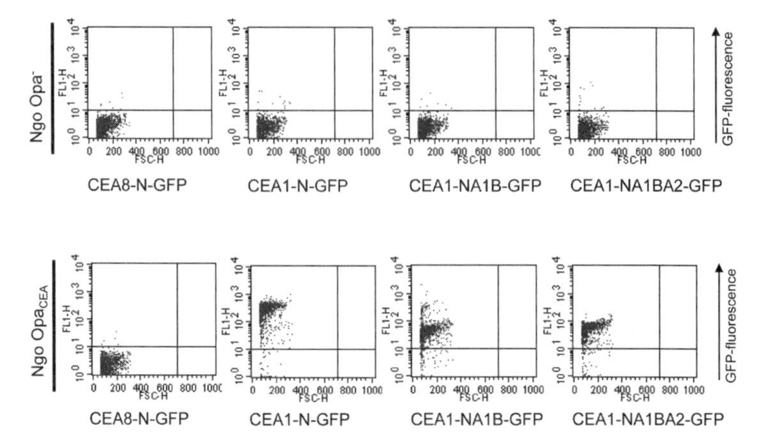

Fig. 5.2. Recognition of neisserial Opa_{CEA} proteins by soluble CEACAM-GFP constructs. *Neisseria gonorrhoea* expressing a CEACAM1-recognizing adhesin (Ngo Opa_{CEA}) or nonopaque gonococci (Ngo Opa^{-}) were incubated with cell culture supernatants containing the indicated GFP-tagged soluble CEACAM constructs. Bacteria were analyzed for CEACAM binding by flow cytometry measuring GFP-derived fluorescence associated with bacteria. Shown are the original dot blots of a representative experiment.

2. Materials

2.1. Construction of Plasmids Encoding for Fluorescent-Tagged Soluble Receptor

1. cDNA encoding extracellular receptor domain(s) (sufficient for adhesin interaction).
2. Material for PCR amplification of the extracellular receptor domain(s) (primers, dNTPs, polymerase, 10x PCR puffer).
3. Eukaryotic expression vector allowing the fusion of the amplified cDNA to the coding sequence of GFP or a variant of GFP (e.g., YFP, CFP, RFP) (e.g., pLPS3'EGFP, pEGFP-N1 [Clontech, Palo Alto, CA], pTurboGFP-N [Evrogen, San Diego, CA], phrGFP II-C [Stratagene, La Jolla, CA]).
4. Appropriate enzymes/cloning kits for cloning the amplified PCR-fragment into the eukaryotic expression vector.

2.2. Materials for Production of Recombinant Fluorescent-Tagged Soluble Receptor

1. Plasmid encoding the fluorescent-tagged soluble receptor.
2. double destillated H_2O.
3. 2x HBS: 280 mM NaCl, 50 mM Hepes, 1.5 mM Na_2HPO_4, pH 7.05.
4. 2.5 M $CaCl_2$ in H_2O; sterile.
5. Human embryonic kidney cell line 293T (293T cells; ATCC CRL-11268).
6. 25 mM chloroquine solution in H_2O; sterile.

7. Cell culture medium I: DMEM, 10% calf serum.
8. Cell culture medium II: OptiMem (Gibco BRL, Paisely, UK).

2.3. Receptor-Binding Assay

1. Bacteria (used for investigating receptor interaction).
2. PBS: 140 mM NaCl, 2.7 mM KCl, 8 mM Na_2HPO_4, 2 mM KH_2PO_4, pH 7.4.
3. PBS^{++}: PBS containing 0.9 mM $CaCl_2$ and 0.5 mM $MgCl_2$.
4. Cell culture supernatant with recombinant fluorescence-tagged soluble receptor.
5. Head-over-head rotator.
6. Flow buffer (PBS/2% FCS).
7. Flow cytometer.

2.4. Materials for the Bacterial Pull-Down Assay

1. Bacteria (used for investigating receptor interaction).
2. PBS: 140 mM NaCl, 2.7 mM KCl, 8 mM Na_2HPO_4, 2 mM KH_2PO_4, pH 7.4.
3. PBS^{++}: PBS containing 0.9 mM $CaCl_2$ and 0.5 mM $MgCl_2$.
4. Cell culture supernatant with recombinant fluorescence-tagged soluble receptor.
5. Head-over-head rotator.
6. 2x SDS sample puffer: 125 mM Tris-HCl (pH 6.8), 10% β-mercaptoethanol, 5% sodium dodecylsulfate, 0.2% bromphenol blue, 10% glycerol.
7. Setup for SDS-PAGE and Western blotting.
8. Antibody against the used fluorescent protein (e.g. anti-GFP antibody).

3. Methods

3.1. Construction of Plasmids Encoding Fluorescent-Tagged Soluble Receptor

The cDNA encoding extracellular receptor domain(s) has to be amplified by PCR using appropriate primers according to standard protocols. It is instrumental to include an amino-terminal signal sequence of the receptor, so that the protein will later be co-translationally imported into the endoplasmic reticulum and secreted into the cell culture supernatant. The PCR-primers have to be designed so that the resulting PCR fragment is cloned in frame with the coding sequence of GFP (or a variant of GFP). Usually, the goal is to have the carboxy-terminus of the receptor fused to the amino-terminus of GFP in the final protein.

3.2. Production of Recombinant Fluorescent-Tagged Soluble Receptor

For production of soluble, fluorescence-tagged receptor, 293T cells are transfected with the appropriate plasmid by a calcium phosphate coprecipitation method (*see* **Note 1**). The fusion protein will be secreted into the cell culture supernatant and can be concentrated or purified from this source.

1. Seed 2 × 10^6 293T cell in a 10-cm cell culture dish using DMEM/10% CS as cell culture medium and incubate the cells at 37°C in a humid atmosphere with 4.5% CO_2.
2. The next day, mix 500 μL double destillated H_2O with 8 μg of plasmid DNA (plasmid encoding fluorescent-tagged soluble receptor or the empty expression vector) (*see* **Note 2**) and 500 μL 2x HBS. Add 50 μL of 2.5M $CaCl_2$ dropwise to the solution during vortexing to achieve instant mixing, and incubate the mixture for 5 min at room temperature.
3. Add 10 μL of chloroquine solution to subconfluent 293T cells in a 10-cm dish (seeded the day before; see **step 1**) and distribute the calcium-DNA mixture dropwise over the dish.
4. Incubate the cells for 8 h at 37°C and 4.5% CO_2. At this point, exchange the medium of the transfected cells with 10 mL of fresh DMEM/10% CS.
5. The next day, replace the culture medium with 10 mL of OptiMem (*see* **Note 3**) and incubate the cells for 2 days at 37°C and 4.5% CO_2.
6. Collect the cell culture supernatants (containing recombinant fluorescence-tagged soluble receptor or control protein) and purify the supernatants from cell debris by centrifugation (2500 × *g*, 4°C, 10 min).
7. Decant each supernatant into a sterile tube and investigate supernatant for secreted fluorescent-tagged receptor by Western blotting according to standard methods using an antibody against the tag (e.g., anti-GFP antibody).
8. Adjust the volume of each supernatant with OptiMem to obtain equal levels of soluble receptor.

Supernatants can be used immediately for receptor-binding assay or can be stored for several weeks at 4°C. For long-term storage, supernatants can be stored at −80°C.

3.3. Receptor-Binding Assay

1. Streak out bacteria (which are used for investigating receptor interaction) on appropriate agar plates and incubate bacteria overnight at the required growth conditions or grow bacteria in liquid media.
2. Scrape bacteria from plates using a sterile cotton swab or harvest from liquid media by centrifugation and resuspend bacteria in PBS.

3. Wash bacteria twice with PBS^{++} and suspend 4×10^6 bacteria (estimated by OD readings according to a standard curve) in 1 mL cell culture supernatant (if the supernatant containing the fluorescence-tagged soluble receptor is used for the first time, *see* **Note 4**).
4. Incubate bacteria with the supernatant (containing fluorescence-tagged soluble receptor or control protein) for 30 min at 20°C with head-over-head rotation (*see* **Note 5**).
5. Wash bacteria twice with PBS^{++} and resuspend bacteria in 1 mL flow buffer.
6. Analyze samples in a flow cytometer by gating on the bacteria based on forward and sideward scatter and by measuring receptor-derived fluorescence associated with bacteria using settings appropriate for the fluorescent protein of choice.

3.4. Bacterial Pull-Down Assay

For the bacterial pull-down assay, perform **steps 1** to **4** of **Section 3.3** and proceed with the following steps:

5. Wash bacteria twice with PBS^{++}.
6. Resuspend bacteria in 50 μL SDS sample buffer and boil the samples at 95°C for 5 min.
7. Analyze samples for receptor binding by SDS-PAGE and Western blotting according to standard procedures using an antibody against the used fluorescent protein (e.g., anti-GFP antibody).

4. Notes

1. Besides using 293T cells, other commonly used cell lines (e.g., CHO, COS) can be transfected by calcium phosphate coprecipitation or various means (e.g., liposome-based transfection or electroporation), which are not described here.
2. As a negative control, use the empty expression vector in order to produce a mock supernatant or, even better, express a soluble receptor-related protein, which is known not to be involved in bacterial adhesion (in our example, CEACAM8).
3. DMEM/10%CS is replaced by OptiMem, a low serum medium, in order to reduce total protein content of the cell culture supernatant. However, changing medium at this step is not necessarily required.
4. To guarantee the sensitivity of the assay, you first have to make sure that GFP-tagged receptor is abundant in the cell culture supernatant. Accordingly, perform the bacterial pull-down assay as described in the protocol, using

stepwise dilutions of the receptor-containing cell culture supernatant, and determine the minimum amount of supernatant needed for sufficient bacterial-receptor interaction. Employing CMV-promoter–based expression vectors in 293T cells, the amount of fluorescence-tagged receptor secreted in 1 mL of cell culture supernatant is usually sufficient for individual samples containing 4×10^6 bacteria.

5. Incubation time and incubation temperature might be increased or decreased at this step depending on the affinity of the fluorescence-tagged receptor for the bacterial adhesin.

Acknowledgment

We thank T.F. Meyer (MPI für Infektionsbiologie, Berlin, Germany) for the *Neisseria* strains used in this study and W. Zimmermann (Universität München, Germany) for CEACAM cDNAs. This work was supported by funds from the DFG (Ha2568/3-2) to C.R.H.

References

1. Finlay BB, Cossart P (1997) Exploitation of mammalian host cell functions by bacterial pathogens. *Science* **276**: 718–725.
2. Lecuit M, Dramsi S, Gottardi C, Fedor-Chaiken M, Gumbiner B, Cossart P (1999) A single amino acid in E-cadherin responsible for host specificity towards the human pathogen *Listeria monocytogenes*. *EMBO J* **18**: 3956–3963.
3. Hauck CR (2006) The role of bacterial adhesion to epithelial cells in pathogenesis. In McCormick BA (ed.), *Bacterial-Epithelial Cell Cross-Talk: Molecular Mechanisms in Pathogenesis*. Cambridge University Press, Cambridge, pp. 158–183.
4. Le Bouguenec C (2005) Adhesins and invasins of pathogenic *Escherichia coli*. *Int J Med Microbiol* **295**: 471–478.
5. Stern A, Brown M, Nickel P, Meyer TF (1986) Opacity genes in *Neisseria gonorrhoeae*: Control of phase and antigenic variation. *Cell* **47**: 61–71.
6. Meyer TF, Gibbs CP, Haas R (1990) Variation and control of protein expression in *Neisseria*. *Annu Rev Microbiol* **44**: 451–477.
7. Elm C, Braathen R, Bergmann S, Frank R, Vaerman JP, Kaetzel CS, Chhatwal GS, Johansen FE, Hammerschmidt S (2004) Ectodomains 3 and 4 of human polymeric Immunoglobulin receptor (hpIgR) mediate invasion of *Streptococcus pneumoniae* into the epithelium. *J Biol Chem* **279**: 6296–6304.
8. Heise T, Dersch P (2006) Identification of a domain in *Yersinia* virulence factor YadA that is crucial for extracellular matrix-specific cell adhesion and uptake. *Proc Natl Acad Sci U S A* **103**: 3375–3380.
9. Pils S, Schmitter T, Neske F, Hauck CR (2006) Quantification of bacterial invasion into adherent cells by flow cytometry. *J Microbiol Methods* **65**: 301–310.
10. Rambukkana A, Yamada H, Zanazzi G, Mathus T, Salzer JL, Yurchenco PD, Campbell KP, Fischetti VA (1998) Role of alpha-dystroglycan as a Schwann cell receptor for *Mycobacterium leprae*. *Science* **282**: 2076–2079.
11. Mengaud J, Ohayon H, Gounon P, Mege RM, Cossart P (1996) E-cadherin is the receptor for internalin, a surface protein required for entry of *L. monocytogenes* into epithelial cells. *Cell* **84**: 923–932.

12. Kuespert K, Pils S, Hauck CR (2006) CEACAMs—their role in physiology and pathophysiology. *Curr Opin Cell Biol* **18**: 565–571.
13. Hauck CR, Meyer TF (2003) "Small" talk: Opa proteins as mediators of *Neisseria*-host cell communication. *Curr Opin Microbiol* **6**: 43–49.
14. Popp A, Dehio C, Grunert F, Meyer TF, Gray-Owen SD (1999) Molecular analysis of neisserial Opa protein interactions with the CEA family of receptors: identification of determinants contributing to the differential specificities of binding. *Cell Microbiol* **1**: 169–181.

Part II
Fungal Pathogens

Chapter 6

Introduction: Fungal Pathogens

Steffen Rupp

Fungal infections represent today a serious health problem in industrialized countries. In particular, multimorbid patients are highly susceptible to life-threatening infections by opportunistic fungi, such as *Candida*, *Aspergillus*, or *Cryptococcus* species. In Europe, fungal infections account for 17% of intensive care unit infections. In addition, common non–life-threatening superficial infections impose significant restrictions on patients resulting in a reduced quality of life. For opportunistic pathogens, like most fungal pathogens, the interplay between both host and pathogen factors is crucial for the emergence of symptoms of disease. This is especially relevant for opportunistic pathogens that also are commensal organisms, such as *Candida albicans*, where in general a quiet colonization state is maintained. The aim of **Part II** of this volume is to provide a selection of methodologies to study the interaction of fungal pathogens with the host on a molecular level. For this purpose, a set of currently used *in vitro* and *in vivo* models are described as well as tools to identify genes, proteins, or carbohydrate structures expressed in fungi especially during infection of the host.

The cell wall of fungi is the major contact site between the host and the pathogen. Not surprisingly, cell wall components like glucans or mannoproteins play a major role in recognition of fungal pathogens by the host and therefore are important virulence factors that also influence the interchange between commensalism/saprophytism and infection. The analysis of these carbohydrate-based structures has been shown to be crucial for understanding some of the virulence properties of fungal pathogens. Wozniak and Levitz (**Chapter 7**) and Martínez-Esparza and colleagues (**Chapter 8**) respectively describe how some of these structures can be

Steffen Rupp, Kai Sohn (eds.), *Host-Pathogen Interactions*, DOI: 10.1007/978-1-59745-204-5_6,

identified or isolated from *Cryptococcus neoformans* and from *Candida albicans.*

One of the first steps of pathogens during infection of the host is their ability to attach to the surface of host tissues. This step in host-pathogen interaction in general is crucial for colonization by the pathogen and for persistence of the pathogen in the host. The development of *in vitro* models of different complexity have been a major tool to gain information about both adhesion and the first steps of invasion of fungal pathogens on a molecular level. These models have been used widely to investigate both the response of fungal pathogens to human epithelia or immune cells and the response of the host cells, which is outlined in **Part IV** of this volume.

Currently, several different models are used representing the different host niches that may be affected by fungal pathogens. Most of them are derived from cell lines that are simple to maintain and grow. However, more complex systems involving primary cells also have been used for this purpose. Ready-to-use models based on cell lines derived from oral or vaginal tissue can also be bought if only limited cell culture facilities are available. Methods regarding how to set up and use a selection of these models based on cell lines or primary cells are described in the chapters by Sohn and Rupp (**Chapter 9**), Hernandez and Rupp (**Chapter 10**), and Bourgeois and colleagues (**Chapter 11**).

The most commonly used *in vivo* model of infection is the mouse model. Several procedures have been developed to generate defined infections in the lung, the skin, vaginal and intraperitoneal areas of mice, as well as systemic infections, which are described in the chapter by Schweizer and Schröppel (**Chapter 12**).

Factors relevant to virulence of a pathogen have to be expressed in the host. Therefore, sensitive methods based on RT-PCR or antibody-mediated technologies have been developed to identify genes/proteins expressed in fungal pathogens during infection of the host. Bok and colleagues (**Chapter 13**) and Clancy and colleagues (**Chapter 14**) respectively describe the methods to identify genes/proteins expressed by *Aspergillus* species in mice or plants and by *Candida albicans* in the human host. By using mass spectrometry, even the entire set of fungal proteins leading to an immune reaction of the host (the immunome)—as accessible by proteomics—can be identified as outlined in detail by Pitarch and colleagues (**Chapter 15**).

The selection of methods presented in **Part II** can be directly applied as described. However, they also may serve to give an idea of the different possibilities for investigation of the factors that are responsible for the pathogenicity of fungal species and may serve as an entry point to choose the most appropriate technology for the question to be addressed.

Chapter 7

Isolation and Purification of Antigenic Components of *Cryptococcus*

Karen L. Wozniak and Stuart M. Levitz

Abstract

The encapsulated fungal pathogens *Cryptococcus neoformans* and *Cryptococcus gattii* are significant agents of life-threatening infections, particularly in persons with suppressed cell-mediated immunity. This chapter provides detailed methodology for the purification of two of the major antigen fractions of *C. neoformans*: glucuronoxylomannan (GXM) and mannoprotein (MP). GXM is the primary component of the polysaccharide capsule, which is the major cryptococcal virulence factor. In contrast, MPs have been identified as key antigens that stimulate T-cell responses. Purification of GXM and MP should assist investigators studying the antigenic, biochemical, and virulence properties of *Cryptococcus* species.

Key words: *Cryptococcus*, glucuronoxylomannan, mannoprotein.

1. Introduction

Cryptococcus neoformans is an opportunistic fungal pathogen that has a predilection to cause disease in persons with compromised T-cell function. Major risk factors for cryptococcosis include acquired immunodeficiency syndrome (AIDS), lymphoid malignancies, and immunosuppressive medications to prevent transplant rejection *(1–4)*. *Cryptococcus* is divided into serotypes A to D, based on the structure of its major capsular component, glucuronoxylomannan (GXM). Serotypes B and C were previously classified as *Cryptococcus neoformans* var. *gattii* and have recently been classified as a new species, *Cryptococcus gattii (5)*. *C. neoformans* can be isolated from environmental sources worldwide, particularly soil contaminated with bird excreta and rotting wood *(6)*. In contrast with *C. neoformans, C. gattii* is found mostly in tropical and subtropical regions, although Vancouver Island, Canada, has emerged

Steffen Rupp, Kai Sohn (eds.), *Host-Pathogen Interactions*, DOI: 10.1007/978-1-59745-204-5_7,

as an endemic focus *(7)*. Environmental niches of *C. gattii* include eucalyptus trees and Douglas fir trees *(8)*. For both species, infections are thought to be typically acquired after inhalation of aerosolized organisms.

The *C. neoformans* capsule has been identified as a major virulence factor of the organism. It is composed primarily of the polysaccharide GXM but also contains galactoxylomannan (GalXM). The capsule inhibits phagocytosis, and it can be shed from the cryptococcal cells into blood, cerebrospinal fluid (CSF), or infected tissues *(9–11)*. In experimental models, upon shedding of the capsule, GXM accumulates in marginal zone macrophages in the spleen *(10)* and in Kupffer cells in the liver *(9)*. In patients suffering from cryptococcosis, GXM circulates in blood and CSF at high concentrations *(11)* and can be detected for months to years after successful antifungal therapy *(12)*.

GXM has many immunomodulatory properties. It has been shown to downregulate proinflammatory cytokine secretion from human monocytes *(13)*, inhibit leukocyte migration *(13)*, impair neutrophil anticryptococcal activity *(14)*, and diminish T-cell responses *(15)*. GXM is recognized by many receptors on immune cells, including CD14, CD11/CD18, TLR2 and TLR4, and can be internalized by monocytes, neutrophils, and macrophages *(16, 17)*. Accumulation of GXM in human monocyte-derived macrophages (MDMs) results in decreased human neutrophil anticryptococcal activity *(14)* as well as modulation of MHC II and costimulatory molecule expression on MDMs *(18)*. Recent studies also show that both human and murine dendritic cells (DCs) can internalize GXM *in vitro (15, 19)*. However, soluble GXM does not impair human DC maturation when examined for the markers MHC I, MHC II, CD40, or CD86 *(19)*. GXM from all serotypes of *C. neoformans* and *C. gattii* directly inhibits T-cell proliferation, without affecting antigen presentation by DCs or macrophages, and without inhibiting T-cell activation *(15)*.

An adaptive Th1-type immune response is required for protection against cryptococcal infection *(20–24)*. The search for protective antigens began when Bennett and colleagues characterized a cryptococcal skin test antigen that caused delayed-type hypersensitivity (DTH) reactions *(25)* and continued when Murphy and colleagues isolated a crude *C. neoformans* culture supernatant (CneF) *(26)* that stimulated DTH reactions and cytokine production in mice *(27, 28)*. After separation of CneF by concanavalin A (Con A) affinity columns, the adherent mannoprotein (MP) fraction was found to be responsible for the DTH reaction *(29)*. Since that finding, both clinical and experimental studies have identified cryptococcal MPs as critical antigens responsible for stimulating T-cell responses *(30–33)*.

Clinically, MPs stimulate lymphoproliferative responses and cytokine production from patients recovered from cryptococcosis

(30, 31). Experimentally, in human monocytes and murine macrophages, MPs have been reported to induce the production of the cytokines TNF-α *(32, 34)*, IL-12, and IFN-γ *(33, 35)*, which are critical for host defenses in murine models of cryptococcosis *(36–39)*. Additionally, in a mouse model of infection, mice vaccinated with MP and then challenged with *C. neoformans* had increased survival, increased TNF-α, IFN-γ, and IL-2 in the brain, and a stronger infiltrate of immune cells into the brain, kidney, and liver compared with nonvaccinated mice *(40)*. Whereas most MPs promote proinflammatory cytokine production, one fraction, termed MP-4, has been shown to inhibit neutrophil migration, downregulate neutrophil expression of L-selectin, and desensitize neutrophils toward chemotactic factors *(41)*.

This chapter provides detailed methods for isolating GXM and MP. The protocol described for isolating GXM is applicable to all strains and serotypes of *C. neoformans* and *C. gattii*. The protocol described for isolation of MP is applicable to the MPs that are >10 kDa but can be easily modified in order to isolate lower-molecular-weight MPs. Moreover, subfractionization of the MPs can be performed to isolate specific antigenic components.

2. Materials

2.1. GXM Preparation

1. *Cryptococcus neoformans* grown on Sabouraud dextrose agar plates (Remel, Lenexa, KS).
2. 10x YNB: Yeast Nitrogen Base with amino acids (Difco, Detroit, MI) 6.7 g, dextrose (D-glucose) 5.0 g, dH_2O to 100 mL. Sterile filter using an 0.22-μm filter bottle, and store at 4°C. Before using, dilute to 1x with dH_2O from a pure water system and add penicillin (50 U/mL) and streptomycin (50 μg/mL).
3. Sodium acetate (powder) and acetic acid (to pH the solution).
4. Hexadecyltrimethyl ammonium bromide (CTAB), make a 0.3% (w/v) solution of CTAB in water at room temperature
5. Ethanol.
6. Phenol, glucose, sulfuric acid, glass test tubes, and 96-well U-bottom plates (for polysaccharide concentration assay).
7. 2 M NaCl.
8. 1 M NaCl.
9. Baked glassware, sterile latex gloves, and nanopure water.
10. Limulus amoebocyte lysate assay (Associates of Cape Cod, East Falmouth, MA).

2.2. MP Preparation

1. YNB medium (described in **Section 2.1**).
2. *Cryptococcus neoformans* acapsular strain Cap67 (ATTC no. 52817) grown on Sabouraud dextrose agar plates (Remel).
3. Filter bottles, 0.45 μm and 0.22 μm (Corning, Lowell, MA).
4. Tangential filtration cassette, 10,000 MW cutoff (Millipore, catalog no. CDUF001LG, Billerica, MA).
5. Veristaltic pump (Manostat, division of Barnant Company, Barrington, IL model 72-310-000).
6. Dulbecco's phosphate-buffered saline (DPBS) with Ca^{+2} and Mg^{+2}.
7. Concanavalin A (Con A) immobilized on 4% cross-linked Sepharose beads (Sigma, St. Louis, MO).
8. 0.2 M methyl α-D manno-pyranoside (Sigma).
9. Slide-a-lyzer dialysis cassettes (10,000 MW cutoff) (Pierce, Rockford, IL).
10. BCA protein assay kit (Pierce).
11. SDS gels (12%), Coomassie blue GelCode Blue Stain reagent (Pierce), Silver stain kit (Biorad, Hercules, CA), and Artisan Periodic Acid–Schiff (PAS) Stain kit (Dako, Carpinteria, CA).

3. Methods

GXM is the major polysaccharide that comprises the capsule of *C. neoformans*, and, as noted above, it has many immunomodulatory functions. These immunomodulatory properties make GXM a useful tool for examining immune responses. Other methods exist for isolating GXM from *C. neoformans*, but the method described herein summarizes the protocol that our laboratory routinely uses to obtain purified GXM. GXM can be isolated from all serotypes of *C. neoformans* and *C. gattii* (A to D). Differences in GXM structure (**Fig. 7.1**) affect virulence, inhibition of neutrophil migration, and tissue accumulation of GXM *(42–44)*. Because of these differences, the GXM isolated from different serotypes may have differing degrees of immunomodulation.

Mannoproteins (MPs) are major T-cell antigenic determinants isolated from *C. neoformans (29, 40)*. MPs are mannosylated proteins that contain both N- and O-linked glycans (**Fig. 7.2**) and can be recognized by mannose receptors on antigen-presenting cells, which results in efficient antigen uptake, processing, and presentation to T cells *(45)*. They are readily purified from the Cap67 acapsular mutant of *C. neoformans*, because this mutant does not have GXM on its surface to interfere with MP purification. However, other laboratories have successfully isolated MP from various other strains, including the encapsulated strains B3501 *(41)* and 184-A *(46, 47)*, as well as other nonencapsulated strains, such as strain 602 *(47)*. The

MP isolation described in this protocol is used to isolate total MP, not individual MPs. Additional purification is necessary to subfractionate MPs or to purify individual MPs. This can be accomplished using standard techniques, including size-exclusion chromatography, ion-exchange chromatography, hydrophobic interaction chromatography, and elution of bands from excised gels *(30, 41, 48)*. MPs have also been subfractionated based on the molar strength of methyl α-D manno-pyranoside required to elute it from Con A beads *(41)*. Once purified, MP (either total MP or specific MPs) can serve as components of candidate vaccines against cryptococcosis. Additionally, because MPs are effectively taken up by mannose receptors on dendritic cells and macrophages, they can be used for *in vitro*

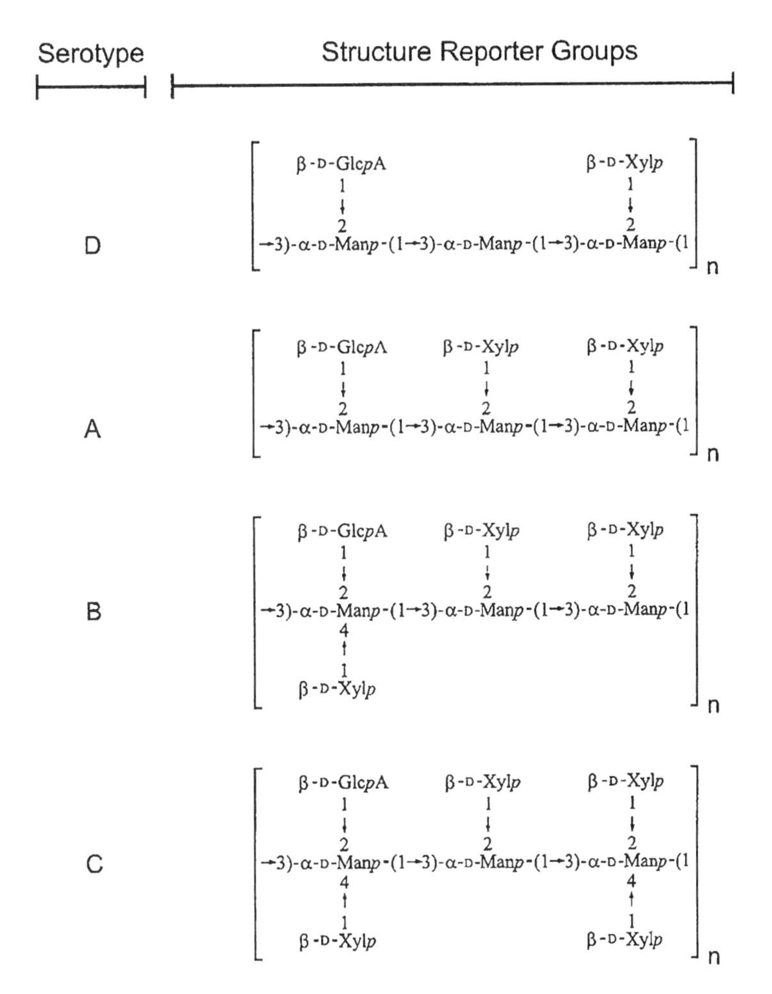

Fig. 7.1. Shown are the different structures for the glucuronoxylomannan (GXM) of *Cryptococcus* serotypes A to D. Serotypes A and D belong to *C. neoformans*, and serotypes B and C belong to *C. gattii*. The mannose backbone of GXM has variable degrees of O-acetylation (not shown). Glc*p*A, glucopyranosyluronic acid; Man*p*, mannopyranan; Xyl*p*, xylopyranosyl. (Reproduced and adapted with permission from Cherniak, R., Valafar, H., Morris, L.C., and Valafar, F. 1998. *Cryptococcus neoformans* Chemotyping by Quantitative Analysis of 1H Nuclear Magnetic Resonance Spectra of Glucuronoxylomannans with a Computer-Simulated Artificial Neural Network. *Clin. Diagn. Lab. Immunol.* **5,** 146–159.)

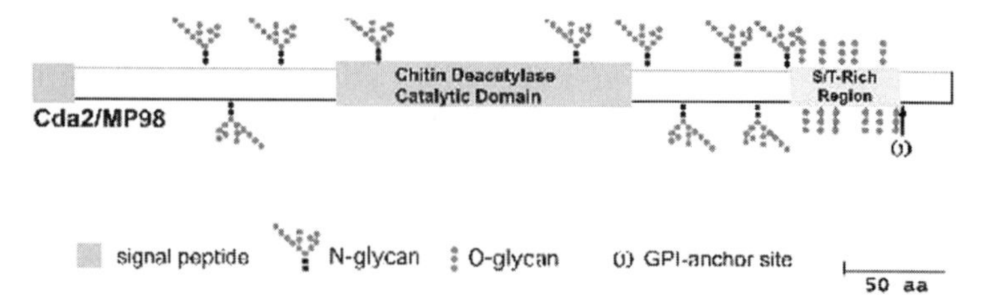

Fig. 7.2. Mannoprotein Cda1/MP-98 *Cryptococcus neoformans* is shown as a model mannoprotein. The major defining characteristics of MPs are their S/T rich region and their GPI anchor. The putative N-glycans are indicated by the branching structures, and the putative O-glycans are indicated by the straight dotted structures in the S/T rich regions. (Reproduced and adapted with permission from Levitz, S. M., and Specht, C. A. 2006. The molecular basis for the immunogenicity of *Cryptococcus neoformans* mannoproteins. *FEMS Yeast Research* **6**, 513-524.)

studies to examine cytokine production, antigen presentation, and T cell activation. Finally, the biochemical properties of MPs can be investigated by assaying for functions such as enzymatic activity.

3.1. GXM Isolation

1. Inoculate a colony of *C. neoformans* grown on a Sabouraud dextrose agar plate into 1 L of 1x YNB culture medium. Shake at 30°C overnight.
2. Spin down *C. neoformans* to pellet fungi. Discard fungal pellet and save supernatant.
3. Precipitate the polysaccharide in the supernatant by slowly adding sodium acetate (add as a powder) to make the solution 10% (w/v). Use a stir bar and adjust the pH to 7.0 with acetic acid as the sodium acetate is being added (*see* **Note 1**). Add 2.5 volumes of EtOH. As the EtOH is added, a precipitate will form on top with the appearance of a "glob of cheese."
4. Leave the solution overnight on the benchtop. By the morning, the polysaccharide will have settled to the bottom and formed a "glaze" (*see* **Note 2**). Decant the supernatant and invert the flask so the EtOH is completely removed. Air dry. Dissolve the polysaccharide in 2 to 3 mL dH_2O. This is the GXM and GalXM.
5. Measure the polysaccharide concentration by the phenol sulfuric method of Dubois *(49)*. First, make a standard curve: Add 1 g glucose to 100 mL dH_2O. Set up six tubes for the standard curve. To each tube add 2 mL dH_2O and 50 μL phenol. Add 0, 5, 10, 20, 40, or 80 μL of glucose solution to the six tubes. For the unknowns, set up tubes with 2 mL dH_2O, 50 μL phenol, and 40 μL of unknown. In a fume hood, add 5 mL sulfuric acid directly into the solution in the tubes (*see* **Notes 3** and **4**). Take 100-μL samples from the glass test tubes after 2 to 3 inversions to mix, place in a U-bottom 96-well plate, and read at 485 nm using a spectrophotometric plate reader. Compare GXM unknowns to the standard curve to determine carbohydrate content.

6. To purify the GXM, adjust the polysaccharide solution to 0.2M NaCl. Make 0.3% (w/v) solution of CTAB in water at room temperature. Add CTAB dropwise and slowly (1 mL/min) to a stirring mixture of polysaccharide. The total amount of CTAB added should be 3 times the total amount of polysaccharide (w/w). The amount of polysaccharide is calculated from the phenol sulfuric measurement (described previously). Thus, if you have 100 mg of polysaccharide, you would add 100 mL of 0.3% CTAB solution (=300 mg of CTAB). A milky precipitate will form, which is the CTAB-GXM. The GalXM should remain in solution.
7. Spin the CTAB-GXM precipitate at 10,000 × *g* for 10 min. Discard supernatant. Wash the CTAB-GXM precipitate with 10% EtOH in ddH20 (v/v). Centrifuge the sample at 10,000 × *g* for 10 min and discard the supernatant.
8. To remove the CTAB: At room temperature, dissolve the CTAB-GXM in 1 M NaCl. The solution should be clear at this point. It cannot be filtered, but it can be centrifuged to clarify if it is not too thick (*see* **Note 5**). Add EtOH dropwise with stirring (about 2 volumes EtOH will be needed). The EtOH will precipitate the GXM while the CTAB will remain in solution. Centrifuge the precipitate at 10,000 × *g* for 10 min at room temperature. Discard the supernatant (*see* **Note 6**).
9. Dissolve precipitate in 2 M NaCl (keep adding NaCl until the pellet dissolves) until a viscous solution forms. Put this viscous solution in a dialysis cassette with a 10,000 MW cutoff and dialyze overnight against 1 M NaCl. Then dialyze against dH_2O for 1 week, changing dialysate every day (*see* **Note 7**).
10. Measure the polysaccharide concentration by the method of Dubois *(49)*, as described above. The expected yield varies greatly depending upon the strain of *C. neoformans* used. Typical yields range from 5 to 200 mg per liter. Lyophilize.
11. Reconstitute the lyophilized GXM in sterile PBS (or tissue culture media) for use. Perform the Limulus amoebocyte lysate assay to determine levels of endotoxin, if desired (*see* **Note 8**).

3.2. MP Preparation

1. Inoculate a single colony of *Cryptococcus neoformans* (Cap 67) from a Sabouraud dextrose agar plate into 15 mL 1x YNB culture medium. Shake at 30°C overnight.
2. Transfer the entire culture into 20 L of 1x YNB medium (*see* **Note 9**). Shake at 30°C for 5 to 6 days.
3. Spin down the yeast cells in 500 mL centrifuge bottles at 5000 × *g* for 20 min (this will require many spins in order to spin the entire 20 L of culture).

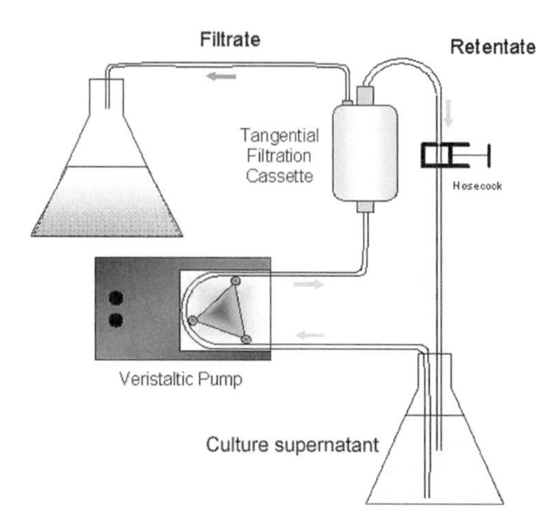

Fig. 7.3. Diagram of the mannoprotein concentration system. Culture supernatant is placed in the lower flask, pumped through a tangential filtration cassette, and concentrated before MP purification. (Figure courtesy of Dr. Michael Mansour.)

4. Filter the supernatant through an 0.45-μm filter bottle, followed by filtering through an 0.22-μm filter bottle.
5. Concentrate the supernatant with a Millipore tangential filtration cassette from 20 L to <100 mL (**Fig. 7.3**) (*see* **Note 10**). Run DPBS (including Ca^{++} and Mg^{++}) through the filtration cassette several times in order to remove any remaining YNB present in the concentrated material (*see* **Note 11**).
6. Run a gravity ConA column. For this step, the running buffer is DPBS and the elution buffer is DPBS with 0.2 M methyl α-D manno-pyranoside (*see* **Note 12**). Collect the eluate in fractions.
7. Take aliquots of the eluate fractions and measure protein using the BCA protein assay (*see* **Note 13**). To measure carbohydrates, perform the DuBois assay (described in **Section 3.1, step 5**).
8. After the protein assay, perform dialysis on the eluate fractions containing protein using a Slide-A-Lyzer dialysis cassette (10,000 MW cutoff), and dialyze against dH_2O at 4°C for 2 days. The dH_2O should be replaced at least 4 times.
9. To analyze MP, run three 12% SDS gels with 20 to 30 μg/well of the purified MP. Stain one gel each with Coomassie blue (to visualize protein), silver (to visualize protein) (*see* **Note 14**), and periodic acid–Schiff (PAS; to visualize carbohydrates), according to the manufacturer's instructions (**Fig. 7.4**).
10. Filter sterilize the dialyzed sample with an 0.22-μm filter and freeze at −80°C in a dry ice–alcohol bath.

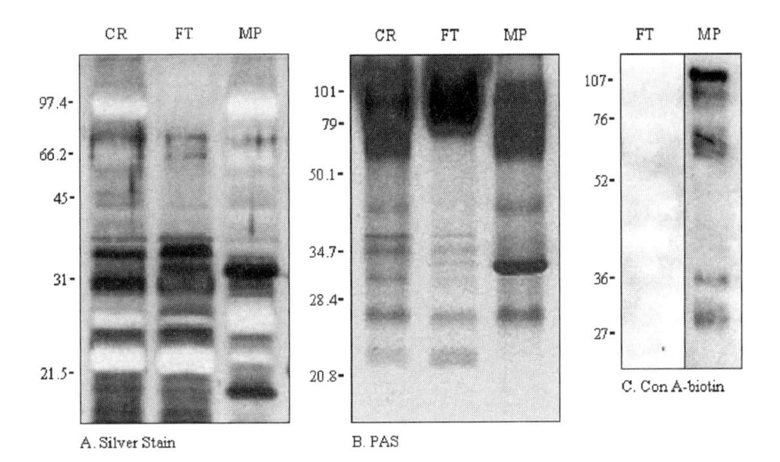

Fig. 7.4. Analysis of *C. neoformans* strain CAP67 supernatant fractions by SDS-PAGE. Crude (CR), flow-through (FT), and MP fractions were resolved by 12% SDS-PAGE and analyzed by (**A**) silver stain, (**B**) PAS, and (**C**) Con A–biotin blot. Migration of commercial molecular mass standards, expressed in kilodaltons, is indicated to the left of the gels. (Reproduced and adapted with permission from Mansour, M. K., Schlesinger, L.S., and Levitz, S. M. 2002. Optimal T cell responses to *Cryptococcus neoformans* mannoprotein are dependent on recognition of conjugated carbohydrates by mannose receptors. *Journal of Immunology* **168**, 2872–2879.)

11. Lyophilize and store the sample at −80°C. Reconstitute the MP in PBS before use. Perform another BCA protein assay on the reconstituted MP to confirm the concentration (*see* **Note 15**).

4. Notes

1. Make sure to monitor and adjust the pH during the precipitation. GXM has varying degrees of O-acetylation, and if the pH is too high, it will destroy the acetyl groups.
2. For some *C. neoformans* strains, the capsular polysaccharide may stay on the top of the solution rather than settle to the bottom.
3. It is important to add the sulfuric acid directly into the solution and not down the sides of the tubes.
4. The reaction is strongly exothermic, and the tubes should be cooled for 30 to 60 min in a water bath at 23°C. The colorimetric reaction is directly proportional to the concentration of carbohydrate.
5. If the NaCl solution is very thick, it is hard to centrifuge, and more NaCl should be added.
6. **Steps 6,** 7, and **8** have to be done at room temperature otherwise the CTAB will precipitate. Make sure the centrifuge is room temperature before adding sample!

7. If the solution becomes cloudy, then the CTAB has not been adequately removed, and the sample needs to be re-dialyzed against NaCl.
8. If the GXM will be used for immunologic experiments, it is important to minimize endotoxin contamination. For all steps, meticulously avoid the introduction of endotoxin by wearing gloves, baking glassware, using endotoxin-free water, and so forth. Our laboratory is able to routinely purify endotoxin-free GXM (<0.03 endotoxin U/mL).
9. The reaction can be scaled up or down depending upon the desired amount of MP.
10. The setup of this device is important for correctly concentrating MP. Incorrect setup can lead to loss of the MP.
11. This step uses a 10-kDa MW cutoff for concentrating the supernatant. This concentration step will lead to loss of lower MW MPs. In order to retain the smaller MPs, a 3.5-kDa MW cutoff filter may be used *(41)*.
12. If retention of the MP-4 fraction is desired, elute MP with 0.4 M methyl α-D manno-pyranoside instead of 0.2 M. Additionally, the flow-through (FT) fraction (which contains non-mannoprotein antigens) can be collected.
13. Perform the BCA protein assay on the eluate fractions prior to dialysis in order to determine which fractions contain MP (these will have measurable protein by the BCA assay). Alternatively, the MP-containing fractions can be determined using an ultraviolet monitor. Then use the MP-containing fractions for the following steps.
14. Coomassie blue is preferentially used to stain the protein portion of MP instead of the normally more sensitive silver staining. MPs that are extensively glycosylated generally will not stain and often appear as a negative band after silver staining (**Fig. 7.4**).
15. A small amount of Con A will inevitably leach off of the affinity column. Because Con A is a potent mitogen, this contamination may have untoward effects in immunologic assays. To biologically inactivate Con A, MP can be boiled for 5 min prior to use. This boiling step will not destroy the antigenic activity of MP.

Acknowledgments

The protocols were developed with the help of Drs. Arturo Casadevall, Robert Cherniak, Michael Mansour, Shuhua Nong, and Lauren Yauch. The research was supported in part by National Institutes of Health grants R01 AI25780 and R01 AI066087.

References

1. Levitz, S. M. (1991) The ecology of *Cryptococcus neoformans* and the epidemiology of cryptococcosis. *Reviews of Infectious Diseases* 13, 1163–1169.
2. Mitchell, T. G. & Perfect, J. R. (1995) Cryptococcosis in the era of AIDS - 100 years after the discovery of *Cryptococcus neoformans*. *Clinical Microbiology Reviews* 8, 515–548.
3. Shoham, S. & Levitz, S. M. (2005) The immune response to fungal infections. *British Journal of Haematology* 129, 569–582.
4. Singh, N., Gayowski, T., Wagener, M. M. & Marino, I. R. (1997) Clinical spectrum of invasive cryptococcosis in liver transplant recipients receiving tacrolimus. *Clinical Transplantation* 11, 66–70.
5. Kwon-Chung, K. J. & Varma, A. (2006) Do major species concepts support one, two or more species within Cryptococcus neoformans? *FEMS Yeast Research* 6, 574–587.
6. Ellis, D. H. & Pfeiffer, T. (1990) Ecology, life cycle, and infectious propagule of Cryptococcus neoformans. *The Lancet* 336, 923–925.
7. Kidd, S. E., Hagen, F., Tscharke, R. L., Huynh, M., Bartlett, K. H., Fyfe, M., MacDougall, L., Boekhout, T., Kwon-Chung, K. J. & Meyer, W. (2004) A rare genotype of Cryptococcus gattii caused the cryptococcosis outbreak on Vancouver Island (British Columbia, Canada). *Proceedings of the National Academy of Sciences of the United States of America* 101, 17258–17263.
8. Duncan, C., Schwantje, H., Stephen, C., Campbell, J. & Bartlett, K. (2006) Cryptococcus gattii in Wildlife of Vancouver Island, British Columbia, Canada. *Journal of Wildlife Diseases* 42, 175–178.
9. Grinsell, M., Weinhold, L. C., Cutler, J. E., Han, Y. M. & Kozel, T. R. (2001) In vivo clearance of glucuronoxylomannan, the major capsular polysaccharide of Cryptococcus neoformans: A critical role for tissue macrophages. *Journal of Infectious Diseases* 184, 479–487.
10. Goldman, D. L., Lee, S. C. & Casadevall, A. (1995) Tissue localization of Cryptococcus-neoformans glucuronoxylomannan in the presence and absence of specific antibody. *Infection and Immunity* 63, 3448–3453.
11. Chuck, S. L. & Sande, M. A. (1989) Infections with Cryptococcus-neoformans in the acquired immunodeficiency syndrome. *New England Journal of Medicine* 321, 794–799.
12. Lu, H., Zhou, Y., Yin, Y., Pan, X. & Weng, X. (2005) Cryptococcal antigen test revisited: significance for cryptococcal meningitis therapy monitoring in a tertiary Chinese hospital. *Journal of Clinical Microbiology* 43, 2989–2990.
13. Ellerbroek, P. M., Walenkamp, A.M., Hoepelman, A.I. & Coenjaerts, F.E. (2004) Effects of the capsular polysaccharides of Cryptococcus neoformans on phagocyte migration and inflammatory mediators. *Current Medical Chemistry* 11, 253–266.
14. Monari, C., Retini, C., Casadevall, A., Netski, D., Bistoni, F., Kozel, T. R. & Vecchiarelli, A. (2003) Differences in outcome of the interaction between Cryptococcus neoformans glucuronoxylomannan and human monocytes and neutrophils. *European Journal of Immunology* 33, 1041–1051.
15. Yauch, L. E., Lam, J. S. & Levitz, S. M. (2006) Direct inhibition of T-cell responses by the cryptococcus capsular polysaccharide glucuronoxylomannan. *PLoS Pathogens* 2, e120.
16. Dong, Z. M. & Murphy, J. W. (1997) Cryptococcal polysaccharides bind to CD18 on human neutrophils. *Infection and Immunity* 65, 557–563.
17. Shoham, S., Huang, C., Chen, J. M., Golenbock, D. T. & Levitz, S. M. (2001) Toll-like receptor 4 mediates intracellular signaling without TNF-alpha release in response to *Cryptococcus neoformans* polysaccharide capsule. *Journal of Immunology* 166, 4620–4626.
18. Monari, C., Pericolini, E., Bistoni, G., Casadevall, A., Kozel, T. R. & Vecchiarelli, A. (2005) *Cryptococcus neoformans* capsular glucuronoxylomannan induces expression of fas ligand in macrophages. *Journal of Immunology* 174, 3461–3468.
19. Vecchiarelli, A., Pietrella, D., Lupo, P., Bistoni, F., McFadden, D. C. & Casadevall, A. (2003) The polysaccharide capsule of *Cryptococcus neoformans* interferes with human dendritic cell maturation and activation. *Journal of Leukocyte Biology* 74, 370–378.

20. Hill, J. O. & Harmsen, A. G. (1991) Intrapulmonary growth and dissemination of an avirulent strain of *Cryptococcus neoformans* in mice depleted of CD4+ or CD8+ T-cells. *Journal of Experimental Medicine* 173, 755–758.
21. Huffnagle, G. B., Lipscomb, M. F., Lovchik, J. A., Hoag, K. A. & Street, N. E. (1994) The role of CD4(+) and CD8(+) T-cells in the protective inflammatory response to a pulmonary cryptococcal infection. *Journal of Leukocyte Biology* 55, 35–42.
22. Huffnagle, G. B., Yates, J. L. & Lipscomb, M. F. (1991) Immunity to a pulmonary *Cryptococcus neoformans* infection requires both CD4+ and CD8+ T-cells. *Journal of Experimental Medicine* 173, 793–800.
23. Huffnagle, G. B., Yates, J. L. & Lipscomb, M. F. (1991) T-cell-mediated immunity in the lung - a *Cryptococcus neoformans* pulmonary infection model using Scid and athymic nude-mice. *Infection and Immunity* 59, 1423–1433.
24. Mody, C. H., Lipscomb, M. F., Street, N. E. & Toews, G. B. (1990) Depletion of CD4+ (L3T4+) lymphocytes in vivo impairs murine host defense to *Cryptococcus neoformans. Journal of Immunology* 144, 1472–1477.
25. Bennett, J. E. (1981) Cryptococcal skin test antigen: preparation variables and characterization. *Infection and Immunity* 32, 373–380.
26. Murphy, J. W. & Cozad, G. C. (1972) Immunological unresponsiveness induced by cryptococcal capsular polysaccharide assayed by hemolytic plaque technique. *Infection and Immunity* 5, 896.
27. Murphy, J. W. (1998) Protective cell-mediated immunity against *Cryptococcus neoformans. Research in Immunology* 149, 373–386.
28. Murphy, J. W. & Moorhead, J. W. (1982) Regulation of cell-mediated immunity in cryptococcosis. I. Induction of specific afferent T suppressor cells by cryptococcal antigen. *Journal of Immunology* 128, 276–283.
29. Murphy, J. W. (1988) Influence of cryptococcal antigens on cell-mediated-immunity. *Reviews of Infectious Diseases* 10, S432–S435.
30. Levitz, S. M., Nong, S. H., Mansour, M. K., Huang, C. & Specht, C. A. (2001) Molecular characterization of a mannoprotein with homology to chitin deacetylases that stimulates T cell responses to *Cryptococcus neoformans. Proceedings of the National Academy of Sciences of the United States of America* 98, 10422–10427.
31. Hoy, J. F., Murphy, J. W. & Miller, G. G. (1989) T-cell response to soluble cryptococcal antigens after recovery from cryptococcal infection. *Journal of Infectious Diseases* 159, 116–119.
32. Chaka, W., Verheul, A. F., Vaishnav, V. V., Cherniak, R., Scharringa, J., Verhoef, J., Snippe, H. & Hoepelman, A. I. (1997) Induction of TNF-alpha in human peripheral blood mononuclear cells by the mannoprotein of Cryptococcus neoformans involves human mannose binding protein. *Journal of Immunology* 159, 2979–2985.
33. Pietrella, D., Cherniak, R., Strappini, C., Perito, S., Mosci, P., Bistoni, F. & Vecchiarelli, A. (2001) Role of mannoprotein in induction and regulation of immunity to Cryptococcus neoformans. *Infection and Immunity* 69, 2808–2814.
34. Chaka, W., Verheul, A. F., Vaishnav, V. V., Cherniak, R., Scharringa, J., Verhoef, J., Snippe, H. & Hoepelman, I. M. (1997) Cryptococcus neoformans and cryptococcal glucuronoxylomannan, galactoxylomannan, and mannoprotein induce different levels of tumor necrosis factor alpha in human peripheral blood mononuclear cells. *Infection and Immunity* 65, 272–278.
35. Pitzurra, L., Cherniak, R., Giammarioli, M., Perito, S., Bistoni, F. & Vecchiarelli, A. (2000) Early induction of interleukin-12 by human monocytes exposed to Cryptococcus neoformans mannoproteins. *Infection and Immunity* 68, 558–563.
36. Hoag, K. A., Lipscomb, M. F., Izzo, A. A. & Street, N. E. (1997) IL-12 and IFN-gamma are required for initiating the protective Th1 response to pulmonary cryptococcosis in resistant C.B-17 mice. *American Journal of Respiratory Cell and Molecular Biology* 17, 733–739.
37. Kawakami, K., Qifeng, X., Tohyama, M., Qureshi, M. H. & Saito, A. (1996) Contribution of tumour necrosis factor-alpha (TNF-alpha) in host defence mechanism against Cryptococcus neoformans. *Clinical and Experimental Immunology* 106, 468–474.
38. Kawakami, K., Tohyama, M., Teruya, K., Kudeken, N., Xie, Q. F. & Saito, A. (1996) Contribution of interferon-gamma

in protecting mice during pulmonary and disseminated infection with Cryptococcus neoformans. *FEMS Immunology and Medical Microbiology* 13, 123–130.

39. Kawakami, K., Tohyama, M., Xie, Q. & Saito, A. (1996) IL-12 protects mice against pulmonary and disseminated infection caused by Cryptococcus neoformans. *Clinical and Experimental Immunology* 104, 208–214.
40. Mansour, M. K., Yauch, L. E., Rottman, J. B. & Levitz, S. M. (2004) Protective efficacy of antigenic fractions in mouse models of cryptococcosis. *Infection and Immunity* 72, 1746–1754.
41. Coenjaerts, F. E. J., Walenkamp, A. M. E., Mwinzi, P. N., Scharringa, J., Dekker, H. A. T., van Strijp, J. A. G., Cherniak, R. & Hoepelman, A. I. M. (2001) Potent inhibition of neutrophil migration by cryptococcal mannoprotein-4-induced desensitization. *Journal of Immunology* 167, 3988–3995.
42. Kozel, T. R., Levitz, S. M., Dromer, F., Gates, M. A., Thorkildson, P. & Janbon, G. (2003) Antigenic and biological characteristics of mutant strains of Cryptococcus neoformans lacking capsular O acetylation or xylosyl side chains. *Infection and Immunity* 71, 2868–2875.
43. Ellerbroek, P. M., Lefeber, D. J., van Veghel, R., Scharringa, J., Brouwer, E., Gerwig, G. J., Janbon, G., Hoepelman, A. I. M. & Coenjaerts, F. E. J. (2004) O-Acetylation of cryptococcal capsular glucuronoxylomannan is essential for interference with neutrophil migration. *Journal of Immunology* 173, 7513–7520.
44. Janbon, G., Himmelreich, U., Moyrand, F., Improvisi, L. & Dromer, F. (2001) Cas1p is a membrane protein necessary for the O-acetylation of the Cryptococcus neoformans capsular polysaccharide. *Molecular Microbiology* 42, 453–467.
45. Mansour, M. K., E. Latz, and S. M. Levitz (2006) *Cryptococcus neoformans* glycoantigens are captured by multiple lectin receptors and presented by dendritic cells. *Journal of Immunology* 176, 3053–3061.
46. Buchanan, K. L. & Murphy, J. W. (1993) Characterization of cellular infiltrates and cytokine production during the expression phase of the anticryptococcal delayed-type hypersensitivity response. *Infection and Immunity* 61, 2854–2865.
47. Dong, Z. M. & Murphy, J. W. (1993) Mobility of human neutrophils in response to Cryptococcus neoformans cells, culture filtrate antigen, and individual components of the antigen. *Infection and Immunity* 61, 5067–5077.
48. Huang, C., Nong, S. H., Mansour, M. K., Specht, C. A. & Levitz, S. M. (2002) Purification and characterization of a second immunoreactive mannoprotein from *Cryptococcus neoformans* that stimulates T-cell responses. *Infection and Immunity* 70, 5485–5493.
49. Dubois, M., K. Gilles, J. K. Hamilton, P. A. Rebers, F. Smith (1951) A colorimetric method for the determination of sugars. *Nature* 168, 167.

Chapter 8

A Method for Examining Glycans Surface Expression of Yeasts by Flow Cytometry

Maria Martínez-Esparza, Aurore Sarazin, Daniel Poulain, and Thierry Jouault

Abstract

Recognition of pathogenic yeasts by host cells is based on components of the yeast cell wall, which are considered part of its virulence attributes. Cell wall glycans play an important role in the continuous interchange that regulates the balance between saprophytism and parasitism and between resistance and infection. Flow cytometry is a useful method for probing surface yeast glycans in order to compare their expression depending on strains and growth conditions. By using different monoclonal or polyclonal antibodies, levels of β- and α-linked mannosides as well as β-glucans can be successfully evaluated by flow cytometry methods. The cytometric method we describe here represents a useful tool to investigate to what extent yeasts are able to regulate their glycan surface expression and therefore modify their virulence properties.

Key words: *Candida albicans*; surface markers; glycans; cell wall.

1. Introduction

Interactions between yeasts and host cells involve the binding of target ligands expressed at the yeast cell surface to receptors present in the host cell membrane. Glycans play an important role in the continuous interchange that regulates the balance between saprophytism and parasitism *(1)*. The outermost layers of the cell wall are made of phosphopeptidomannan (**Fig. 8.1**), a polymer of O- and N-linked mannose residues commonly referred to as mannan. The O-linked mannose residues consist of short chains of α-1,2- and α-1,3-linked mannose. The N-linked part consists of a backbone of α-1,6-linked mannopyranose residues with branches composed of α-1,2- and α-1,3-linked mannopyranose

Steffen Rupp, Kai Sohn (eds.), *Host-Pathogen Interactions*, DOI: 10.1007/978-1-59745-204-5_8,

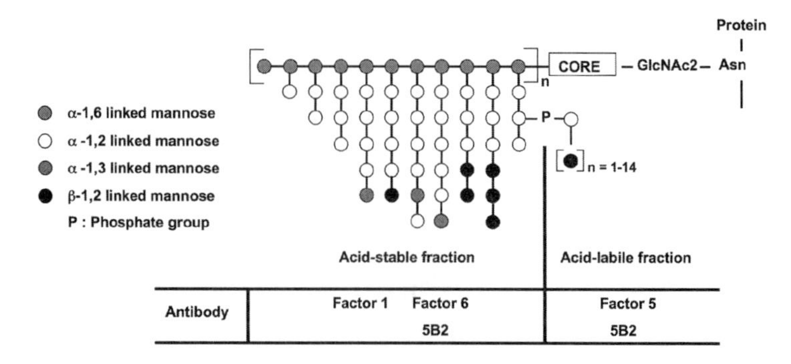

Fig. 8.1. Scheme of *C. albicans N*-linked mannan. The Man core group is attached to protein through two *N*-acetyl-glucosamine residues ($GlcNAc_2$). The acid-stable fraction, recognized by factor 1, consists of the outer chain with α-1,2- and α-1,3-linked oligomannoside branches. Some β-1,2-mannosides linked to the α-1,2-mannosides are recognized by factor 6 and mAb 5B2. Acid-labile fraction made of β-1,2-linked oligomannosides recognized by factor 5 and mAb 5B2 is linked to the acid-stable fraction by phosphodiester bond.

units. N-linked mannan can be resolved by mild acid hydrolysis into acid-stable fraction, consisting of all proximal mannose residues, and acid-labile fraction, which corresponds in *Candida* with phosphomannan made of β-1,2-linked mannose chains *(2)*. In *Candida albicans*, additional linked β-1,2 mannosides are present as terminal mannosides of acid-stable fraction side chains and constitute the *C. albicans* serotype A *(3)*. The serologic classification of Tsuchiya et al. *(4)*, resulting from the serologic study on a variety of yeasts based on their agglutination abilities with a series of rabbit polyclonal antibodies (Iatron kit), has led to description of 10 antigenic factors more or less specific for the different strains of *Candida*. These factors have been used as tool to identify medically important *Candida* species in clinical specimens. Immunochemical investigations, carried out together with structural analysis of mannan from different strains of the genus *Candida*, have resulted in the definition of the epitopic structure of those antigenic factors *(5)*. Among them, antigenic factor 5 consisted in the homopolymers of β-1,2-linked mannose present in the acid-labile fraction of *C. albicans* mannan. Antigenic factor 6 corresponds with the β-linked mannose residues at the nonreducing end of the α-1,2-linked lateral chains specific for the mannan acid-stable fraction of *C. albicans* serotype A. The more ubiquitous antigenic factor 1 consists of α-1,2-linked mannosides present both in the O-linked and N-linked chains of *C. albicans* and *Saccharomyces cerevisiae* (for a description of the expression and accessibility of these antigenic factors, *see* **Figs. 8.2** and **8.3**).

The yeast cell wall is also composed of β-1,3- or β-1,6-linked glucans, which are expressed more deeply in the cell wall *(6)* and are poorly accessible at the surface. They can be exposed after

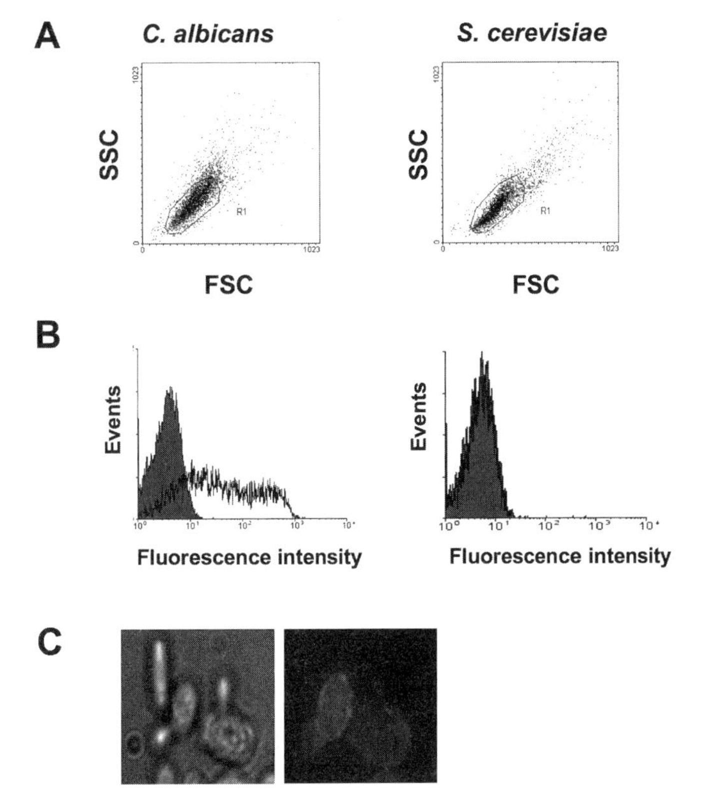

Fig. 8.2. Representative results obtained after staining of *C. albicans* and *S. cerevisiae* yeasts with 5B2 monoclonal antibody. **(A)** Analysis of yeast morphology by flow cytometry. Histogram plots the cell size (SSC) on the vertical axis against cellular complexity (FSC). The gate used for acquisition and data analysis is shown. **(B)** Histogram of fluorescence intensity obtained after staining of both yeasts with mAb 5B2 (unfilled area) compared with controls (filled area). **(C)** A detail of *C. albicans* yeast cells stained on slides with 5B2 mAb and imaged by fluorescence microscopy.

heating of the yeast *(7, 8)* (an example of such treatment and the resulting accessibility of these epitopes is shown in **Fig. 8.4**). They are nonetheless accessible during budding and at the levels of birth scars of yeast, where mother-daughter cell separation occurs *(9)*.

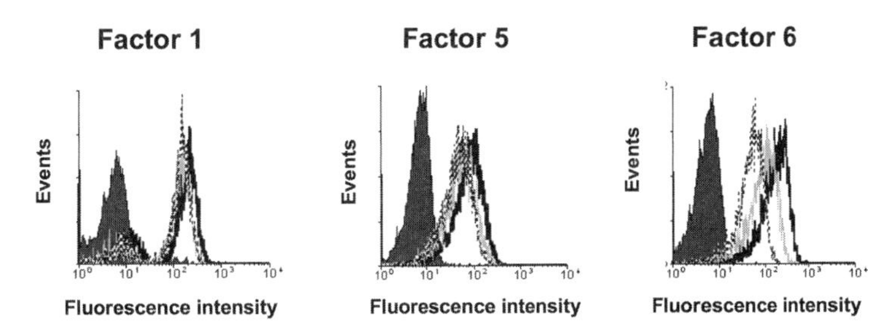

Fig. 8.3. Expression of antigenic factors on *C. albicans* cell wall surface. *C. albicans* strain was probed with 1/100 (solid line), 1/200 (thin line), or 1/300 (dashed line) antibody against factors 1, 5, and 6 and compared with controls (filled line) by flow cytometry.

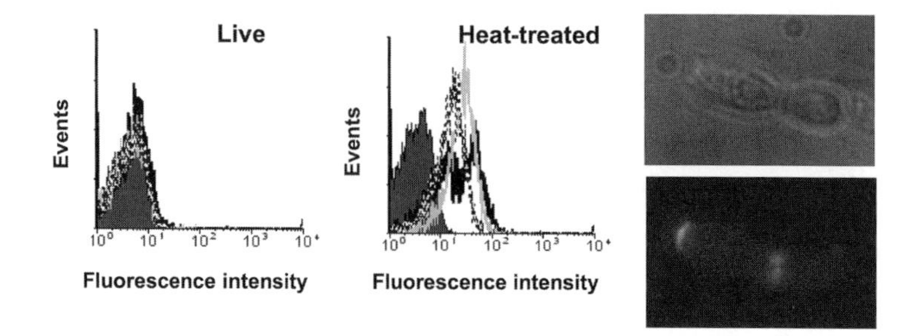

Fig. 8.4. Effect of heat treatment on β-glucan surface accessibility. Live or heat-killed *C. albicans* cells were probed with 20 μg/mL (solid line), 10 μg/mL (thin line), or 4 μg/mL (dashed line) of anti-β-1,3-glucan mAb and compared with unstained control (filled) by flow cytometry. On the right is shown the microscopic examination of β-1,3-glucan distribution on live *C. albicans* yeast after staining with the mAb.

The most representative macrophage receptor for the α-linked mannosides is the mannose receptor *(10)*. It is mainly involved in the uptake and phagocytosis of microbes presenting these mannosides among which are the yeasts. DC-SIGN [dendritic cell-specific ICAM (intercellular adhesion molecule)-grabbing nonintegrin] has also been shown to be involved in recognition of yeasts through α-mannosides *(11)*, but it is poorly phagocytic. Recently, the participation of dectin-2 in the recognition of α-mannosides has been evidenced *(12, 13)*. In contrast with α-linked mannosides, recognition of β-1,2-mannosides by macrophages involves a non–C lectin, galectin-3 *(14, 15)*. Whereas α-mannosides induced cytokine production through a TLR-4-dependent pathway *(16)*, β-1,2-oligmanosides stimulate proinflammatory cytokine production by cells in a TLR-2–dependent manner in association with galectin-3 *(7)*. β-linked glucans are recognized by dectin-1, which mediates binding to macrophage membrane and initiates inflammatory responses in a TLR-2–dependent manner *(9, 17)*. Recognition of yeast cell wall glycans by immune cells through these different receptors and lectins is followed by the induction of a variety of immunologic mechanisms *(18)*. Thus, identification of antigens expressed by the yeast that may interact with macrophage membrane receptors is of great importance for understanding the host-yeast interaction.

Among the different methods for such investigation, flow cytometry represents an easy way for exploring modulation and relative expression of the different antigens expressed at the cell surface. In parallel, genetic approaches have led to the availability of different mutants in glycosyl transferases that differently express such epitopes *(19)*. A large range of antibodies, either monoclonal or polyclonal, specific for some important *C. albicans* epitopes as well as specific lectins, are available. When involved in flow cytometry, they provide an easy way for exploring expression of glycans known to interact with innate immune receptors.

2. Materials

2.1. Yeast Cultures

1. The reference strains *C. albicans* (serotype A) SC5314 and *S. cerevisiae* S288C were used throughout this study.
2. YPD (1% yeast extract, 2% peptone, 2% dextrose) and YPD-agar medium (*see* **Note 1**).
3. Incubator with shaker.
4. Erlenmeyer flasks and Petri plates.
5. Spectrophotometer with 600-nm filter.

2.2. Antibodies (see Note 2)

1. The anti-β-1,2-mannoside monoclonal antibody 5B2 is a rat IgM and has been developed in our laboratory. The anti-β-1,3-glucan, a mouse monoclonal IgG, was provided by Biosupplies (Parkville, Australia).
2. Rabbit polyclonal antibodies against antigenic factors 1, 5, and 6 were from Iatron (Tokyo, Japan).
3. Fluorescein isothiocyanate (FITC)- and phycoerythrin (PE)-conjugated anti-rat IgM, anti-mouse IgG, or anti-rabbit IgG were obtained from Southern Biotechnology Laboratories (Birmingham, AL).

2.3. Flow Cytometry

1. Phosphate-buffered saline (PBS): Prepare 10X stock with 1.37 M NaCl, 27 mM KCl, 100 mM Na_2HPO_4, 18 mM KH_2PO_4 (adjust to pH 7.4 with HCl if necessary), and autoclave before storage at room temperature. Prepare working solution by dilution of one part with nine parts water.
2. Paraformaldehyde 0.4%. Use paraformaldehyde either diluted from commercial stock solution (37%) or prepare 0.4% (w/v) solution in PBS fresh for each experiment. In this case, the solution may need to be carefully heated (use a stirring hot-plate in a fume hood) to dissolve, and then cool to room temperature for use.
3. Fetal calf serum (FCS) is inactivated by heating at 56°C for 30 min.
4. Antibody dilution buffer and washing buffer: PBS containing 2% inactivated FCS.

3. Methods

Although based on similar technical approaches, flow cytometry and immunofluorescence staining bring complementary results. For each experiment, a microscopic examination should be performed before cytometry analysis. This allows one to verify the morphology and integrity of yeast cells and to determine the antigen distribution among individual cells, which is not possible with flow cytometry. We thus describe both methods giving examples of results obtained with both in order to allow comparison.

3.1. Preparation of Yeasts

1. Yeast cells are maintained at 4°C in YPD-agar medium. Before the experiments, cells are transferred onto fresh YPD and incubated in an Erlenmeyer flask by shaking at 37°C. Cells are harvested in exponential (OD_{600} = 0.2 to 0.6, 3 to 4 h in culture) or stationary phase of growth (OD_{600} = 1.5 to 3, 14 to 16 h in culture).
2. Cells are washed once with PBS.
3. To estimate the concentration of yeast, the optical density of the yeast cultures are measured in a spectrophotometer at 600 nm or number of yeast cells are counted in a Neubauer hemacytometer chamber (1 OD is equivalent to 2×10^7 yeast cells); 10^6 blastoconidia cells are used per sample.
4. To obtain heat-killed yeasts, cells are warmed for 20 min at 90°C and then washed once in PBS.

3.2. Flow Cytometry Assay

1. Yeast cells are washed with PBS.
2. Samples are incubated at 4°C for 15 min with 20 μL of first antibody in Eppendorf tubes (*see* **Note 3**) or with the same volume of antibody dilution buffer for negative controls (*see* **Notes 4** and **5**).
3. The primary antibody is washed twice with 500 μL of washing solution. The washings consisted in resuspending the cells in washing solution, centrifugation at 1000 × *g* at 4°C for 6 min, and discarding the supernatant.
4. The secondary antibody (20 μL) is incubated at 4°C for 15 min in Eppendorf tubes for all samples (*see* **Note 6**).
5. The secondary antibody is washed twice with 500 μL of washing solution and once in PBS, as described previously (*see* **Note 7**).
6. After washing, cells are fixed with 0.4% paraformaldehyde and examined at the moment by fluorescence-activated cell sorter (FACS). The samples could also be maintained at 4°C up to 24 h before examination (*see* **Note 8**).

3.3. Flow Cytometry Acquisition and Analysis

1. Flow cytometry is performed using an EPICS XLMCL4 (Beckman Coulter, High Wycombe, UK) equipped with an argon ion laser with an excitation power of 15 mW at 488 nm.
2. Forward scatter (FSC) and side scatter (SSC) are analyzed on linear scales, whereas green (FL1) and red fluorescence intensity (FL2) are analyzed on logarithmic scales.
3. Analysis gates are set around debris and intact cells on an FSC versus SSC dot plot. The fluorescence histograms of 5000 cells are generated using the gated data.
4. Data analysis is performed using WINMDI software (available from http://facs.scripps.edu) (*see* **Note 9**).

3.4. Microscopic Examination and Immunofluorescent Staining

1. Yeast cells are plated onto 10-well microscopy slides (10^6 yeast in 50 μL per well) and allowed to dry overnight under hood at 20°C.
2. Dried preparation is delicately washed with 50 μL of PBS per well and then aspirated dry.
3. Fifty microliters per well of PBS containing 1% BSA is added for blocking for 20 min at room temperature and then aspirated.
4. Cells are then incubated for 1 h at 37°C with 50 μL of the primary antibody (generally 1/50 dilution) in PBS–1% BSA.
5. Unbound antibody is washed out by addition of 50 μL of PBS–1% BSA and aspiration. This step is performed twice.
6. Fluorescent secondary antibody is then incubated for 1 h at room temperature. Generally, 50 μL of 1/50 dilution of the corresponding FITC-conjugated secondary antibody is used.
7. Wells are then washed 5 times as above and allowed to dry.
8. The samples are then ready to be mounted. Ten microliters of mounting medium (such as Moviol) is then added to each well and recovered with a coverslip and sealed with nail polish (*see* **Note 10**). The sample can be examined immediately after the polish is dried or can be stored in the dark at 4°C for subsequent examination.
9. The slides are viewed under phase contrast microscopy (to locate the cells and identify the focal plane) and under fluorescence microscopy.

4. Notes

1. YPD medium should be supplemented when mutant strains are used.
2. Primary antibodies may be replaced by specific lectins conjugated either directly with fluorophores or biotinylated (both

are available). In this case, the use of conjugated streptavidin will allow staining.

3. Flow cytometric assay could be performed in 96 V-bottom well plates instead of Eppendorf tubes. In this case, the washings consist in resuspending the cells with 180 μL of washing solution, centrifugation of the plates at 800 × *g* at 4°C for 6 min, and discarding the supernatant.
4. Homogenous labeling of yeasts with antibodies requires that yeasts are carefully mixed with antibody. This is performed by pipetting the suspension.
5. Incubation steps should be performed under agitation. If using microtiter plates, this requires the use of a special apparatus.
6. All steps with conjugated secondary antibodies should be performed with a cover to avoid light.
7. Double staining may be realized for subsequent analysis either by flow cytometry or by microscopy. For this, if primary antibodies used are not conjugated, it is necessary that they have been obtained from two different species (mice and rat, for example) and to use different corresponding secondary antibodies (PE conjugated anti-mouse and FITC conjugated anti-rat antibodies, for example).
8. Fluorescence is generally stable. However, analysis at the end of the experiment always gives optimal results. If not possible, samples in PFA can be stored at 4°C in the dark for 24 h.
9. After data acquisition, mean fluorescence intensities are obtained by subtracting values for negative controls from the values given by each antigen.
10. Air bubbles are undesirable in the mounting medium, and slow, careful application of the top layer minimizes their appearance.

Acknowledgments

This work was supported in part by the European Community (Feder fund, program Interreg IIIA). María Martínez-Esparza was supported by the Fundación Séneca, Spain. Aurore Sarazin was supported by Region Nord-Pas de Calais.

References

1. Poulain, D., and Jouault, T. (2004) *Candida albicans* cell wall glycans, host receptors and responses: elements for a decisive crosstalk. *Curr Opin Microbiol* 7, 342–349.
2. Shibata, N., Kobayashi, H., Takahashi, S., Okawa, Y., Hisamichi, K., Suzuki, S., and Suzuki, S. (1991) Structural study on a phosphorylated mannotetraose obtained from the phosphomannan of Candida albicans NIH B-792 strain by acetolysis. *Arch Biochem Biophys* **290**, 535–542.
3. Kobayashi, H., Shibata, N., Osaka, T., Miyagawa, Y., Ohkubo, Y., and Suzuki, S.

(1992) Structural study of cell wall mannan of a *Candida albicans* (serotype A) strain. *Phytochemistry* **31**, 1147–1153.

4. Tsuchiya, T., Fukazawa, Y., Taguchi, M., Nakase, T., and Shinoda, T. (1974) Serologic aspects on yeast classification. *Mycopathol Mycol Appl* **53**, 77–91.
5. Suzuki, S. (1997) Immunochemical study on mannans of genus *Candida*. I. Structural investigation of antigenic factors 1, 4, 5, 6, 8, 9, 11, 13, 13b and 34. *Curr Top Med Mycol* **8**, 57–70.
6. Kapteyn, J. C., Hoyer, L. L., Hecht, J. E., Muller, W. H., Andel, A., Verkleij, A. J., Makarow, M., Van Den Ende, H., and Klis, F. M. (2000) The cell wall architecture of *Candida albicans* wild-type cells and cell wall-defective mutants. *Mol Microbiol* **35**, 601–611.
7. Jouault, T., El Abed-El Behi, M., Martinez-Esparza, M., Breuilh, L., Trinel, P. A., Chamaillard, M., Trottein, F., and Poulain, D. (2006) Specific recognition of *Candida albicans* by macrophages requires galectin-3 to discriminate *Saccharomyces cerevisiae* and needs association with TLR2 for signaling. *J Immunol* **177**, 4679–4687.
8. Martinez-Esparza, M., Sarazin, A., Jouy, N., Poulain, D., and Jouault, T. (2006) Comparative analysis of cell wall surface glycan expression in *Candida albicans* and *Saccharomyces cerevisiae* yeasts by flow cytometry. *J Immunol Methods* **314**, 90–102.
9. Gantner, B. N., Simmons, R. M., and Underhill, D. M. (2005) Dectin-1 mediates macrophage recognition of *Candida albicans* yeast but not filaments. *EMBO J* **24**, 1277–1286.
10. Linehan, S. A., Martinez-Pomares, L., and Gordon, S. (2000) Mannose receptor and scavenger receptor: two macrophage pattern recognition receptors with diverse functions in tissue homeostasis and host defense. *Adv Exp Med Biol* **479**, 1–14.
11. Cambi, A., Gijzen, K., de Vries, J. M., Torensma, R., Joosten, B., Adema, G. J., Netea, M. G., Kullberg, B. J., Romani, L., and Figdor, C. G. (2003) The C-type lectin DC-SIGN (CD209) is an antigen-uptake receptor for *Candida albicans* on dendritic cells. *Eur J Immunol* **33**, 532–538.
12. McGreal, E. P., Rosas, M., Brown, G. D., Zamze, S., Wong, S. Y., Gordon, S., Martinez-Pomares, L., and Taylor, P. R. (2006) The carbohydrate-recognition domain of Dectin-2 is a C-type lectin with specificity for high mannose. *Glycobiology* **16**, 422–430.
13. Sato, K., Yang, X. L., Yudate, T., Chung, J. S., Wu, J., Luby-Phelps, K., Kimberly, R. P., Underhill, D., Cruz, P. D., Jr., and Ariizumi, K. (2006) Dectin-2 is a pattern recognition receptor for fungi that couples with the Fc receptor gamma chain to induce innate immune responses. *J Biol Chem.* **281**, 38854–38866.
14. Fradin, C., Poulain, D., and Jouault, T. (2000) beta-1,2-linked oligomannosides from *Candida albicans* bind to a 32-kilodalton macrophage membrane protein homologous to the mammalian lectin galectin-3. *Infect Immun* **68**, 4391–4398.
15. Kohatsu, L., Hsu, D. K., Jegalian, A. G., Liu, F. T., and Baum, L. G. (2006) Galectin-3 induces death of *Candida* species expressing specific beta-1,2-linked mannans. *J Immunol* **177**, 4718–4726.
16. Tada, H., Nemoto, E., Shimauchi, H., Watanabe, T., Mikami, T., Matsumoto, T., Ohno, N., Tamura, H., Shibata, K., Akashi, S., Miyake, K., Sugawara, S., and Takada, H. (2002) *Saccharomyces cerevisiae*- and *Candida albicans*-derived mannan induced production of tumor necrosis factor alpha by human monocytes in a CD14- and Toll-like receptor 4-dependent manner. *Microbiol Immunol* **46**, 503–512.
17. Brown, G. D., Herre, J., Williams, D. L., Willment, J. A., Marshall, A. S., and Gordon, S. (2003) Dectin-1 mediates the biological effects of beta-glucans. *J Exp Med* **197**, 1119–1124.
18. Romani, L., Bistoni, F., and Puccetti, P. (2002) Fungi, dendritic cells and receptors: a host perspective of fungal virulence. *Trends Microbiol* **10**, 508–514.
19. Odds, F. C., Brown, A. J., and Gow, N. A. (2004) *Candida albicans* genome sequence: a platform for genomics in the absence of genetics. *Genome Biol* **5**, 230.

Chapter 9

Human Epithelial Model Systems for the Study of *Candida* Infections *In Vitro*: Part I. Adhesion to Epithelial Models

Kai Sohn and Steffen Rupp

Abstract

Adhesion to host tissue represents one of the first steps during the early phase of fungal infections. In order to mediate pathogenesis in the infected host, this process is crucial for colonization and subsequent penetration of the respective tissue. *In vivo* analyses of the adhesion process in whole organisms are limited because of difficulties in providing reproducible and comparable conditions in the host environment. Therefore, *in vitro* assays provide the opportunity to study such processes under more defined conditions thus allowing for the analysis of events that are involved in more detail. Here we describe an *in vitro* adhesion assay making use of human epithelial cell lines to study fungal associations with host epithelia. This assay not only is suited to determine the rate of adhesion in a time-dependent manner but also facilitates global transcriptional profiling in order to determine the fungal response during adhesion at the molecular level.

Key words: adhesion assay, *Candida*, Caco-2, A-431, transcriptional profiling, DNA microarrays.

1. Introduction

Pathogenic fungi have evolved many strategies to colonize different niches of the respective host. In addition to their ability to withstand different types of stresses when exposed to the host environment *(1)*, one of the key properties is their ability to adhere to a variety of different epithelia and tissues in a very tight manner *(2)*. Adhesion to cellular supports not only is a prerequisite for superficial colonization but also is necessary for the penetration of organs and tissues during pathogenesis. In general, many different pathogenic fungi, including *Histoplasma capsulatum* or, for example, aspergilli are able to adhere to specific epithelia *(2)*, but in particular *Candida albicans* is able to associate with a

Steffen Rupp, Kai Sohn (eds.), *Host-Pathogen Interactions*, DOI: 10.1007/978-1-59745-204-5_9,

wide range of different cell types *(3–5)*. This might be one of the reasons why *C. albicans* is able to colonize host niches as different as the human gut, the vaginal tract, or the oral mucosa making this organism both a commensal and the most successful fungal pathogen in humans.

In most cases, adhesion of *C. albicans* to the respective substrate is mediated by cell surface adhesins, including proteins of the fungal cell wall like Hwp1p or members of the Als-family *(6, 7)*. Hwp1p is a surface protein that is specifically expressed in the hyphal growth form of *C. albicans* and that is covalently linked to the cell wall via a GPI-anchor *(8)*. It serves as a substrate for host transglutaminases thus facilitating a covalent attachment of the fungal cell to the epithelium *(8)*. This form of attachment represents one of the most rigid associations with host tissue, providing the basis for a successful colonization even under unfavorable environmental conditions. In fact, mutants lacking Hwp1p are less adherent, have defects in biofilm formation, and are less virulent compared with wild-type cells *(8, 9)*.

The characteristic properties of fungal cells with respect to adhesion and the signaling pathways involved in this process can be commonly analyzed by *in vitro* test assays *(3–5)*. Using such assays, clinical isolates or laboratory mutant strains can be conveniently analyzed under defined conditions and directly compared with each other in order to identify and characterize proteins with putative functions as adhesins or that are implicated in the regulation of colonization *(3, 5)*. In contrast with *in vivo* models, *in vitro* assays offer the advantage to control the conditions during infection more accurately and to define the substrates that are analyzed. Such *in vitro* assays are not restricted to the analysis of fungal adhesion, rather more they are also employed for the study of other organisms such as, for example, *E. coli* or *E. histolytica* as well *(10, 11)*. Here we describe an *in vitro* adhesion assay that is suitable to study fungal adhesion to different substrates, including human epithelia or abiotic plastic surfaces (**Fig. 9.1**). It has been successfully applied to study *C. albicans* wild-type and mutant strains *(3, 4)* but might also be suitable for other pathogenic fungi or bacteria. These models are also used to study invasion as described in **Chapter 10**. The assay described here not only facilitates study of the extent and the time-dependent kinetics of adhesion in a quantitative manner but also allows an up-scaled format to analyze the global response of the fungal cell on the transcriptional level *(4)*.

For this purpose, the colorectal carcinoma cell line Caco-2 *(12, 13)* or vulvo-vaginal epidermoid A-431 cells *(14)* are grown to confluence in 24-well cell culture plates, and subsequently different *C. albicans* strains are applied for defined periods of time to

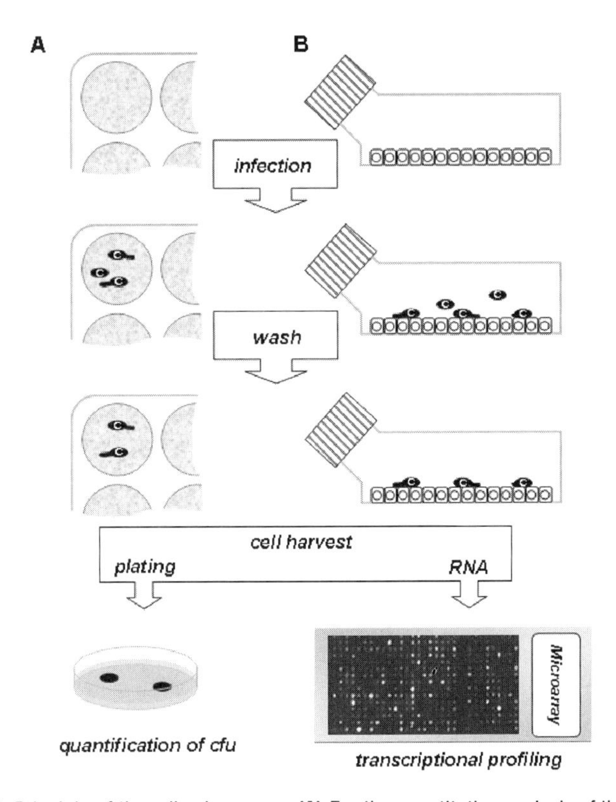

Fig. 9.1. Principle of the adhesion assay. **(A)** For the quantitative analysis of the adhesion kinetics, the *in vitro* adhesion assay is performed in 24-well cell culture plates. Human epithelial cells are grown to confluence and are subsequently infected with *Candida* strains. Afterwards, unbound cells are removed by washing, and the adhering cells are detached and plated on agar plates for the quantitation of the total number of adhering cells. **(B)** For the analysis of the transcriptional response during adhesion, the assays are performed in cell culture flasks under similar conditions as described for the 24-well assay. After harvesting *Candida* cells bound to the human epithelia, total RNA is isolated to perform microarray analysis, quantitative real-time PCR, or Northern blotting.

let the cells adhere to the respective surfaces. Adhesion of fungal cells to the different substrates is accompanied by the induction of morphogenesis as well as by tight interactions of pathogen with the surface of the epithelium (**Fig. 9.2A, B**). Afterwards, unbound *Candida* cells are removed by washing, and the attached cells are harvested to quantify the ratio between adhering and nonadhering cells (**Fig. 9.2C**). In an up-scaled format using standard cell culture flasks, adherent cells can be cultured under similar conditions providing sufficient amounts of RNA to allow for global transcriptional profiling using DNA microarrays. In this context, further amplification of RNA is not necessary, thus circumventing any problems related to the introduction of additional bias to the transcriptome analysis.

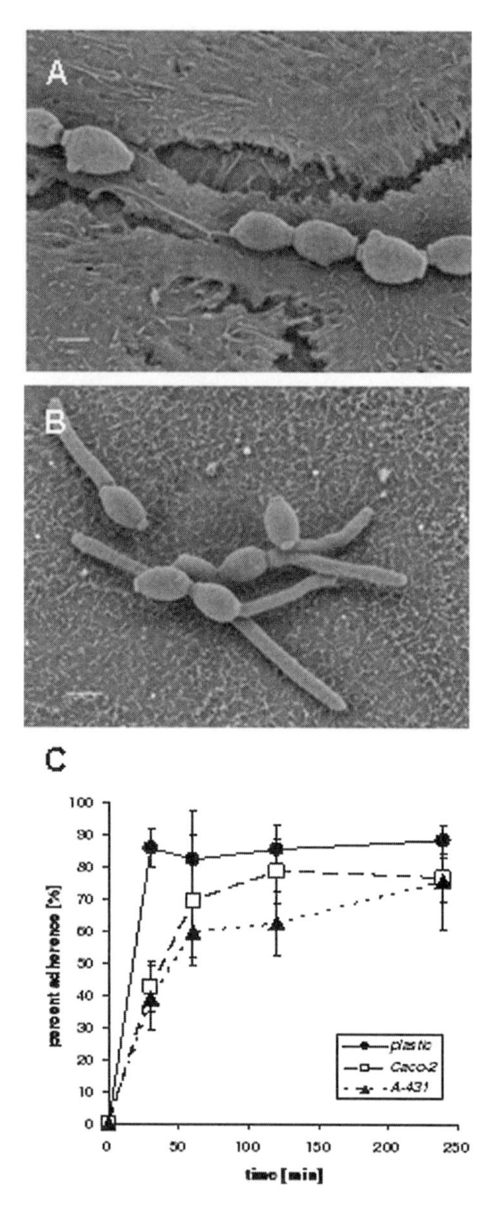

Fig. 9.2. Adhesion of *a Candida albicans* wild-type strain to different human epithelia. *C. albicans* (SC5314) adhering to **(A)** A-431 cells for 30 min or to **(B)** Caco-2 cells for 60 min are analyzed by scanning electron microscopy. Bars represent **(A)** 1 μm or **(B)** 2 μm, respectively. **(C)** Quantitation of adhesion kinetics of *C. albicans* wild-type cells bound to A-431 cells (triangles), to Caco-2 cells (open squares), or directly onto the polystyrene surface (circles).

2. Materials

2.1. Strains and Cell Lines

1. *Candida albicans* wild-type strain (SC5314).
2. Colorectal carcinoma cell line Caco-2 (ATCC: HTB-37).
3. Epidermoid vulvo-vaginal cell line A-431 (ATCC: CRL-1555).

2.2. Media and Solutions

1. Dulbecco's Modified Eagle's Medium (DMEM/Invitrogen Gibco, Karlsruhe, Germany) supplemented with 10% heat inactivated fetal calf serum (FCS/Gibco) and 1 mM sodium pyruvate (Gibco).
2. YPD medium: 10 g/L (Becton-Dickinson, Heidelberg, Germany) yeast extract, 20 g/L (Becton-Dickinson) bacto-peptone, 2% glucose, 0.1 mM uridine.
3. YPD agar plates: 10 g/L (Becton-Dickinson) yeast extract, 20 g/L (Becton-Dickinson) bacto-peptone, 2% glucose, 2% bacto agar.
4. PBS^-: 137 mM NaCl, 2.7 mM KCl, 1.5 mM KH_2PO_4, 8.1 mM Na_2HPO_4, 1 mM EDTA, pH 7.2.
5. Trypsin/EDTA: 0.25% trypsin in PBS^- (Invitrogen Gibco).
6. RNA-isolation kit (e.g., RNeasy Midi kit, Qiagen, Hilden, Germany).
7. RNase-free water.
8. 4 M LiCl in DEPC-treated water.
9. 70% EtOH in DEPC-treated water.

2.3. Disposables

1. 15 and 50 mL tubes (Greiner bio-one, Frickenhausen, Germany).
2. 24-well polystyrene cell culture plate (Nunc GmbH, Thermo Fisher Scientific, Langenselbold, Germany).
3. 250-mL cell culture flask T-75 (Greiner Bio-One).
4. Cell scraper (Sarstedt Inc., Newton, USA).
5. Syringe (21 gauge) (Braun, Melsungen, Germany).

2.4. Equipment

1. Incubator (30°C; HAT Infors AG).
2. Tabletop centrifuge (Heraeus Biofuge fresco, Heraeus Instruments, Hanau, Germany).
3. Cell culture centrifuge (Heraeus Megafuge 1.0).
4. Retsch mill (Retsch MM200).

3. Methods

3.1. Quantitative Analysis of Fungal Adhesion

For the quantitative analysis of adhesion kinetics, the assays are performed in the 24-well cell culture plate format.

3.1.1. Preparation of Confluent Human Monolayers of Epithelial Cells (Start 2 to 3 Days Before the Experiment)

For detailed description of cell culture methods, see **Chapter 10**.

1. Apply 1.5 mL of supplemented DMEM to each well of the 24-well cell culture plate.
2. Add 2×10^5 Caco-2 or A-431 cells per well (*see* **Note 1**).
3. Cultivate at 37°C and 5% CO_2 for 2 to 3 days until 100% confluence is reached (*see* **Note 2**).

3.1.2. Preculture of Candida albicans Strains (Start 1 Day Before the Experiment)

1. Inoculate *C. albicans* in 10 mL of YPD in an Erlenmeyer flask.
2. Culture overnight at 30°C and 180 rpm in an incubator.
3. Inoculate 10 mL fresh YPD with the overnight culture to an optical density (OD_{600}) of 0.1 to 0.2.
4. Cultivate at 30°C and 180 rpm in an incubator until OD_{600} of 1.0 is reached.
5. Pellet the cells for 3 min at $3000 \times g$ at RT.
6. Resuspend the pellet in 10 mL supplemented DMEM (equivalent to 3×10^7 cells per mL; *see* **Note 3**).
7. Dilute the cell suspension to 6×10^3 cells per mL in supplemented DMEM (1:5000 dilutions).

3.1.3. Analytical Adhesion Assay

1. Remove DMEM from 24-well plates containing the epithelial monolayers.
2. Add 250 μL fresh and prewarmed (37°C) supplemented DMEM to each well.
3. Apply 50 μL of diluted *Candida* cell suspension per well (equivalent to 300 cells per well).
4. Employ two wells per time point to analyze adhesion in duplicates (use at least 10 wells per strain when you analyze time points, e.g., 0, 30, 60, 120, and 240 min).
5. For time point 0 min, immediately remove the medium containing the *Candida* cells after application to the well (*see* **Note 4**).
6. Plate the supernatant on YPD agar plates and incubate them at 30°C overnight (= nonadherent cells).
7. Wash the well with 300 μL PBS^-.
8. Add another 300 μL PBS^- and scratch off the adherent cells using a 1-mL pipette.
9. Plate the scratched-off cells on YPD agar plates and incubate at 30°C overnight (= adherent cells).

10. For the other time points, proceed similarly as described for **steps 5** to **9**.
11. As a control, plate in duplicate 50 μL of diluted *Candida* cell suspension directly on YPD agar plates.
12. Count the colonies on all YPD agar plates and calculate the ratio of adherent to the total number of cells for each time point as follows:

$$\text{adherence (\%)} = \left(\frac{\text{adherent}}{\text{adherent} + \text{nonadherent}}\right) \times 100$$

Analogously, calculate the ratio of nonadherent to the total number of cells for each time point as follows:

$$\text{nonadherence (\%)} = \left(\frac{\text{adherent}}{\text{adherent} + \text{nonadherent}}\right) \times 100$$

Plot percent adherence versus time in an x/y diagram (**Fig. 9.2C**).

3.2. Assay Format for the Analysis of the Global Transcriptional Response

For the analysis of the transcriptional response, the *in vitro* adhesion assay is up-scaled using 250-mL cell culture flasks. Isolated RNA from such preparative assays can be used to subsequently generate fluorescently labeled cDNA probes for DNA-microarray analyses.

3.2.1. Preparation of Confluent Human Monolayers of Epithelial Cells (Start at Least 1 Week Before the Experiment; All Media Should Be Prewarmed)

1. Let Caco-2 or A-431 cells grow to 80% confluence in supplemented DMEM at 37°C and 5% CO_2.
2. Wash the cells with PBS^-.
3. Detach the cells with 2.5 mL trypsin/EDTA.
4. Add 10 mL of supplemented DMEM and pellet the cells.
5. Resuspend the cells in 10 mL of fresh supplemented DMEM and split them in three to five 250-mL cell culture flasks.
6. Cultivate at 37°C and 5% CO_2 approximately for 3 days until 100% confluence is reached (*see* **Note 2**).

3.2.2. Preculture of Candida albicans Strains (Start 1 Day Before the Experiment)

1. Inoculate *C. albicans* in 10 mL of YPD in an Erlenmeyer flask.
2. Culture overnight at 30°C and 180 rpm in an incubator.
3. Inoculate 10 mL YPD with the overnight culture to an optical density (OD_{600}) of 0.1 to 0.2.
4. Cultivate at 30°C and 180 rpm in an incubator until OD_{600} of 1.0 is reached.
5. Pellet the cells for 3 min at 3000 × *g* at RT.
6. Resuspend the pellet in an appropriate volume (approximately 9 mL) of fresh supplemented DMEM to get a cell concentration equivalent to 3×10^7 cells per mL (*see* **Note 3**).

3.2.3. Preparative Adhesion Assay

1. Remove used DMEM from the cell culture flasks containing the epithelial monolayers.
2. Wash the monolayer with 10 mL PBS^-.

3. Add 9 mL fresh and prewarmed (37°C) supplemented DMEM to each well.
4. Apply 1 mL of *Candida* cell suspension to each flask (equivalent to 3×10^7 cells).
5. Employ at least two flasks per time point to analyze transcriptional profiles using DNA microarrays (*see* **Note 5**).
6. At different time points, remove the supernatant with unbound *Candida* cells.
7. Wash with 10 mL PBS^-.
8. Add 2.5 to 5 mL PBS^- and scratch off the cells using a cell scraper.
9. Collect the cells in a 50-mL tube and pellet them at $3000 \times g$ for 3 min at 4°C and discard the supernatant.
10. Resuspend cells in 5 mL of remaining buffer and break human cells by drawing them up 10 to 20 times in a syringe (21 gauge).
11. Pellet the *Candida* cells at $3000 \times g$ for 3 min at 4°C.
12. Resuspend the cells in 1 to 2 mL of PBS^-.
13. Transfer resuspended cells drop by drop in a 50 mL tube filled with liquid nitrogen to freeze.
14. Store the frozen cell beads at −70°C up to 6 months.

3.2.4. Preparation of Total RNA

1. Cells of the frozen beads are broken using a Retsch mill.
2. Two cell beads are transferred to the precooled (liquid nitrogen) Teflon-container of the Retsch mill (*see* **Note 6**).
3. Apply a shaking frequency of 30 per second for 2 minutes to disintegrate the *Candida* cells.
4. Transfer frozen powder to a 15-mL tube.
5. Add the respective buffer of the RNA-isolation kit (RNeasy-kit from Qiagen or similar) and proceed according to the instruction protocol of the manufacturer.
6. Elute RNA from the provided column of the RNA-isolation kit with 2×250 μL RNase-free water.
7. Add 500 μL 4 M LiCl.
8. Incubate at least 2 h at −20°C to precipitate the RNA (*see* **Note 7**).
9. Centrifuge at $13{,}000 \times g$ and 4°C for 30 min to pellet the RNA.
10. Wash the pellet with 1 mL 70% EtOH-DEPC.
11. Centrifuge at $13{,}000 \times g$ and 4°C for 10 min to pellet the RNA.
12. Repeat the washing step.
13. Dry the pellet and solubilize it with 30 to 50 μL RNase-free water.

14. Determine RNA concentration and purity by measuring the optical density at 260 and 280 nm.

15. Check RNA integrity by TAE agarose gel electrophoresis.

Isolated RNA can be used to generate cDNA probes for hybridization of DNA microarrays, for quantitative real-time PCR, or Northern blotting. For global transcriptional profiling using fluorescently labeled cDNA hybridized to DNA microarrays, about 10 to 25 μg of total RNA is commonly needed. Such amounts can be provided by the assay format described above. For experimental details on how to perform the microarray analysis, we refer to the individual protocols of your DNA microarray provider.

4. Notes

1. For the preparation of confluent monolayers, other cell types like the buccal epithelial cell line TR146 can also be used to study fungal adhesion to oral mucosa. Alternatively, 24-well plates can also be used without any epithelial monolayer to study adhesion to abiotic polystyrene surfaces.
2. It is important to carefully check for 100% confluence as uncovered polystyrene surfaces might have an impact on the overall amount of adhesion.
3. The correlation between optical density and actual cell number can vary among different *Candida* strains (e.g., strains with opaque-like phenotype exhibit lower optical densities at similar cell numbers). Check the exact correlation before by plating various dilutions of defined optical densities on YPD agar plates and count the resulting colonies to make sure that you are using the correct number of cells.
4. Adherence of *Candida* cells to the surface is very rapid. Thus, remove the supernatant at time point 0 min quickly.
5. At later time points of adhesion (60, 120, and 240 min), one to two cell culture flasks will yield sufficient amounts of total RNA to perform a global transcriptional profiling experiment. For earlier time points (30 min), several flasks (up to four) have to be processed simultaneously to yield enough material to perform transcriptional profiling without prior amplification.
6. Particular care is necessary to avoid any contamination with RNAses. Therefore, any parts of the Teflon container should be pretreated with 0.5% sodium hypochlorite for 10 min. Afterwards, wash with 70% ethanol and blow dry using nitrogen.

7. RNA yield after precipitation using LiCl depends on RNA concentration and precipitation time. Highly concentrated RNA solutions precipitated overnight at −20°C give best results.

Acknowledgments

We would like to thank all the members of the department of Molecular Biotechnology of Fraunhofer IGB (Stuttgart, Germany) and the Cell Systems Department of Fraunhofer IGB (Stuttgart, Germany) (Prof. Dr. Heike Mertsching) for kindly providing their installations, technical support, and advice during the histology processing. This work was supported by a grant of the DFG to S. Rupp (Ru608/4–1,2).

References

1. Moye-Rowley W.S. (2002) Transcription factors regulating the response to oxidative stress in yeast. *Antioxid Redox Signal* **4,** 123–140.
2. Mendes-Giannini M.J., Soares C.P., da Silva J.L., Andreotti P.F. (2005) Interaction of pathogenic fungi with host cells: Molecular and cellular approaches. *FEMS Immunol Med Microbiol* **45,** 383–394.
3. Dieterich C., Schandar M., Noll M., Johannes F.J., Brunner H., Graeve T., Rupp S. (2002) In vitro reconstructed human epithelia reveal contributions of *Candida albicans EFG1* and *CPH1* to adhesion and invasion. *Microbiology* **148,** 497–506.
4. Sohn K., Senyurek I., Fertey J., Konigsdorfer A., Joffroy C., Hauser N., Zelt G., Brunner H., Rupp S. (2006) An in vitro assay to study the transcriptional response during adherence of *Candida albicans* to different human epithelia. *FEMS Yeast Res* **6,** 1085–1093.
5. Wiesner S.M., Bendel C.M., Hess D.J., Erlandsen S.L., Wells C.L. (2002) Adherence of yeast and filamentous forms of *Candida albicans* to cultured enterocytes. *Crit Care Med* **30,** 677–683.
6. Hoyer L.L. (2001) The *ALS* gene family of *Candida albicans. Trends Microbiol* **9,** 176–180.
7. Sundstrom P. (1999) Adhesins in *Candida albicans. Curr Opin Microbiol* **2,** 353–357.
8. Staab J.F., Bradway S.D., Fidel P.L., Sundstrom P. (1999) Adhesive and mammalian transglutaminase substrate properties of *Candida albicans* Hwp1. *Science* **283,** 1535–1538.
9. Nobile C.J., Nett J.E., Andes D.R., Mitchell A.P. (2006) Function of *Candida albicans* adhesin Hwp1 in biofilm formation. *Eukaryot Cell* **5,** 1604–1610.
10. Darfeuille-Michaud A., Aubel D., Chauviere G., Rich C., Bourges M., Servin A., Joly B. (1990) Adhesion of enterotoxigenic *Escherichia coli* to the human colon carcinoma cell line Caco-2 in culture. *Infect Immun* **58,** 893–902.
11. Rigothier M.C., Coconnier M.H., Servin A.L., Gayral P. (1991) A new *in vitro* model of *Entamoeba histolytica* adhesion, using the human colon carcinoma cell line Caco-2: scanning electron microscopic study. *Infect Immun* **59,** 4142–4146.
12. Fogh J., Fogh J.M., Orfeo T. (1977) One hundred and twenty-seven cultured human tumor cell lines producing tumors in nude mice. *J Natl Cancer Inst* **59,** 221–226.
13. Hidalgo I.J., Raub T.J., Borchardt R.T. (1989) Characterization of the human colon carcinoma cell line (Caco-2) as a model system for intestinal epithelial permeability. *Gastroenterology* **96,** 736–749.
14. Giard D.J., Aaronson S.A., Todaro G.J., Arnstein P., Kersey J.H., Dosik H., Parks W.P. (1973) *In vitro* cultivation of human tumors: establishment of cell lines derived from a series of solid tumors. *J Natl Cancer Inst* **51,** 1417–1423.

Chapter 10

Human Epithelial Model Systems for the Study of *Candida* infections *In Vitro*: Part II. Histologic Methods for Studying Fungal Invasion

Rosa Hernandez and Steffen Rupp

Abstract

Although the role of invasion in the virulence of *Candida albicans* has been demonstrated, the mechanism that governs fungal invasion is not fully understood. Among the tools that exist to fill these gaps in knowledge, *in vitro* tissue models based on reconstituted human epithelia (RHE) have already been developed. Such models are designed to study more reproducably the fungus-host relationship, as they eliminate the complexity and variability found *in vivo*. Herein we describe the preparation of these RHE and their application in study of the invasion properties of *C. albicans* by further histologic processing and microscopic observation. For this purpose, different epithelial cell lines are grown on a collagen gel to build up models of intestinal (Caco-2 cell line), vaginal (A431 cell line), and oral (TR146 cell line) mucosa. The use of these *in vitro* models applied to test the invasiveness of *C. albicans* strains (clinical isolates or gene deleted mutants) and to identify changes in gene expression during the invasion of the RHE will help to advance our knowledge of pathogenesis and to study specific mechanisms used by *C. albicans* to adapt to changing environments present in different epithelia. Furthermore, because these models are useful to study the host response during the challenge with the pathogen, they will also offer important new insights into host cell biology and identify new targets for treatment.

Key words: *Candida albicans*, invasion, reconstituted human epithelium, Caco-2, A431, TR146, collagen gel, histology.

1. Introduction

Candida albicans infections are normally established by cells that typically grow as a harmless commensal on the skin and mucosal surfaces of healthy individuals. As an opportunistic pathogen, under certain circumstances usually linked to a compromised host immune system, *C. albicans* causes infections restricted to

Steffen Rupp, Kai Sohn (eds.), *Host-Pathogen Interactions*, DOI: 10.1007/978-1-59745-204-5_10,

the mucosa or, in severe cases, life-threatening systemic infections *(1)*. Several virulence mechanisms are known to contribute to the ability of this fungus to colonize and invade host cells and tissues like dimorphic and phenotypic switching *(2, 3)*, expression of host recognition molecules or adhesins *(4, 5)*, thigmotropism *(6)*, and secretion of hydrolytic enzymes such as lipases, phospholipases, and secreted aspartyl proteinases *(7–10)*.

One of the main factors affecting the pathogenesis of candidiasis is the ability of *C. albicans* to penetrate into epithelial and endothelial cell layers and invade the host tissues. One mechanism of invasion used by this fungus already described is the production of lytic enzymes, such as secreted aspartyl proteinases (SAPs), able to digest the surface of the epithelial cell and intercellular junctions *(11)*. Another mechanism of *Candida* invasion proposed is passive and induced endocytosis by epithelial cells *(12–14)*. Nevertheless, although the role of invasion in the virulence of *C. albicans* has been demonstrated and different mechanisms have been proposed, many aspects related to this process still remain to be determined.

For the study of host-pathogen interaction, several models of experimental candidiasis in animal species have been applied *(15–17)*, but these models incompletely mimic the situation in humans and present a complex system in which many variables cannot be controlled *(18)*. Therefore, *in vitro* reconstituted human epithelia (RHE) have been developed in order to characterize the cellular and molecular mechanisms of epithelial-fungus interactions at a relevant biological surface. Several models of superficial candidiasis based on these RHE—cutaneous *(19, 20)*, oral *(21, 22)*, esophageal *(23)*, intestinal *(19)*, and vaginal *(24, 25)*—have already been used to further characterize the contribution of individual factors to the invasion of *C. albicans* into human tissue. In this respect, *in vitro* models have offered important new insights into the invasion process through studies on gene expression of *C. albicans* (DNA microarrays, RT-PCR) *(14, 22, 26, 27)* (*see* **Chapter 9**) and microscopic analysis of the extent of fungal invasion of *C. albicans* mutant strains. These studies have shown the importance of several factors during the invasion of different epithelia such as SAPs *(28, 29)*, phospholipases *(30)*, GPI anchored proteases *(31)*, protein mannosyltransferases *(21)*, the siderophore iron transporter Sit1/Arn1 *(32)*, the high-affinity cAMP phosphodiesterase Pde2 *(33)*, and genes implicated in hyphae formation such as the transcription factor Efg1 *(19)*, among others.

On the other side, as the ability to stimulate a strong inflammatory response has been shown to correlate well with the severity of infection, the study of the host response is of prime importance *(23, 34)*. Therefore, the use of these *in vitro* models of candidiasis to determine which genes of the host are expressed

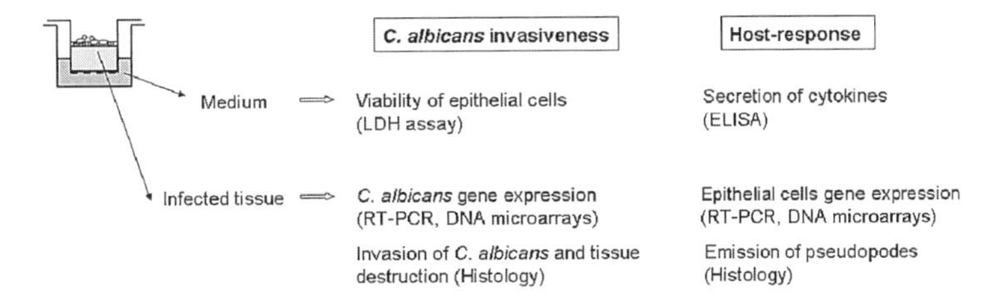

Fig. 10.1. Examples of applications of *in vitro* RHE to the study of host-pathogen interaction. After incubation of the tissue model with *C. albicans*, the culture medium and infected tissue can be used to analyze the viability of the tissue (LDH release), the pattern of secreted cytokines (ELISA, RT-PCR), changes in gene expression (RT-PCR, DNA microarrays), and for histology processing to visualize invasion process.

during the invasion of the fungus and the response in secretion of cytokines *(35–38)* or antimicrobial peptides *(39)* will help to achieve new insights into the pathogenesis of *Candida* infections (**Fig. 10.1**). Furthermore, the use of RHE also appears to be a good basis to test tolerability of new antifungal agents simultaneously with their antifungal activity, studying specific alterations of the epithelium and reduction in the invasion of the mucosa by fungal cells *(19, 40, 41)* (**Fig. 10.2**).

A Drug-testing

\- +

B Characterization of mutant strains and clinical isolates of *C. albicans*

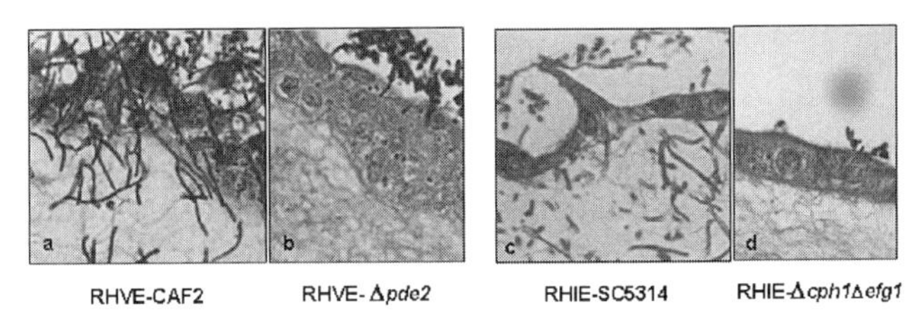

Fig. 10.2. Application of histologic examination of RHE models in antifungal drug-testing assays and in the characterization of mutant strains and clinical isolates of *C. albicans*. **(A)** Testing of tolerability and antifungal activity is done by administration of the drug to the RHE together with the infection with *C. albicans* (SC5314). The picture shows the complete invasion and destruction of the epithelial monolayer after 20 h of interaction when no drug is applied (−) versus no marked morphologic alteration in the tissue and the inability of *C. albicans* to grow and invade when the drug is present (+). **(B)** Characterization of mutant strains of *C. albicans*. Differences in invasive properties of different deletion mutant strains (*efg1cph1* and *pde2*) **(b, d)** when compared with wild-type strains **(a, c)** after 8 h of interaction with the RHE can be clearly observed using histologic methods.

Herein we describe the preparation of model systems of oral, vaginal, and intestinal candidiasis using *in vitro* reconstituted tissues for the study of events related to the invasion of human mucosa by *C. albicans.* These models also can be used to study adhesion as described in **Chapter 9**. One of the models is based on the Caco-2 cell line derived from human colorectal adenocarcinoma *(19, 42)*, as the gastrointestinal tract is considered a natural reservoir of *C. albicans* in healthy individuals and a frequent source of hematogenous candidiasis *(43, 44)*. The other models are prepared from the epidermoid vulvo-vaginal cell line A431 *(45)* or from the buccal epithelial cell line TR146 *(46)*, as the most frequently reported superficial diseases caused by this fungus are vulvovaginal and oral candidiasis *(47, 48)*.

To start the preparation of the RHE, a collagen solution is gelified on an inert supporting membrane, and afterwards the selected cell line is cultured as the basis for the *in vitro* tissue equivalent. The use of the collagen gel as supporting surface for the epithelial cell lines enables us to visualize the invasion processes in more detail, to facilitate the handling of the tissues for histology, and it can be used as a matrix to embed other cell types to complete the mucosa model. Four days after seeding the collagen gel with the epithelial cell line, the RHE is ready for further studies.

The infection of the *in vitro* tissue equivalent is conducted by application of yeast cells, and after the desired time of interaction (at 37°C and 5% CO_2), the progress of infection is monitored by histologic studies. For this purpose, the tissue model is fixed, infiltrated, and embedded in paraffin and cut with a microtome. Finally, the histologic sections are examined under the microscope after being stained sequentially with the periodic acid–Schiff (PAS) method of McManaus and the method of Papanicolaou *(49, 50)* (**Fig. 10.2**).

The model described here represents a simple, easy to prepare and handle *in vitro* tissue model. Further improvements of the system to mimic the *in vivo* situation are possible including fibroblasts or any other cell type embedded in the collagen matrix *(19)* or even integrating a third cell type from the immune system to perform more deep studies of the host response during the infection (e.g., neutrophils, dendritic cells) *(51)*. To study the effect of bacterial colonization in mucosal surfaces, bacteria from the normal human flora can also be added to the model during infection with *C. albicans (52)*.

The multiple applications of these *in vitro* systems hopefully will add significant new insights into the nature of the invasion of *C. albicans* and other pathogens in host mucosal tissue.

2. Materials

2.1. Preparation of Oral, Vaginal, and Intestinal Mucosa Models

2.1.1. Cell Culture

1. Human cell lines: colorectal carcinoma cell line Caco-2 (ATCC HTB-37), epidermoid vulvo-vaginal cell line A431 (ATCC CRL-1555), and neck node metastasis of a human buccal carcinoma cell line TR146 (kindly provided by the Imperial Cancer Research Technology, London, UK), for the intestinal, vaginal, and oral candidiasis models, respectively (*see* **Note 1**).
2. Cell culture medium for Caco-2 and A431 cell lines: High glucose (4.5 g/L) Dulbecco's Modified Eagle's Medium (DMEM) with L-glutamine supplemented with sodium pyruvate (1 mM), gentamicin (100 μg/mL), and 10% of heat-inactivated fetal bovine serum (FBS) (for preparation, *see* **Notes 2** and **3**). All the components are from Invitrogen Gibco (Karlsruhe, Germany).
3. Cell culture medium for TR146 cell line: High glucose (4.5 g/L) DMEM with L-glutamine supplemented with penicillin (100 IU/mL), streptomycin (100 μg/mL), and 10% heat-inactivated FBS (for preparation, *see* **Notes 2, 4,** and **5**). All the components are from Invitrogen Gibco.
4. Trypsin/EDTA solution (0.25% Trypsin, 1 mM EDTA) (Invitrogen Gibco) (*see* **Note 5**).
5. CMF-PBS (calcium-magnesium–free phosphate-buffered solution): 137 mM NaCl, 2.7 mM KCl, 1.5 mM KH_2PO_4, 8.1 mM Na_2HPO_4, 1 mM EDTA, pH 7.2.
6. Cell culture flask, 75 cm^3 (Greiner Bio-One, Frickenhausen, Germany).
7. Sterile glass pipettes (5, 10, 25 mL) and pipetting aid.
8. Pasteur pipettes with a vacuum pump for medium aspiration.
9. Electronic cell counter and analyzer (Casy; Schaerfe System, Reutlingen, Germany) or Neubauer counting chamber (Carl Roth, Karlsruhe, Germany) and 0.4% Trypan blue solution.
10. CO_2 incubator (Espec BNA-311; Tabai Espec Corp., Osaka, Japan) and laminar flow hood (Heraeus Instruments, Hanau, Germany).

2.1.2. Collagen Gel

1. Frozen tails from rats (Charles River Laboratories, Sulzfeld, Germany).
2. PBS sterile: 137 mM NaCl, 2.7 mM KCl, 1.5 mM KH_2PO_4, 8.1 mM Na_2HPO_4.
3. Acetic acid 0.1 M.
4. DMEM, powder (high glucose) (Invitrogen Gibco).

5. Sodium bicarbonate 7.5%, liquid (Invitrogen Gibco).
6. HEPES Buffer solution 3 M, liquid (Invitrogen Gibco).
7. Heat-inactivated-FBS (Invitrogen Gibco) (*see* **Note 2**).
8. Sodium hydroxide (NaOH) 10 M.
9. Ethanol 70%.
10. Tweezers and scalpel.
11. Sterile glass bottle and petri dish.
12. Stir bar and magnetic stirrer.

2.1.3. Preparation of Different RHE

1. Collagen solution (*see* **Section 3.1.2**).
2. Neutralizing solution (*see* **Section 3.1.2**).
3. Cell culture medium free of antibiotics (*see* **Note 6**).
4. Culture of the cell lines that will be used for the preparation of the tissue model (*see* **Section 3.1.1**).
5. Inserts: ThinCert 12 mm diameter, with microporous PET membrane, 0.4-μm pore size (Greiner Bio-One) (*see* **Note 7**).
6. 24-well polystyrene plate (Greiner Bio-One).
7. Tweezers.
8. CO_2 incubator (Espec BNA-311; Tabai Espec Corp.) and laminar flow hood (Heraeus Instruments).

2.1.4. Infection of the RHE

1. Clinical isolate of *C. albicans* SC5314 *(53)*.
2. YPD liquid medium: 10 g/L yeast extract (Becton Dickinson, Heidelberg, Germany), 20 g/L bacto-peptone (Becton Dickinson), and 2% (w/v) glucose in ddH_2O.
3. Sterile PBS.
4. Cell culture medium free of antibiotics (*see* **Note 7**).
5. Sterile Erlenmeyer flasks.
6. Neubauer counting chamber (Carl Roth).
7. Incubation shaker at 30°C.
8. CO_2 incubator (Espec BNA-311; Tabai Espec Corp.) and laminar flow hood (Heraeus Instruments) used only for infection purposes.

2.2. Histology Processing

2.2.1. Fixation

1. Bouin's solution (Sigma, Taufkirchen, Germany).
2. Pasteur pipettes and rubber bulb.
3. Nitrile gloves.
4. Fume extraction hood.

2.2.2. Infiltration and Embedding in Paraffin

1. Tissue Embedding Cassettes (Labonord, Templemars, France).
2. Filter papers for embedding cassettes (Labonord).
3. Tissue Processor Shandon Citadel 1000 (Thermo Scientific, Frankfurt, Germany).
4. Stainless Steel Base Molds (Labonord).
5. Paraffin wax melting point 54°C to 56°C for histology embedding (Labonord).
6. Paraffin dispenser, 60°C (Medax Nagel, Kiel, Germany).
7. Heating plate (Medax Nagel).
8. Incubator 65°C (Heraeus Instruments).

2.2.3. Microtome Slicing and Tissue Section Mounting

1. Cooling plate −20°C (Medax Nagel).
2. Paraffin tissue floating bath, 37°C (Medax Nagel).
3. Drying table, 42°C (Medax Nagel).
4. Leica RM2145 rotary microtome and microtome blades (Leica Microsystems, Wetzlar, Germany).
5. Twin frosted slides (Labonord).
6. Tweezers and small paint brush.

2.2.4. PAS–Papanicolaou Staining

1. Roticlear (Carl Roth).
2. Ethanol: 50%, 70%, 96% and 100% (Carl Roth).
3. Isopropanol (Carl Roth).
4. 0.1% (w/v) Periodic acid solution (Sigma-Aldrich, Taufkirchen, Germany) (*see* **Note 8**).
5. Sodium Bisulfite Solution: 1100 mL ddH_2O + 55 mL HCl 1M + 6.6 mL sodium bisulfite 38% to 40% (Riedel- de Häen, Seelze, Germany) (*see* **Notes 8** and **9**).
6. 0.03 M hydrochloric acid (HCl).
7. Schiff's reagent (Sigma-Aldrich).
8. Mayers Hematoxylin (Merck, Darmstadt, Germany).
9. Polychromic staining solution EA 50 (Sigma-Aldrich).
10. Rectangular glass staining dish with lid.
11. Removable glass tray with slots (20-slide capacity) and a nickel, spring-wire holder for lifting tray out of staining solution.
12. Fume extraction hood.

2.2.5. Slides Coverslipping

1. Aquamount (Labonord).
2. Coverslip 20 × 60 mm (Labonord).
3. Pasteur pipettes and rubber bulb.

3. Methods

3.1. Preparation of Intestinal and Vaginal Mucosa Models (see Note 1)

Before preparation of the model, the different cell lines are grown in 75-cm^2 tissue-culture flasks by standard procedures *(54, 55)* to be up to 80% to 90% of confluence for the day of the preparation of the RHE. From these flasks, cells are detached and distributed in the infection models as described below.

3.1.1. Cell Culture

1. The cell culture medium is removed from the flask with a sterile Pasteur pipette and CMF-PBS is added (5 mL to a 75-cm^2 flask) to remove any residual FBS that may inhibit the action of trypsin. Rinse the CMF-PBS over the cells and discard it (*see* **Note 10**).
2. Trypsin/EDTA solution is added to completely cover the monolayer surface (3 mL to a 75-cm^2 flask), and the flasks are incubated at 37°C in the CO_2 incubator until the cells detach (spherical shape) (3 min for Caco-2 cell line and 6 min for A431 and TR146 cell lines) (*see* **Note 11**).
3. As soon as cells are detached, serum or medium containing serum is added to inhibit further trypsin activity that might damage the cells. Afterwards, the cells are pelleted by centrifugation at 800 to 1000 $\times$ *g* for 3 min.
4. The cell pellet is resuspended in fresh cell culture medium and live cells counted using an electronic cell counter (*see* **Note 12**).
5. The cell suspensions are diluted to the appropriate seeding concentration for the inserts (1 $\times$ 10^6 cells/mL) by adding fresh cell culture medium.

3.1.2. Collagen Gel

Type I collagen is extracted from rat tail tendons and processed in acetic acid solution to obtain sterile soluble collagen. Then neutralizing solution is added for gelling the collagen.

1. Four rat tails (stored at −20°C) are thawed by soaking in 70% ethanol for 15 min.
2. Starting from the tip, the skin of each tail is removed with sterile tweezers.
3. The tendon fibers attached to the distal section of the tail are drawn out, cut with a scalpel, and collected in a preweighed Petri dish.
4. The fibers are weighted and soaked in sterile PBS in a sterile bottle and washed again with fresh sterile PBS for 15 min with continuous stirring.
5. The fibers are then soaked in 70% ethanol for 30 min to sterilize them.
6. Sterilized tendon fibers are then added to 0.1 M acetic acid (8 g of fibers/L) and stirred continuously for 72 h at 4°C for solubilization of the collagen.

7. The solution is afterwards clarified by centrifugation at 4000 × *g* at 4°C for 1 h. The supernatant is decanted into a sterile bottle (*collagen stock solution*) and stored at 4°C.
8. To prepare 100 mL of *neutralizing solution*: 2.7 g of DMEM powder is dissolved in 60 mL of ddH_2O and sterilized by filtering. Then 10 mL of 7.5% (w/v) sodium bicarbonate, 7.5 mL of 3 M HEPES, 20 mL of heat-inactivated FBS, and 2 mL of 10 M NaOH are added. Keep stored at 4°C.
9. Any new stock of collagen solution is needed to titrate with the neutralizing solution for the formation of a collagen gel. For that purpose, 1 mL of collagen solution is placed in several cavities of a 24-well plate and using the neutralizing solution, the volume that must be added to form a gel of suitable consistency is determined (*see* **Note 13**).

3.1.3. Preparation of Different RHE

The three-dimensional *in vitro* model of intestinal, vaginal, or oral epithelium is composed of human epithelial cells grown on a collagen matrix over a 4-day period as described next (**Fig. 10.3**).

1. The tissue culture inserts are removed from the sterile package with sterile tweezers and placed in the cavities of a 24-well plate.
2. The collagen solution is mixed with the neutralizing solution in a Petri dish in the proportion found during the titration process (*see* **Section 3.1.2, step 9**).
3. After mixing by pipetting, 200 μL of this formulation is placed into each insert and incubated for 15 min at 37°C to allow the gel to solidify (*see* **Note 14**).
4. The collagen gel is then seeded with oral, vaginal, or intestinal epithelial cells by adding 100 μL of a cell suspension of 1×10^6 cells/mL (*see* **Section 3.1.1**). Afterwards, cell culture medium free of antibiotics (600 μL) is added in the outer part of the insert, and the 24-well plates are incubated at 37°C with 5% of CO_2 (saturated humidity).
5. Cells are allowed to adhere to the collagen surface for 24 h, and then the culture is submerged in fresh cell culture medium (approximately 2 mL per well).
6. Within 4 days from the seeding of cells, the models of *in vitro* reconstituted tissue are ready for further studies (*see* **Note 15**):
   - Reconstituted Human Intestinal Epithelium (RHIE), age day 4, human colorectal carcinoma cell line Caco-2.
   - Reconstituted Human Vaginal Epithelium (RHVE), age day 4, human vulvo-vaginal epidermoid carcinoma cell line A431.
   - Reconstituted Human Oral Epithelium (RHOE), age day 4, neck node metastasis of a human buccal carcinoma cell line TR146.

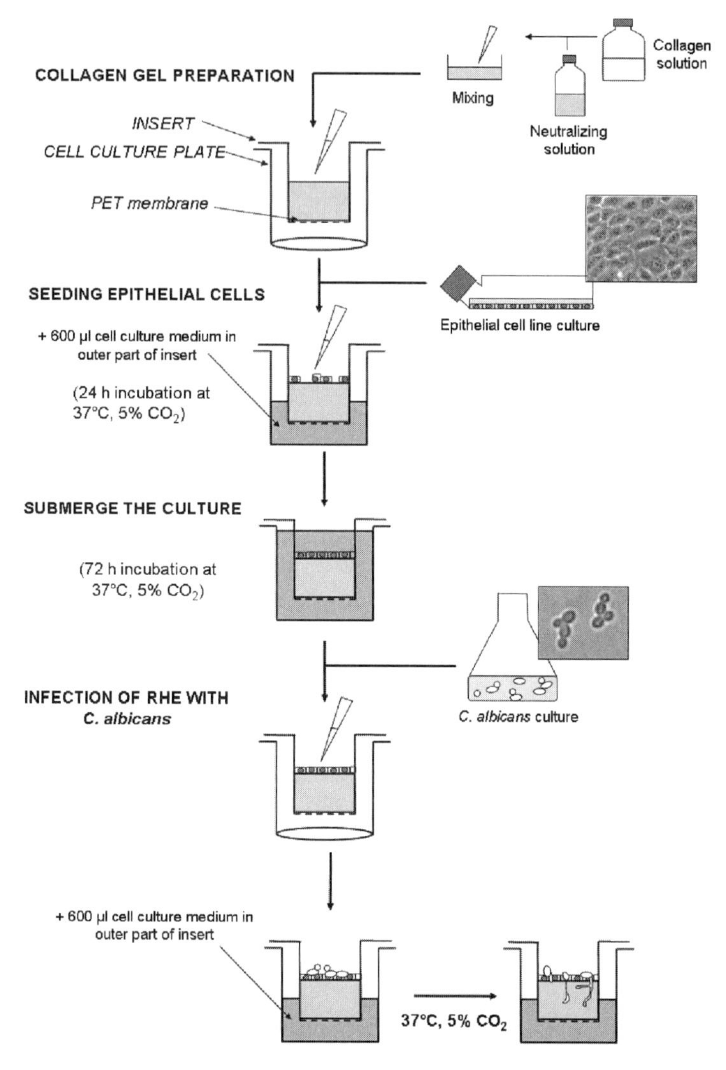

Fig. 10.3. Scheme for the preparation of the *in vitro* RHE and infection with *C. albicans*. First the collagen solution is gelified in the insert, and afterwards the epithelial cells are seeded on top of the gel as the basis for the different *in vitro* tissue equivalents. After 4 days of submerged culture, the tissue model is ready and can be infected by application of the pathogen and incubation at 37°C and 5% of CO_2.

3.1.4. Infection of the RHE

3.1.4.1. *C. albicans* Culture

1. The day before the infection assay, the *C. albicans* strain is inoculated in liquid YPD medium.
2. Culture overnight in an incubator at 30°C with continuous shaking.
3. 5 mL of fresh YPD medium is inoculated with the overnight culture to an optical density (OD_{600}) of 0.2 to 0.3 and incubated at 30°C with continuous shaking until the culture reaches OD_{600} of 1.
4. The cells are pelleted by centrifugation at 2000 × *g* for 3 min.

5. The pellet is resuspended in sterile PBS and adjusted to the desired density (10^8 yeast cells/mL), after cell counting with Neubauer chamber, for the inoculation of the RHE.

3.1.4.2. Infection of the RHE

1. Before the infection, the RHE age 4 days (*see* **Section 2.1**) is taken from the CO_2 incubator, and the medium from the inner and outer part of the inserts is removed (*see* **Note 16**).
2. Cell culture medium free of antibiotics (600 μL) is added in the outer part of the insert to feed the cells only through the basolateral surface. Afterwards, the RHE is infected by pipetting 20 μL of the suspension of *C. albicans* in PBS (2×10^6 yeast cells) (*see* **Note 17**). Control of noninfected RHE contains 20 μL of PBS alone.
3. Infected and noninfected RHE are incubated in a CO_2 incubator at 37°C with 5% of CO_2 at 100% of humidity for 8, 24, and 48 h. After the first 24 h, the medium of the outer part of the inserts is replaced with fresh medium.

3.2. Histologic Processing

The progress of infection in this *in vitro* model of candidiasis is monitored by microscopic examination of histologic sections. Here are described the fixation, infiltration with paraffin, embedding in paraffin, microtome slicing, and staining processes prior to the observation under the microscope (**Fig. 10.4**).

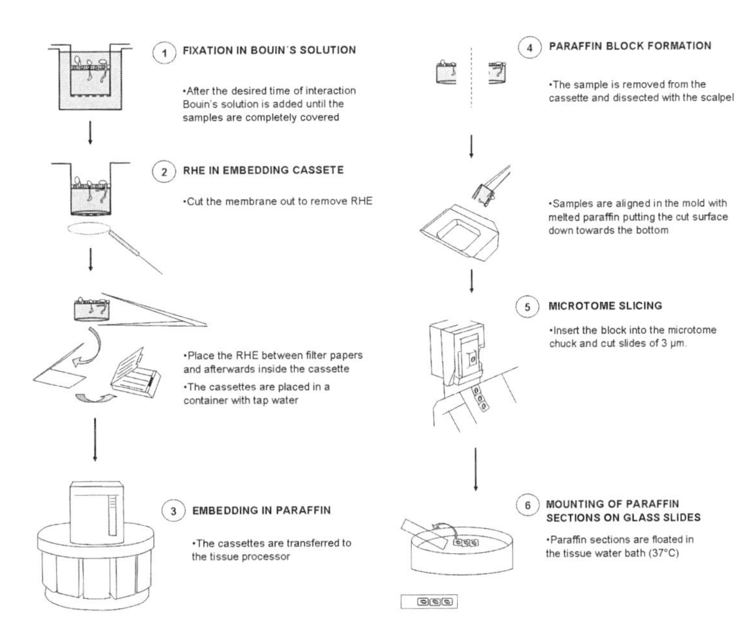

Fig. 10.4. Summary of main steps during the histologic processing. After the desired time of infection, the sample is fixed in Bouin's solution for 1 h **(1)** and placed in embedding cassettes for washing **(2)**. The cassettes are transferred to the tissue processor for the embedding in paraffin **(3)**, and afterwards the sample is manually put into molds **(4)**. The resulting paraffin blocks can be cut in the microtome **(5)** and the 3-μm sections mounted on glass slides in the tissue floating bath **(6)**. Histologic sections are now ready for staining procedures.

3.2.1. Fixation

1. After the desired time of infection, the inserts are transferred into a new 24-well plate.
2. Bouin's solution is added until the samples are completely covered (*see* **Note 18**).
3. After 1 h of fixation, the Bouin's solution is removed. Then, with the help of tweezers go around the edge of the insert and cut the membrane free.
4. Carefully, the sample is placed between filter papers in the embedding cassette and transferred to a container with tap water for washing for 1 h (*see* **Note 19**).

3.2.2. Infiltration and Embedding in Paraffin

1. The cassettes are transferred to the Tissue Processor for the infiltration with paraffin. This instrument features a carousel design with 12 reagent stations: Water I, Water II, 70% Ethanol, 90% Ethanol, 96% Ethanol, Isopropanol I, Isopropanol II, Xylol:Isopropanol (1:1), Xylol I, Xylol II, Paraffin I, Paraffin II. In a 16-h process, the tissues will be moved around through the various agents on a preset timescale. Alternatively, samples can be manually processed.

After the infiltration, the samples are removed from the tissue processor and manually put in molds for the embedding in paraffin.

1. All the instruments (glass container with paraffin and cassettes, molds, scalpel, and tweezers) must be placed in the heating plate (70°C).
2. The sample is removed from the cassette and dissected in two halves with the scalpel on the heating plate.
3. One mold is filled halfway with melted paraffin and chilled in the cold plate only until the bottom has turned white.
4. Using the heated tweezers, the pieces of the sample are aligned in the mold with the cut surface down toward the bottom (*see* **Note 20**).
5. When the tissue is in the desired orientation, the labeled part of the embedding cassette is placed on top of the mold as a backing, and the mold is filled all the way with melted paraffin.
6. Carefully, the mold is cooled in the cold plate and set aside until the paraffin has solidified. Afterwards, the mold is transferred to 4°C for several hours for complete solidification.

3.2.3. Microtome Slicing and Paraffin Section Mounting

The paraffin blocks of tissue are cut using a rotatory microtome in 3-μm sections that will be mounted on glass slides.

1. Turn on the cold plate (−20°C), drying table (42°C), and tissue floating bath (37°C). The tissue floating bath should be filled with tap water about 1 inch from the top.
2. The paraffin blocks are popped out of the molds by scoring the four corners and then pulling each side apart.

3. Then the blocks to be sectioned are placed face down on the cooling plate (−20°C) for 30 min.
4. The paraffin block is inserted into the microtome chuck so the paraffin faces the blade and is aligned in the vertical plane.
5. To start sectioning, the dial is set to cut 15-μm sections. The first few sections will probably be only fragments of paraffin; they are discarded until the tissue can be seen coming through.
6. After that, the dial is set to cut 3-μm sections, and when several slices have been sectioned, they are removed from the blade with tweezers or a fine paint brush (*see* **Note 21**).
7. Then the sections are floated in the tissue water bath (37°C) to remove the wrinkles and are picked up on a glass microscope slide (*see* **Note 22**).
8. The glass slides are then placed in the drying plate to remove the excess of water and to help the section to adhere to the slide.

3.2.4. PAS–Papanicolaou Staining

The slides are placed in the tray for staining and kept in the incubator at 65°C for 30 min to bind the tissue to the glass. Afterwards, the tray is transferred sequentially through different solutions as it is described in the following (*see* **Note 23**):

1. Deparaffinization in Roticlear for 10 min.
2. Second solution of Roticlear for 3 min.
3. Then start the progressive hydration by transferring to decreasing concentrations of ethanol: 2 min in ethanol 96%, then 2 min in a second solution of ethanol 96%, 2 min in ethanol 70%, 2 min in ethanol 50%, and double-distilled water (ddH_2O).
4. 1% periodic acid for 5 min.
5. Rinse in ddH_2O.
6. Schiff's reagent for 15 min. Watch until tissue turns to deep magenta color.
7. Sodium bisulfite solution for 6 min changing for new solution every 2 min.
8. Wash in running tap water for 5 min.
9. Mayer's hematoxylin for 15 s.
10. Rinse in HCl 0.03 M.
11. Wash in running tap water for 10 min. The washing will be more effective if the water is hand warm.
12. Start the progressive dehydration by transferring to increasing concentrations of ethanol: 2 min in ethanol 50%, then 2 min in ethanol 70%, and finally 2 min in ethanol 96%.

13. Polychromic staining solution EA50 for 1 min.
14. Rinse twice in ethanol 96% (3 times, any of them change for new solution).
15. Ethanol 100% for 30 s.
16. Clearing in isopropanol for 5 min.
17. Second isopropanol for 5 min.

3.2.5. Slides Coverslipping

The slides are removed from the isopropanol bath and coverslipped before they get dry:

1. A small drop of mounting solution is applied on every tissue sample and a glass coverslip placed over the slide.
2. Finally, the tissue sections are studied and documented using an optical microscope with adapted digital camera as shown in **Fig. 10.2**.

4. Notes

1. All the solutions, media, and material used must be sterile and maintained in sterile conditions by handling under a laminar flow hood.
2. The serum must be inactivated by high temperature before using it to supplement the cell culture medium. As the storage of commercial FBS is at −20°C, thaw the serum in a water bath at 37°C and afterwards transfer the bottle to another waterbath at 56°C for 30 min. The serum is now heat-inactivated and can be aliquoted into 50-mL sterile tubes and stored at −20°C.
3. To prepare the cell culture medium for the Caco-2 and A431 cell lines, open the bottle of commercial DMEM under sterile conditions and add 50 mL of heat-inactivated FBS, 5 mL of 100 mM sodium pyruvate, and 5 mL of 10 mg/mL gentamicin.
4. To prepare the cell culture medium for the TR146 cell line, open the bottle of commercial DMEM under sterile conditions and add 50 mL of heat-inactivated FBS and 5 mL of 100x (10,000 IU/mL – 10 mg/mL) penicillin-streptomycin.
5. To avoid activity loss by repeated freeze-thaw cycles, make small aliquots of trypsin-EDTA and penicillin-streptomycin. Thaw a bottle of trypsin-EDTA and a bottle of penicillin-streptomycin by placing in the 37°C water bath (~1 h). In the hood, sterilely aliquot into 15-mL sterile tubes. Each tube should have 5 mL. Store the labeled aliquots at −20°C until needed.

6. The construction of the model and the infection assay are performed in medium free from any antibiotic. Cell culture medium should be prepared for the infection assays as previously explained (*see* **Notes 3** and **4**) except for the addition of gentamicin or penicillin-streptomycin.
7. If bigger amounts of tissue are required for further analysis, cell culture inserts for 6- and 12-well plates can also be used to prepare the RHE.
8. The solution of periodic acid and of sodium bisulfite must be prepared fresh before starting the staining protocol.
9. The sodium bisulfite is an irritant to the respiratory tract and should always be handled in a fume extraction hood. The amount prepared of sodium bisulfite should be enough for the three changes needed during the staining protocol.
10. This step is designed to remove traces of serum that would inhibit the action of trypsin. A buffered salt solution free of Ca^{2+} and Mg^{2+} should be used to wash cells, as divalent cations can cause cells to stick together.
11. Tap the bottom of the flask to dislodge cells, and the monolayer should slide down the surface. Check the culture with an inverted microscope to be sure that cells have a spherical shape and are detached from the surface. Do not leave the flasks longer than necessary but, on the other hand, if cells are not sufficiently detached, return plate to 37°C incubator for an additional minute or two. Do not force the cells to detach before they are ready to do so, or else clumping may result.
12. Before counting, the cell suspension is pipetted up and down a few times carefully to avoid mechanical damage of the cells and the creation of foam. The degree of pipetting required will vary from one cell line to another. Cells can be counted using a Neubauer counting chamber and 0.4% Trypan blue alternatively if an electronic cell counter is not available.
13. Each batch of collagen solution will vary, depending on source and age of rats, and must be titrated for producing consistent gels. Between 450 and 650 μL of neutralizing solution is added to each well with 1 mL of collagen to find the minimum amount necessary to produce the firmest gel. Place the plate at 37°C for 15 min to solidify and choose the amount added that produces a gel that remains firm in the multiwell when inverted.
14. This procedure must be fast. To avoid an early gelification, solutions must be kept on ice before mixing, and the distribution in the inserts must be done with a multipette. To allow

the formation of a homogeneous gel and to avoid clump formation, place the plate in the CO_2 incubator at 37°C for 15 min and do not move during the gel formation.

15. Further improvement of the system to mimic the *in vivo* situation more closely can be done by the addition of fibroblasts or any other cell type embedded in the collagen matrix *(19)*, colonization with bacteria from the normal human flora *(52)*, or even to integrate a third cell type, such as neutrophils *(51)* or dendritic cells.
16. The medium is carefully removed from the inner part of the insert to avoid any damage of the RHE by using a manual micropipette with tip. To be able to aspirate the medium slowly, do not use the Pasteur pipette connected to a vacuum flask.
17. Avoid touching the RHE with the tip of the pipette and be sure that the inoculum is equally distributed over the surface of the epithelium.
18. Bouin's solution has harmful components like picric acid and formaldehyde. It should be handled with care using nitrile gloves and always in the fume extraction hood. The material in contact with the solution should be left under the extraction hood for 24 h for evaporation before disposal. Consult your safety officer for handling and disposal of the fixative.
19. Tissue embedding cassettes are used to hold tissue specimens during the washing, embedding, sectioning, and storage. The cassettes must be labeled with the characteristics of the sample using a pencil and a double filter paper that will hold the sample inserted in the cassette.
20. Alignment of the tissue in the direction of cutting with the microtome also is necessary for excellent ribboning.
21. If the block is not ribboning well, then place it back on the cooling plate or change the blade of the microtome for a new one.
22. Slides should be labeled using a pencil to avoid loss of information during the staining protocol.
23. During the staining process, wear gloves and avoid contact and inhalation with components working under the extraction hood. The steps during the staining protocol are performed by transferring the glass tray with the slides through different solutions. The amount of every solution should be enough to completely cover the slides (in our case, 250 mL) and can be recycled at least 3 times, except for periodic acid, sodium bisulfite solution, and HCl 0.03 M. After the staining, the toxic solutions (Schiff's reagent and polychromic staining solution) must be placed in a special disposal unit for hazardous wastes.

Acknowledgments

We would like to thank all the members of the department of Molecular Biotechnology of Fraunhofer IGB (Stuttgart, Germany) and the Cell Systems Department of Fraunhofer IGB (Stuttgart, Germany) (Prof. Dr. Heike Mertsching) for kindly providing their installations, technical support, and advice during the histology processing. This work was supported by the European Union through the funding of the EC Marie Curie Training Network *"CanTrain"* (CT-2004-512481).

References

1. Calderone, R.A. and Fonzi, W.A. (2001) Virulence factors of *Candida albicans*. *Trends Microbiol.* 9, 327.
2. Braun, B.R. and Johnson, A.D. (1997) Control of filament formation in *Candida albicans* by the transcriptional repressor TUP1. *Science* 277, 105.
3. Slutsky, B., Staebell, M., Anderson, J., Risen, L., Pfaller, M. and Soll, D.R. (1987) "White-opaque transition": a second high-frequency switching system in *Candida albicans*. *J. Bacteriol.* 169, 189.
4. Fu, Y., Ibrahim, A.S., Sheppard, D.C., Chen, Y.C., French, S.W., Cutler, J.E., Filler, S.G. and Edwards, J.E., Jr. (2002) *Candida albicans* Als1p: an adhesin that is a downstream effector of the EFG1 filamentation pathway. *Mol. Microbiol.* 44, 61.
5. Sheppard, D.C., Yeaman, M.R., Welch, W.H., Phan, Q.T., Fu, Y., Ibrahim, A.S., Filler, S.G., Zhang, M., Waring, A.J. and Edwards, J.E., Jr. (2004) Functional and structural diversity in the Als protein family of *Candida albicans*. *J. Biol. Chem.* 279, 30480.
6. Davies, J.M., Stacey, A.J. and Gilligan, C.A. (1999) *Candida albicans* hyphal invasion: thigmotropism or chemotropism? *FEMS Microbiol. Lett.* 171, 245.
7. Ghannoum, M.A. (2000) Potential role of phospholipases in virulence and fungal pathogenesis. *Clin. Microbiol. Rev.* 13, 122.
8. Ibrahim, A.S., Mirbod, F., Filler, S.G., Banno, Y., Cole, G.T., Kitajima, Y., Edwards, J.E., Jr., Nozawa, Y. and Ghannoum, M.A. (1995) Evidence implicating phospholipase as a virulence factor of *Candida albicans*. *Infect. Immun.* 63, 1993.
9. Naglik, J.R., Challacombe, S.J. and Hube, B. (2003) *Candida albicans* secreted aspartyl proteinases in virulence and pathogenesis. *Microbiol. Mol. Biol. Rev.* 67, 400.
10. Schaller, M., Borelli, C., Korting, H.C. and Hube, B. (2005) Hydrolytic enzymes as virulence factors of *Candida albicans*. *Mycoses* 48, 365.
11. Filler, S.G. and Sheppard, D.C. (2006) Fungal invasion of normally non-phagocytic host cells. *PLoS. Pathog.* 2, e129.
12. Chiang, L.Y., Sheppard, D.C., Bruno, V.M., Mitchell, A.P., Edwards, J.E., Jr. and Filler, S.G. (2007) *Candida albicans* protein kinase CK2 governs virulence during oropharyngeal candidiasis. *Cell Microbiol.* 9, 233.
13. Park, H., Myers, C.L., Sheppard, D.C., Phan, Q.T., Sanchez, A.A., Edwards, E. and Filler, S.G. (2005) Role of the fungal Ras-protein kinase A pathway in governing epithelial cell interactions during oropharyngeal candidiasis. *Cell Microbiol.* 7, 499.
14. Zakikhany, K., Naglik, J.R., Schmidt-Westhausen, A., Holland, G., Schaller, M. and Hube, B. (2007) *In vivo* transcript profiling of *Candida albicans* identifies a gene essential for interepithelial dissemination. *Cell Microbiol.* 9, 2938–54.
15. Allen, C.M. (1994) Animal models of oral candidiasis. A review. *Oral Surg. Oral Med. Oral Pathol.* 78, 216.
16. de Repentigny, L. (2004) Animal models in the analysis of *Candida* host-pathogen interactions. *Curr. Opin. Microbiol.* 7, 324.
17. McMillan, M.D. and Cowell, V.M. (1985) Experimental candidiasis in the hamster cheek pouch. *Arch. Oral Biol.* 30, 249.
18. Ekenna, O. and Sherertz, R.J. (1987) Factors affecting colonization and

dissemination of *Candida albicans* from the gastrointestinal tract of mice. *Infect. Immun.* 55, 1558.

19. Dieterich, C., Schandar, M., Noll, M., Johannes, F.J., Brunner, H., Graeve, T. and Rupp, S. (2002) *In vitro* reconstructed human epithelia reveal contributions of *Candida albicans* EFG1 and CPH1 to adhesion and invasion. *Microbiology* 148, 497.
20. Korting, H.C., Patzak, U., Schaller, M. and Maibach, H.I. (1998) A model of human cutaneous candidosis based on reconstructed human epidermis for the light and electron microscopic study of pathogenesis and treatment. *J. Infect.* 36, 259.
21. Rouabhia, M., Schaller, M., Corbucci, C., Vecchiarelli, A., Prill, S.K., Giasson, L. and Ernst, J.F. (2005) Virulence of the fungal pathogen *Candida albicans* requires the five isoforms of protein mannosyltransferases. *Infect. Immun.* 73, 4571.
22. Schaller, M., Schafer, W., Korting, H.C. and Hube, B. (1998) Differential expression of secreted aspartyl proteinases in a model of human oral candidosis and in patient samples from the oral cavity. *Mol. Microbiol.* 29, 605.
23. Bernhardt, J., Herman, D., Sheridan, M. and Calderone, R. (2001) Adherence and invasion studies of *Candida albicans* strains, using *in vitro* models of esophageal candidiasis. *J. Infect. Dis.* 184, 1170.
24. Cheng, G., Wozniak, K., Wallig, M.A., Fidel, P.L., Jr., Trupin, S.R. and Hoyer, L.L. (2005) Comparison between *Candida albicans* agglutinin-like sequence gene expression patterns in human clinical specimens and models of vaginal candidiasis. *Infect. Immun.* 73, 1656.
25. Schaller, M., Bein, M., Korting, H.C., Baur, S., Hamm, G., Monod, M., Beinhauer, S. and Hube, B. (2003) The secreted aspartyl proteinases Sap1 and Sap2 cause tissue damage in an *in vitro* model of vaginal candidiasis based on reconstituted human vaginal epithelium. *Infect. Immun.* 71, 3227.
26. Green, C.B., Cheng, G., Chandra, J., Mukherjee, P., Ghannoum, M.A. and Hoyer, L.L. (2004) RT-PCR detection of *Candida albicans* ALS gene expression in the reconstituted human epithelium (RHE) model of oral candidiasis and in model biofilms. *Microbiology* 150, 267.
27. Jayatilake, J.A., Samaranayake, Y.H. and Samaranayake, L.P. (2005) An ultrastructural and a cytochemical study of candidal invasion of reconstituted human oral epithelium. *J. Oral Pathol. Med.* 34, 240.
28. Korting, H.C., Hube, B., Oberbauer, S., Januschke, E., Hamm, G., Albrecht, A., Borelli, C. and Schaller, M. (2003) Reduced expression of the hyphal-independent *Candida albicans* proteinase genes SAP1 and SAP3 in the efg1 mutant is associated with attenuated virulence during infection of oral epithelium. *J. Med. Microbiol.* 52, 623.
29. Schaller, M., Korting, H.C., Schafer, W., Bastert, J., Chen, W. and Hube, B. (1999) Secreted aspartic proteinase (Sap) activity contributes to tissue damage in a model of human oral candidosis. *Mol. Microbiol.* 34, 169.
30. Hube, B. and Naglik, J. (2001) *Candida albicans* proteinases: resolving the mystery of a gene family. *Microbiology* 147, 1997.
31. Albrecht, A., Felk, A., Pichova, I., Naglik, J.R., Schaller, M., de Groot, P., Maccallum, D., Odds, F.C., Schafer, W., Klis, F., Monod, M. and Hube, B. (2006) Glycosylphosphatidylinositol-anchored proteases of *Candida albicans* target proteins necessary for both cellular processes and host-pathogen interactions. *J. Biol. Chem.* 281, 688.
32. Heymann, P., Gerads, M., Schaller, M., Dromer, F., Winkelmann, G. and Ernst, J.F. (2002) The siderophore iron transporter of *Candida albicans* (Sit1p/Arn1p) mediates uptake of ferrichrome-type siderophores and is required for epithelial invasion. *Infect. Immun.* 70, 5246.
33. Wilson, D., Tutulan-Cunita, A., Jung, W., Hauser, N.C., Hernandez, R., Williamson, T., Piekarska, K., Rupp, S., Young, T. and Stateva, L. (2007) Deletion of the high-affinity cAMP phosphodiesterase encoded by PDE2 affects stress responses and virulence in *Candida albicans. Mol. Microbiol.* 65, 841.
34. Villar, C.C., Kashleva, H., Mitchell, A.P. and Dongari-Bagtzoglou, A. (2005) Invasive phenotype of *Candida albicans* affects the host proinflammatory response to infection. *Infect. Immun.* 73, 4588.
35. Fidel, P.L., Jr. (2007) History and update on host defense against vaginal candidiasis. *Am. J. Reprod. Immunol.* 57, 2.

36. Saegusa, S., Totsuka, M., Kaminogawa, S. and Hosoi, T. (2004) *Candida albicans* and *Saccharomyces cerevisiae* induce interleukin-8 production from intestinal epithelial-like Caco-2 cells in the presence of butyric acid. *FEMS Immunol. Med. Microbiol.* 41, 227.
37. Schaller, M., Korting, H.C., Borelli, C., Hamm, G. and Hube, B. (2005) *Candida albicans*-secreted aspartic proteinases modify the epithelial cytokine response in an *in vitro* model of vaginal candidiasis. *Infect. Immun.* 73, 2758.
38. Schaller, M., Mailhammer, R., Grassl, G., Sander, C.A., Hube, B. and Korting, H.C. (2002) Infection of human oral epithelia with *Candida* species induces cytokine expression correlated to the degree of virulence. *J. Invest. Dermatol.* 118, 652.
39. Lu, Q., Jayatilake, J.A., Samaranayake, L.P. and Jin, L. (2006) Hyphal invasion of *Candida albicans* inhibits the expression of human beta-defensins in experimental oral candidiasis. *J. Invest. Dermatol.* 126, 2049.
40. Bernhardt, J., Bernhardt, H., Knoke, M. and Ludwig, K. (2004) Influence of voriconazole and fluconazole on reconstituted multilayered oesophageal epithelium infected by *Candida albicans*. *Mycoses* 47, 330.
41. Schaller, M., Laude, J., Bodewaldt, H., Hamm, G. and Korting, H.C. (2004) Toxicity and antimicrobial activity of a hydrocolloid dressing containing silver particles in an ex vivo model of cutaneous infection. *Skin Pharmacol. Physiol.* 17, 31.
42. Frank, C.F. and Hostetter, M.K. (2007) Cleavage of E-cadherin: a mechanism for disruption of the intestinal epithelial barrier by *Candida albicans*. *Transl. Res.* 149, 211.
43. Cole, G.T., Halawa, A.A. and Anaissie, E.J. (1996) The role of the gastrointestinal tract in hematogenous candidiasis: from the laboratory to the bedside. *Clin. Infect. Dis.* 22 (Suppl 2), S73.
44. Andrutis, K.A., Riggle, P.J., Kumamoto, C.A. and Tzipori, S. (2000) Intestinal lesions associated with disseminated candidiasis in an experimental animal model. *J. Clin. Microbiol.* 38, 2317.
45. Giard, D.J., Aaronson, S.A., Todaro, G.J., Arnstein, P., Kersey, J.H., Dosik, H. and Parks, W.P. (1973) *In vitro* cultivation of human tumors: establishment of cell lines derived from a series of solid tumors. *J. Natl. Cancer Inst.* 51, 1417.
46. Rupniak, H.T., Rowlatt, C., Lane, E.B., Steele, J.G., Trejdosiewicz, L.K., Laskiewicz, B., Povey, S. and Hill, B.T. (1985) Characteristics of four new human cell lines derived from squamous cell carcinomas of the head and neck. *J. Natl. Cancer Inst.* 75, 621.
47. Sobel, J.D. (2007) Vulvovaginal candidosis. *Lancet* 369, 1961.
48. Zaremba, M.L., Daniluk, T., Rozkiewicz, D., Cylwik-Rokicka, D., Kierklo, A., Tokajuk, G., Dabrowska, E., Pawinska, M., Klimiuk, A., Stokowska, W. and Abdelrazek, S. (2006) Incidence rate of *Candida* species in the oral cavity of middle-aged and elderly subjects. *Adv. Med. Sci.* 51 (Suppl 1), 236.
49. McManaus, G.N. (1989) Darstellung von paraplasmatischen substanzen, PAS reaktion nach McManaus. In: Boeck P., ed. *Romeis, Mikroskopische Technik.* Muenchen, Wien, Baltimore: Urban and Schwarzenberg, p. 394.
50. Papanicolaou, J.F.A. (1989) Faerbetechniken der zytodiagnostik, faerbung nach Papanicolaou. In: Boeck P., ed. *Romeis, Mikroskopische Technik.* Muenchen, Wien, Baltimore: Urban and Schwarzenberg, p. 646.
51. Schaller, M., Boeld, U., Oberbauer, S., Hamm, G., Hube, B. and Korting, H.C. (2004) Polymorphonuclear leukocytes (PMNs) induce protective Th1-type cytokine epithelial responses in an in vitro model of oral candidosis. *Microbiology* 150, 2807.
52. Kaewsrichan, J., Peeyananjarassri, K. and Kongprasertkit, J. (2006) Selection and identification of anaerobic *lactobacilli* producing inhibitory compounds against vaginal pathogens. *FEMS Immunol. Med. Microbiol.* 48, 75.
53. Gillum, A.M., Tsay, E.Y. and Kirsch, D.R. (1984) Isolation of the *Candida albicans* gene for orotidine-5′-phosphate decarboxylase by complementation of *S. cerevisiae* ura3 and *E. coli* pyrF mutations. *Mol. Gen. Genet.* 198, 179.
54. Freshney, R.I. (1993) Culture of animal cells. In: *A Manual of Basic Techniques*, 3rd ed. New York: Wiley-Liss.
55. Phelan, M.C. (2003) Basic techniques for mammalian cell tissue culture. In: Bonifacio J.S., Dasso M., Harford J.B., Lippincott-Scwartz J., Yamada K.M., eds. *Current Protocols in Cell Biology*, Hoboken, NJ: John Wiley & Sons. Chapter 1, Unit 1.1, Page 1.

Chapter 11

In Vitro Systems for Studying the Interaction of Fungal Pathogens with Primary Cells from the Mammalian Innate Immune System

Christelle Bourgeois, Olivia Majer, Ingrid Frohner, and Karl Kuchler

Abstract

The incidence of invasive fungal diseases has increased over the past decades, particularly in relation with the increase of immunocompromised patient cohorts (e.g., HIV-infected patients, transplant recipients, immunosuppressed patients with cancer). Opportunistic fungal pathogens such as *Candida* spp. are most often associated with serious systemic infections. Currently available antifungal drugs are rather unspecific, often with severe side effects. In some cases, their prophylactic use has favored emergence of resistant fungal strains. Major antifungal drugs target the biosynthesis of lipid components of the fungal plasma membrane or the assembly of the cell wall. For a more specific and efficient treatment and prevention of fungal infection, new therapeutic strategies are needed, including strengthening or stimulation of the residual host immune response. Achieving such a goal requires a better understanding of factors important for the defense and the survival of the host combating *Candida* spp. Where possible, primary cultures of mammalian immune cells of the innate immune system constitute a better suited model than transformed cell lines to study host-pathogen response and virulence. Hence, *in vitro* primary cell culture systems are a good strategy for a first screening of mutant strains of *Candida* spp. to identify virulence traits with regard to host cell response and pathogen invasion.

Key words: primary cell culture, bone marrow–derived macrophages, myeloid dendritic cells, *Candida* spp., host-pathogen interaction, cell signaling, MAPK, cytokines.

1. Introduction

Candida albicans (C.a) and other *Candida* spp. are harmless commensals in most healthy people. However, they cause both superficial infections and life-threatening systemic candidiasis in immunocompromised patients. Cells of the innate immune system such as dendritic cells, macrophages, or neutrophils comprise the first line of defense against microbial pathogens. *Candida* and other

Steffen Rupp, Kai Sohn (eds.), *Host-Pathogen Interactions*, DOI: 10.1007/978-1-59745-204-5_11,

fungi are detected and recognized by the innate immune system through pattern recognition receptors (e.g., *toll*-like receptors, mannose receptor) and coactivators (dectin1, CD14), which recognize pathogen-associated molecular patterns (PAMPs) found in the fungal cell wall.

Mouse models lacking the TLR2,4 or dectin-1 genes indicate a role for these recognition molecules in detecting fungal pathogens and triggering the adaptive cytokine response, which in turn leads to an efficient activation of the acquired immune system. The cytokine response, a consequence of the various combinations of signaling pathways activated by these surface receptors (e.g., mitogen-activated protein kinase [MAPK] pathway, NF-κB activation) drive the host response and determine the outcome of infection (for review, see Refs. *1* and *2*). In the case of *Candida* infections, the balance between the production of inflammatory cytokines (e.g., TNF-α), which promote activation of the immune system and destruction of the pathogen, and the release of anti-inflammatory cytokines (e.g., IL-10), which limits the extent of tissue damage induced by inflammation and activates the adaptive immune response *(3–6)*, appears to be particularly important. To counteract the host response, microbial pathogens have developed escape strategies. In the case of *C.a.*, modulation of the activation of the MAPK/extracellular regulatory kinase (ERK) and p38 pathways may be one of the mechanism by which fungi modulate the cytokine response to its advantage *(7–9)*.

In vitro cell culture models are interesting tools to unravel dynamic changes of signaling activities as they allow for following the initial host attack, with the further goal of identifying downstream factors important for the defense and the survival of the host innate immune cells facing fungal pathogens in general and in particular *Candida* spp. They can be a good compromise for a first screening of virulence properties of fungal mutant strains lacking potential pathogenicity genes affecting host response or pathogen invasiveness even before the use of animal models for *in vivo* studies. Here we describe highly standardized primary cell culture models suitable to study early stages of innate immune cell–*Candida* interaction (e.g., pathogen phagocytosis, MAPK activation, cytokine production) and signaling events driving fungal invasion.

2. Materials and Media Components

2.1. Primary Culture of Bone Marrow–Derived Macrophages

1. High glucose (4.5 g/L) Dulbecco's Modified Eagle's Medium (DMEM) with L-glutamine, without pyruvate (PAA, Vienna, Austria).

2. Sterile PBS.
3. Colony-stimulating-factor 1 (CSF-1)-producing L929 cell line (ATCC no. CCL-1).
4. Bone marrow–derived macrophage (mMP) culture medium, high-glucose DMEM with L-glutamine supplemented with 10% fetal calf serum (FCS), 100 U/mL penicillin, 100 μg/mL streptomycin (Invitrogen, Carlsbad, CA), and 15% to 20% L-conditioned medium, as source of CSF-1 (for preparation, *see* **Notes 1** and **2**).
5. 10 × 10 cm square sterile Petri dishes (nontreated for cell culture; Barloworld Scientific, Stone, UK).
6. Soft-rubber spatula (Deutsch & Neumann, Berlin, Germany).

2.2. Primary Culture of Myeloid Dendritic Cells

1. Red blood-cell lysis buffer, 8.29 g/L NH_4Cl, 1 g/L $KHCO_3$, 0.0372 g/L EDTA, pH 7.2 to 7.4. Adjust pH if necessary, sterile filtrate through 0.2-μm membrane filter, store at 4°C.
2. Granulocyte-macrophage colony-stimulating factor (GM-CSF)-producing X-63 cell line *(10)*.
3. Myeloid DC (mDC) culture medium, high-glucose DMEM with glutamine supplemented with 10% FCS, 100 U/mL penicillin, 100 μg/mL streptomycin, and 5–10% X-conditioned medium as source of GM-CSF (for preparation, *see* **Notes 2** and **3**).
4. Cell-culture treated 24-well plates (NUNC, Roskilde, Denmark).

2.3. Cell Characterization by FACS Analysis

1. FACS buffer, PBS containing 2 g/L sodium azide and 2 g/L BSA, sterile-filtrated through an 0.2-μm filter and stored at 4°C.
2. Anti-mouse antibodies CD16/CD32, CD11b-FITC, CD11c-APC (BD Bioscience, Clontech, Palo Alto, CA).

2.4. Host Cell/Fungi Interaction

1. Laminar hood and 37°C incubator with 5% CO_2, 95% humidity, used only for infection purposes.
2. High glucose (4.5 g/L) DMEM without phenol red (Invitrogen), supplemented with 4 mM L-glutamine.
3. SC5314, clinical isolate of *Candida albicans (11)*.
4. UV-treated *Candida albicans* are prepared by treating an aliquot of the *Candida* infection suspension with 999 μJ/cm^2 in a Stratalinker (Stratagene, La Jolla, CA).
5. YPD agar plates, YPD liquid media for growing and culturing fungi.
6. 2 μM Cytochalasin D (Sigma, St. Louis, MO).
7. Cell scrappers (Becton Dickinson Labware, Franklin Lakes, NJ).

2.5. Microscopic Internalization Assay

1. Autoclaved, 12-mm-diameter glass coverslips, distributed in a 24-well plate.
2. *Candida albicans* strain expressing GFP intrinsically *(12)*.
3. 5 mM Calcofluor White M2R solution (Molecular Probes, Invitrogen, Carlsbad, CA).
4. Nonhardening fluorescence mounting media (Dako, Glostrup, Denmark).

2.6. Protein Extract and Immunoblotting

1. Protein lysis buffer (Frackelton buffer) *(13)*, 10 mM Tris pH 7.5, 50 mM NaCl, 1% Triton X-100, 1 mM PMSF, protease inhibitor cocktail (complete no EDTA; Roche, Basel, Switzerland), phosphatase inhibitors (30 mM NaPPi, 50 mM NaF, 0.1 mM sodium vanadate). Prepare fresh for each experiment and chill on ice.
2. 4x Sample Buffer (SBF), 200 mM Tris pH 6.8, 40% glycerol, 8% SDS, 0.002% bromophenol blue. Add 4% (v/v) β-mercaptoethanol just before use.
3. 1x TBST buffer, 3 g/L Tris-HCl, 8 g/L NaCl, 0.2 g/L KCl, 0.1% (v/v) Tween-20 pH 7.4.
4. Anti-mouse panERK (BD Transduction Laboratories, Palo Alto, CA), anti-mouse phospho-ERK1/2 and anti-mouse p38 (Santa Cruz Biotech Inc., Santa Cruz, CA), anti-mouse phospho-p38 (Cell Signaling Technologies Inc., Danvers, MA).
5. Horseradish peroxidase–coupled secondary antibodies (Merck, Whitehouse Station, NJ).
6. ECL reagents for immunodetection (Pierce, Rockford, IL).

2.7. RNA Extraction Procedure and Real-Time PCR

1. Spin column–based RNA extraction kit (BD Bioscience, Clontech, Palo Alto, CA, or Promega, Madison, WI).
2. First strand cDNA synthesis kit (Fermentas, Hanover, MD).
3. Real-time PCR mix, 75 mM Tris-HCl pH 8.8, 20 mM $(NH_4)_2SO_4$, 0.01% (v/v) Tween-20, 2.5 mM $MgCl_2$, 0.2 mM dNTPs, 300 nM Forward primer, 300 nM Reverse primer, 200 nM SYBR green (Biorad, Hercules, CA), 1 U recombinant Taq DNA polymerase (5 U/μL; Fermentas).
4. Mouse tumor necrosis factor-α (TNF-α); primers used, *forward* 5′-CATCTTCTCAAAATTCGAGTGACAA-3′; and *reverse* 5′-TGGGAGTAGACAAGGTACAACCC-3′ *(14)*.
5. Mouse interleukin 10 (IL-10) primers used: *forward* 5′-GGTTGCCAAGCCTTATCGGA-3′; and *reverse* 5′-ACCTGCTCCACTGCCTTGCT-3′ *(14)*.
6. Mouse GAPDH primers used: *forward* 5′-CATGGCCTTCCGTGTTCCTA-3′; and *reverse* 5′-GCGGCACGTCAGATCCA-3′ (RTPrimerDB, the real-time PCR primer and probe database http://medgen.ugent.be/rtprimerdb/index.php) *(15, 16)*.

3. Methods

3.1. Primary Culture of Bone Marrow–Derived Macrophages (mMPs)

This method is adapted from a protocol published earlier *(17)*.

1. On day 1, dissect mouse tibias and femurs from a 6- to 8-week-old animal in a hood, quickly rinse bones in 70% ethanol, and place in 15 mL ice-cold sterile PBS. When all the limbs are collected, transfer them in fresh 15 mL ice-cold sterile PBS and keep on ice (*see* **Note 4**).
2. To flush out the bone marrow, separate femur from tibia at the knee joint. Holding the bone with forceps above a sterile dish, cut one extremity of the bone and using a 20-mL syringe with a 27GX3/4 needle, flush DMEM with 10% FCS, 100 U/mL penicillin and 100 μg/mL streptomycin, into the medullary cavity until no more cells are coming out.
3. Collect bone marrow suspension and keep it on ice until all bones have been processed. Bone marrow flushing should be performed under semisterile conditions as required for cell culture.
4. To prepare mMPs, centrifuge the collected bone marrow at 300 × *g* for 5 min and resuspend the pellet in 44 mL mMP medium. Distribute the cell suspension equally in four 10 × 10 cm Petri dish (or seven 10-cm-diameter Petri dishes) and transfer to a 37°C incubator with a 5% CO_2, 95% humidity atmosphere (*see* **Notes 5** and **6**).
5. On day 2, add 6 to 8 mL of mMP medium; control for cell density every day.
6. On days 4 to 5, when cells in the plate reach confluency, aspirate the media containing nonadherent cells, gently collect cells by scrapping the plates with a soft rubber spatula, and re-plate at a ratio of 1:2 to 1:3 in square 10 × 10 cm Petri dish (*see* **Note** 7).
7. Let the cells grow for another couple of days, change medium completely every 2 to 3 days.
8. After 9 to 10 days of culture, mMP cell surface markers should be tested before performing interaction experiments (*see* **Section 3.3**).

3.2. Primary Culture of Myeloid Dendritic Cells (mDCs)

This method is based on a method described earlier *(18)* using X-conditioned media as source of GM-CSF.

1. For mouse bone marrow isolation, proceed as described above (*see* **Section 3.1, steps 1** to **3**).
2. To prepare mDCs, centrifuge bone marrow cell suspension at 300 × *g* for 7 min.

3. Resuspend the pellet in 1 mL of room-temperature red cell lysis buffer, and immediately stop the lysis with 1 mL of mDC medium (*see* **Note 5**).
4. Centrifuge cell suspension at 300 × *g* for 7 min. Resuspend pellet in 5 mL mDC medium.
5. Count cells and distribute in a tissue culture–treated 24-well plate, to obtain 10^6 cells/well in 1 mL. Transfer to a 37°C incubator with a 5% CO_2, 95% humidity atmosphere.
6. On day 4, aspirate 0.5 mL of medium from each well with a Gilson pipette and add 1 mL of fresh mDC medium. Under the microscope, loosely attached nodules of mDCs will start appearing as darker mass on the bright layer of adherent cells.
7. On day 7, collect these mDC aggregates by flushing medium against the well wall with a 1 mL Gilson pipette set at 800 μL, in order not to lift up too many of the strongly adherent cells.
8. Pool the cells suspension of each well and re-plate cells at the desired cell density to perform an experiment the next day. Myeloid mDC cell-surface markers should be checked before performing interaction experiments (*see* **Section 3.3**). Myeloid mDCs should be used within 8 days of their preparation.

3.3. Characterization of Cell Markers by FACS Analysis

1. Prepare a 4 × 10^7 cells/mL suspension in FACS buffer; distribute 12.5 μL (0.5 × 10^6 cells) in 3 microcentrifuge tubes.
2. Block nonspecific Fc-binding sites with 12.5 μL of CD36/CD32 antibodies diluted 1/25 in FACS buffer.
3. After a 5-min incubation at room temperature, add 25 μL of anti-mouse CD11b-FITC diluted at 1/25 in FACS buffer, or 25 μL of anti-mouse CD11c-APC diluted at 1/25 in FACS buffer, or 25 μL FACS buffer alone for the negative control (*see* **Note 8**).
4. After 15 to 20 min on ice, wash with 800 μL FACS buffer and centrifuge at 300 × *g* for 10 min at 4°C (low-speed centrifugation is important to prevent cell damage). Repeat washing step once.
5. After the second wash, resuspend cell pellet in 500 μL FACS buffer (or less if less cells) and transfer to FACS tube for analysis (*see* **Note 9**).

3.4. Interaction Experiments In Vitro with Candida spp.

1. One day prior to the interaction assay, plate mMPs or mDCs at a density of 1.0 × 10^5 to 1.25 × 10^5 cells/cm^2 in a volume of cell culture medium of of 0.2 to 0.4 mL/cm^2 and place them at 37°C in a 5% CO_2, 95% humidity atmosphere.
2. Grow *C.a.* to saturation overnight in 25 mL 1X YPD with continuous shaking at 30°C. The next morning, dilute to 0.2

to 0.3 OD_{600} in 25 mL YPD liquid medium and incubate with continuous shaking at 30°C until the culture reaches 1 OD_{600}.

3. Collect the fungal cells by centrifugation at 1200 × *g* for 5 min at room temperature and rinse the pellet in 50 mL sterile, room temperature H_2O or PBS.
4. Centrifuge again, resuspend the fungal pellet in 1 mL sterile PBS, and determine the fungus counts/mL.
5. Dilute the fungal cell suspension in prewarmed (37°C) high-glucose DMEM without phenol red (*see* **Note 10**) so that fungal-mammalian cell coculture is performed at a ratio of 2 fungal cells per 1 cell mMP or mDC (multiplicity of infection [MOI] 2:1; *see* **Note 11**).
6. Proceed to **step 7** for infection of mMPs or to **step 8** for infection of mDCs.
7. Aspirate the media from mMP culture dishes, replace it with high-glucose DMEM without phenol red with or without *Candida*. Typically, interaction *C.a.*-mMPs are carried out at a 2:1 MOI, either in a 2-mL volume/6-cm dishes or in a 0.5 mL/well volume in 24-well plates. Dishes are maintained at 37°C in a 5% CO_2, 95% humidity atmosphere for a 20 min "infection pulse." Then, media are discarded and replaced with fresh high-glucose DMEM without phenol red and dishes are further incubated at 37°C in a 5% CO_2, 95% humidity atmosphere until collection time.
8. Infection of mDCs is performed as described above for mMPs (**step 7**), except that, because of the poor adhesion properties of inactive mDCs, the *Candida* cell suspension is simply added to the mDC media in a 200-μL volume for 6-cm dishes or in 100 μL/well for 24-well plates. After the 20 min "infection pulse," media of each plate is not discarded but collected and centrifuged at 700 × *g* for 7 min to collect floating cells. After addition of fresh high-glucose DMEM without phenol red, infection plates are further incubated at 37°C in a 5% CO_2, 95% humidity atmosphere until collection time. The cell pellets, kept on ice until collection time, are pooled with the corresponding cell samples.

3.5. Microscopic Internalization Assay (see Note 12)

1. The interaction experiment is performed as described in **Section 3.4**, except that mDCs or mMPs are plated on 12-mm-diameter sterile glass coverslips in 24-well plates 1 day prior to the infection, and a green fluorescent *Candida albicans* strain is used.
2. Terminate infection by carefully transferring each glass coverslip in a new 24-well plate prepared with 0.5 mL ice-cold PBS/well using clamps and guiding with a syringe

needle. Wash gently two more times with 0.5 mL ice-cold sterile PBS.

3. Fix the cells with 200 μL of 1% buffered paraformaldehyde for 5 min on ice. Discard paraformaldehyde solution and wash 3 times with ice-cold sterile PBS.
4. On ice, stain the cell wall of noninternalized fungi with 200 to 300 μL of an ice-cold 15 μM Calcofluor White solution for 5 min in the dark to stain the cell wall. Wash then 3 times with ice-cold sterile PBS.
5. Carefully invert the coverslips onto a drop of mounting medium for fluorescence on a microscopy slide. Observe slides using contrast phase, fluorescein (excitation 485 nm/emission 535 nm) and DAPI (excitation 355 nm/emission 460 nm) filters (**Fig. 11.1**).

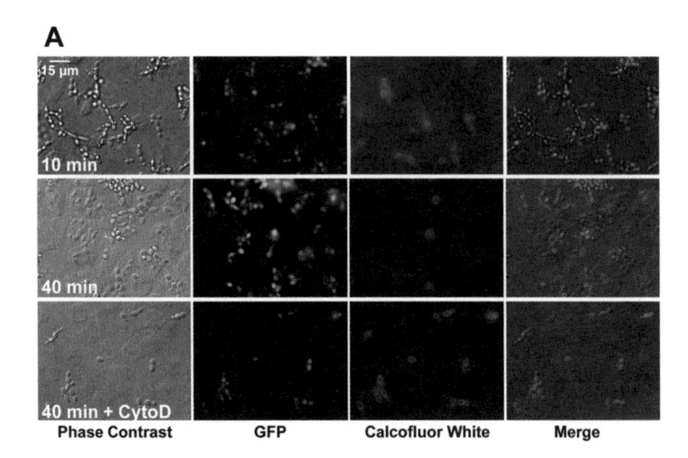

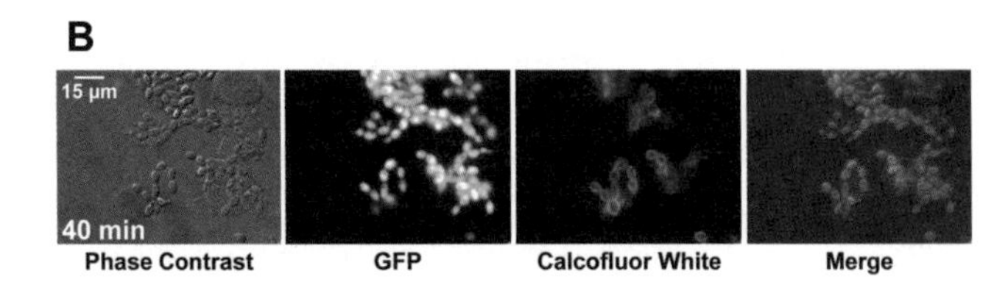

Fig. 11.1. *In vitro* phagocytosis of GFP-labeled *Candida albicans* by mMPs and mDCs. **(A)** mMPs were infected for the indicated time with GFP-*C.a.* at a MOI of 2:1, with or without 2 μM cytochalasin D (cytoD), and processed for Calcofluor White staining (*see* **Section 3.5**). Pictures were obtained on a Zeiss Axioplan2 microscope using a 63× oil-immersion lens, fluorescein and DAPI filters, and a Visitron Imaging System. Phase contrast pictures of the same field at the same magnification are also shown. After a 40-min incubation with GFP-*C.a.*, more "green-only" *C.a.* (internalized) are observed than after 10 min (see merged pictures). As a control for the uptake assay, pretreatment with cytochalasin D is performed as this completely blocks phagocytosis; hence, only double-labeled *C.a.* (noninternalized) are observed. **(B)** representative results of phagocytosis of GFP-*C.a.* by mDCs after infecting for 40 min.

6. For quantification purpose, percentage of internalized fungi can be expressed as (number of Calcofluor White stained fungi)/total of fungi observed in fluorescein channel) × 100.

3.6. Monitoring Activation of Cellular Signaling Pathways

1. Terminate the infection by placing the cell culture dishes on ice and remove media (discard in the case of mMPs, keep and process as described above in **step 8** for mDCs).
2. Scrap the cell layer in 80 μL ice-cold protein lysis buffer, collect in a microcentrifuge tube, and centrifuge at 15,000 × *g* at 4°C for 10 min.
3. Transfer supernatants in fresh tubes containing 30 μL of 4X Sample Buffer.
4. Mix and heat at 95°C for 5 to 7 min, cool on ice, and store at −20°C until further use.
5. After thawing proteins sample at 37°C for 5 min, analyze by SDS-PAGE a 15-μL aliquot on a 10% acrylamide mini gel (0.75 cm) and transfer onto nitrocellulose membrane.
6. Block membranes in 1X TBST containing 10% nonfat dry milk for 1 to 2 h.
7. After a short wash in 1X TBST, probe blots with the primary antibody diluted in 1X TBST with 2% BSA under continuous agitation, at 4°C, overnight (*see* **Note 13**).
8. The next day, wash blots in 1X TBST, and incubate blots with the secondary antibody diluted in 1X TBST with 2% bovine serum albumin (BSA) at room temperature under continuous agitation for 45 min.
9. After 3 to 4 washings in 1X TBST, detect immune complexes using an ECL substrate according to the manufacturer's instructions (**Fig. 11.2**).

3.7. Monitoring Cytokine Gene Expression

1. For RNA isolation from such small amount of mammalian cells, centrifugation column-based kits give very good results. Scraping cells directly in the provided lysis buffer yields a better RNA recovery and better RNA quality.
2. Centrifuge samples at 11,000 × *g* for 8 min and collect supernatants. If too viscous, samples should be passed 4 to 6 times through a syringe fitted with a 20-gauge needle before centrifugation.
3. At that stage, samples can then be kept frozen at −80°C or extraction is pursued according to the manufacturer's instructions.
4. Total RNA samples are eluted in 50 μL RNase-free sterile water.

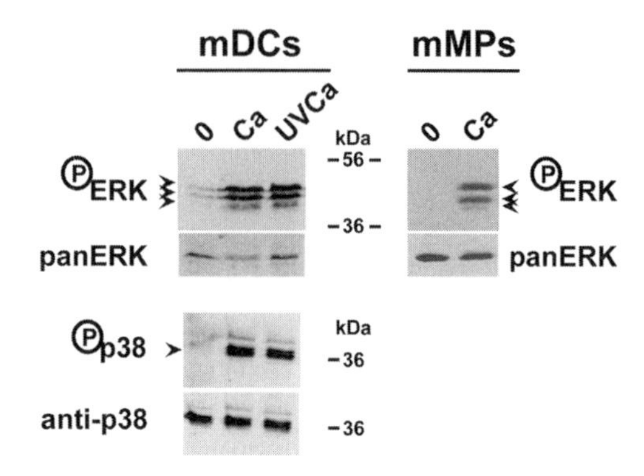

Fig. 11.2. Activation of MAPK in mDCs/mMPs infected with *Candida albicans in vitro*. Cell extracts of mDCs or mMPs incubated with *C.a.* strain SC5314 were collected after 20 and 45 min, respectively, and processed as described in **Section 3.6**. After infecting for 20 min with live SC5314, ERK and p38 activation through phosphorylation was detected in mDCs using phospho-specific antibodies. A similar pattern of MAPK activation was triggered by the UV-killed SC5314 cells, suggesting that early stimulation of these MAPK does not require live fungal cells. ERK phosphorylation is also observed after infecting mMPs for 45 min.

5. Quality control of the samples should include electrophoresis separation of a 5-μL aliquot on a 1% agarose urea-TBE gel for integrity assessment and OD_{260}/OD_{280} measurement of RNA concentration on a 1/20 dilution in Tris 10 mM, pH 7.5.
6. Reverse-transcription is performed on 0.5 to 2 μg total RNA using oligo-dT primers in final volume of 20 μL. The final reverse transcription products are diluted 1:5 with water and stored at −20°C until further use.
7. For the real-time PCR amplification, 5 μL of the diluted cDNAs are added to 20 μL of real-time PCR mix; reactions are submitted to cycling using the following conditions: initial denaturation 95°C for 4 min, followed by 40 cycles (each at 95°C for 10 s, 60°C for 15 s, 72°C for 15 s, and 80°C for 10 s; during these steps, the increase of the fluorescence is recorded); melting curve analysis is done from 60°C to 95°C for 30 min (*see* **Note 14**).
8. For relative quantification, data are analyzed according to the ΔΔCt method and are expressed as the fold-expression (R) of the gene of interest (GOI) versus the expression of a house-keeping gene (GAPDH) in treated (t) versus untreated (ut) conditions. The equation used is $R = 2^{\Delta\Delta Ct}$, where $\Delta\Delta Ct = (\Delta Ct_{GOI}t - \Delta Ct_{GAPDH}t) - (\Delta Ct_{GOI}ut - \Delta Ct_{GAPDH}ut)$ (**Fig. 11.3**).

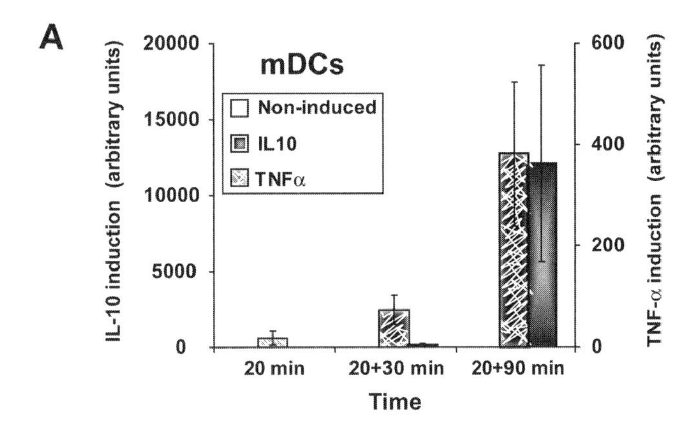

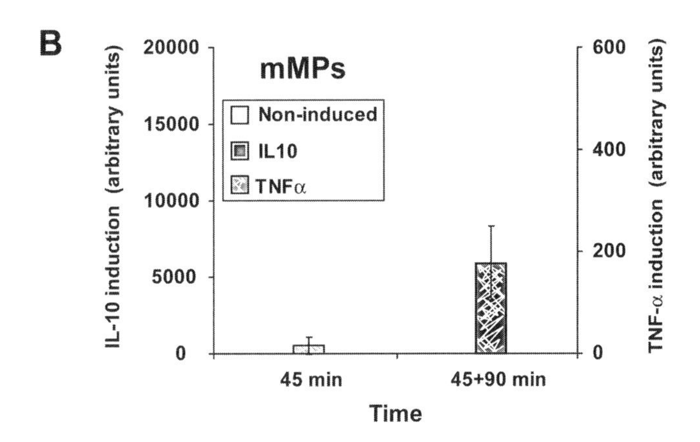

Fig. 11.3. *Candida albicans* triggers TNF-α and IL-10 production in mDCs/mMPs *in vitro*. Cell extracts of mDCs or mMPs incubated with *C.a.* strain SC5314 for a 20-min "pulse" were collected at the indicated times and processed as described in **Section 3.7**. In response to infection, TNF-α mRNA transcription was rapidly activated in both mDCs and mMPs and increased with time. IL-10 mRNA transcription was also induced to a very high extent in mDCs only but not significantly in mMPs.

4. Notes

1. To produce L-conditioned medium, divide 10 confluent 10-cm-diameter dishes of L929 cells (ATCC no. CCL-1) into 20 175-cm^2 flasks with 50 mL/flask of high-glucose DMEM supplemented with 10% FCS without antibiotics. After 36 to 48 hours, when cells are approximately 70% confluent, aspirate the medium and replace it with 100 mL/flask of starving medium (high-glucose DMEM without FCS and antibiotics). After 10 days, collect and filter the conditioned

media on Steritop 0.22-μm GP express PLUS membrane (Millipore, Billerica, MA) to prevent membrane clogging; store 250- to 500-mL aliquots at −20°C. Keep a small aliquot at 4°C for testing (*see* **Note 2**).

2. To test the potency of the L-conditioned media, flush the bones of one mouse as described in **Section 3.1**, divide the cells into five 100 × 100 square Petri dishes. Grow in 11 mL of mMPs cell medium supplemented with 0 to 20% of the fresh batch of L-conditioned medium. As a positive control, also grow two plates in mMPs cell medium supplemented with the optimal concentration of an old batch of L-conditioned medium. After 1 day, add 6 mL of medium. Completely renew the medium every 2 to 3 days. After 5 to 6 days, count the cells at each concentration, and split cells 1:3. Let them grow until confluency and count again. Deduce from the cell count the optimal L-conditioned medium concentration to be used.
3. To produce X-conditioned medium, grow GM-CSF–producing X-63 cell line *(10)* (nonadherent cells) in 10 mL high-glucose DMEM supplemented with 10% FCS, 100 U/mL penicillin, and 100 μg/mL streptomycin in a 75-cm^2 flask until the cell suspension is dense. Dilute the cell suspension at 1:2 with fresh high-glucose DMEM supplemented with 10% FCS, 100 U/mL penicillin, and 100 μg/mL streptomycin. When cells start to become dense again, add 30 mL of starvation media (high-glucose DMEM only) and grow until cells are confluent. Then, transfer into a 250-mL flask and add 50 mL of fresh starvation media. When cells are dense again, add 100 mL of starvation media. After 8 to 10 days, when a large amount of cells are dead, collect X-conditioned medium, centrifuge at 1000 × *g* for 5 min to remove the X-cells, and proceed as described above in **Notes 1 and 2** for sterile-filtration, storage, and testing.
4. Older mice can be used, but the number of bone marrow cells recovered will be smaller.
5. All given volumes are for the limbs of one mouse and should be multiplied according to the number of limbs.
6. Only mMPs will adhere to the non–tissue culture treated plastic, allowing for their separation from other cell types, the latter being eliminated when the media is changed.
7. Alternatively, 4-day-old mMPs can be frozen in FCS with 10% DMSO and stored in liquid nitrogen until use. They should be regrown for about 6 to 7 days in mMPs media before use.
8. To spare cells, double labeling can be performed. In that case, prepare 23 μL FACS buffer + 1 μL from each antibody, and incubate for 15 to 20 min on ice.

9. Typically, mMPs are $CD11^{+}$ positive and $CD11c^{-}$ negative. With the above-described method, one can expect to obtain at least 95% pure mMPs in 9- to 10-day-old primary cultures. In our conditions, the expression of mMP cell surface markers is stable up to 14 days after isolation. Myeloid DCs are both $CD11b^{+}$ positive and $CD11c^{+}$ positive; this method yields routinely cell preparations of 50% to 70% pure mDCs.
10. Host-pathogen interaction experiments are carried out in DMEM without serum and phenol red to slow hyphae formation of dimorphic fungi (i.e., *C.a.*). For the same reason, the dilution in DMEM should be performed just prior to starting the coculture with mammalian cells, as DMEM rapidly induces hyphae formation.
11. To check the *Candida* suspension used for infection, serial dilutions of the fungal cell suspension are plated on YPD agar plates on the day of the infection assay and incubated at 30°C. After 1 to 2 days, colony-forming units (CFU) are counted in order to control the actual MOI of the *in vitro* infection assay and verify the absence of contaminating microbes.
12. This method uses Calcofluor White (CW), a fluorescent dye (excitation 355 nm/emission 460 nm), which specifically binds to nascent fibrils of chitin in the fungal cell wall *(19)*. CW specifically stains fungal cells but fails to penetrate into mammalian cells. After staining with this dye, cell walls of noninternalized fungi appear as a bright blue ring when inspected by microscopy using appropriate filters.
13. Adding 0.05% sodium azide to primary antibody dilutions and performing incubations at 4°C will allow one to re-use primary antibody dilutions several time if stored at 4°C. Sodium azide should *not* be added to the dilution of horseradish peroxidase–coupled secondary antibodies and washing buffers, as it inhibits the enzyme activity.
14. Better reproducibility is achieved with the real-time PCR analysis when using only 1 to 2 μg of total RNA in the reverse transcription reactions. Each real-time PCR assay data point should be performed at least in triplicate. Recording of the fluorescence increase "in real-time" is performed during the "10 s at 80°C" step to favor the dissociation of possible nonspecific products, leading to a more accurate measurement of the amplicon.

Acknowledgments

We are very grateful to Alexander Johnson for kindly providing us with the GFP-labeled *Candida albicans* strain. We thank the laboratory of Thomas Decker (University of Vienna, Vienna,

Austria) for their technical support and advice. In particular, we acknowledge the invaluable advice and encouragements from Tilo Materna, Sylvia Stockinger, Andreas Pilz, and Katrin Ramsauer. We also thank all lab members for the help with culturing fungal pathogens. This work was supported by the "Wiener Wissenschafts Forschungs- und Technologiefonds" (HOPI-LS133-WWTF) and by the European Union through the funding of the EC Marie Curie Training Network *"CanTrain"* (CT-2004-512481). Additional funds came from the Herzfelder Foundation.

References

1. Roeder A, Kirschning CJ, Rupec RA, Schaller M, Korting HC. *Toll*-like receptors and innate antifungal responses. Trends Microbiol 2004;12:44–9.
2. Netea MG, Van der Graaf C, Van der Meer JW, Kullberg BJ. Recognition of fungal pathogens by *Toll*-like receptors: *Toll*-like receptors and the host defense against microbial pathogens: bringing specificity to the innate-immune system. Eur J Clin Microbiol Infect Dis 2004;23:672–6.
3. Mencacci A, Cenci E, Del Sero G, et al. Defective co-stimulation and impaired Th1 development in tumor necrosis factor/lymphotoxin-alpha double-deficient mice infected with *Candida albicans*. Int Immunol 1998;10:37–48.
4. Vazquez-Torres A, Jones-Carson J, Wagner RD, Warner T, Balish E. Early resistance of interleukin-10 knock-out mice to acute systemic candidiasis. Infect Immun 1999;67:670–4.
5. Farah CS, Hu Y, Riminton S, Ashman RB. Distinct roles for interleukin-12p40 and tumour necrosis factor in resistance to oral candidiasis defined by gene-targeting. Oral Microbiol Immunol 2006;21: 252–5.
6. Vonk AG, Netea MG, van Krieken JH, van der Meer JW, Kullberg BJ. Delayed clearance of intraabdominal abscesses caused by *Candida albicans* in tumor necrosis factor-alpha- and lymphotoxin-alpha-deficient mice. J Infect Dis 2002;186:1815–22.
7. Ibata-Ombetta S, Jouault T, Trinel PA, Poulain D. Role of extracellular signal-regulated protein kinase cascade in macrophage killing of *Candida albicans*. J Leukoc Biol 2001;70:149–54.
8. Zhong B, Jiang K, Gilvary DL, et al. Human neutrophils utilize a Rac/Cdc42-dependent MAPK pathway to direct intracellular granule mobilization toward ingested microbial pathogens. Blood 2003;101:3240–8.
9. Choi JH, Choi EK, Park SJ, et al. Impairment of p38 MAPK-mediated cytosolic phospholipase A(2) activation in the kidneys is associated with pathogenicity of *Candida albicans*. Immunology 2007; 120:173–81.
10. Zal T, Volkmann A, Stockinger B. Mechanisms of tolerance induction in major histocompatibility complex class II-restricted T cells specific for a blood-borne self-antigen. J Exp Med 1994;180: 2089–99.
11. Gillum AM, Tsay EY, Kirsch DR. Isolation of the Candida albicans gene for orotidine-5′-phosphate decarboxylase by complementation of *S. cerevisiae* ura3 and *E. coli pyrF* mutations. Mol Gen Genet 1984;198:179–82.
12. Hull CM, Johnson AD. Identification of a mating type-like locus in the asexual pathogenic yeast *Candida albicans*. Science 1999;285:1271–5.
13. Kovarik P, Stoiber D, Novy M, Decker T. Stat1 combines signals derived from IFN-gamma and LPS receptors during macrophage activation. EMBO J 1998;17: 3660–8.
14. Overbergh L, Giulietti A, Valckx D, Decallonne R, Bouillon R, Mathieu C. The use of real-time reverse transcriptase PCR for the quantification of cytokine gene expression. J Biomol Tech 2003;14:33–43.
15. Pattyn F, Speleman F, De Paepe A, Vandesompele J. RTPrimerDB: the Real-Time PCR primer and probe database. Nucl Acids Res 2003;31:122–3.

16. Pattyn F, Robbrecht P, De Paepe A, Speleman F, Vandesompele J. RTPrimerDB: the real-time PCR primer and probe database, major update 2006. Nucl Acids Res 2006;34:D684–8.
17. Hume DA, Gordon S. Optimal conditions for proliferation of bone marrow-derived mouse macrophages in culture: the roles of CSF-1, serum, Ca^{2+}, and adherence. J Cell Physiol 1983;117: 189–94.
18. Inaba K, Inaba M, Romani N, et al. Generation of large numbers of dendritic cells from mouse bone marrow cultures supplemented with granulocyte/macrophage colony-stimulating factor. J Exp Med 1992; 176:1693–702.
19. Herth W. Calcofluor white and Congo red inhibit chitin microfibril assembly of *Poterioochromonas*: evidence for a gap between polymerization and microfibril formation. J Cell Biol 1980;87:442–50.

Chapter 12

Experimental Infection of Rodent Mammals for Fungal Virulence Testing

Anja Schweizer and Klaus Schröppel

Abstract

Invasive fungal infections comprise a group of serious and life-threatening diseases affecting immunocompromised patients. Molecular analysis of fungal virulence involves the deletion of genes that are suspected for contributing to fungal pathogenesis. Phenotypic analysis of the generated mutants includes *in vivo* infection experiments in order to assign a function during fungal disease to a gene of interest.

Key words: *C. albicans*, candidiasis, fungal infection, fungal pathogenesis, virulence, animal model.

1. Introduction

Life-threatening fungal infection of human hosts has become a frequent complication of immunocompromising disorders such as hematooncologic neoplasia or of immunosuppression therapy after solid organ transplantation. Molecular approaches to gain a better understanding of the physiology of fungal microorganisms and their role as pathogens have addressed several biochemical pathways *(1, 2)* and developmental processes such as the synthesis of cell wall components *(3, 4)* and secreted proteins *(5, 6)* or the regulation of hyphal formation *(2, 7)*. The genetics of the two major fungal pathogens, *Candida albicans* and *Aspergillus fumigatus*, have been analyzed, their genome has been sequenced *(8, 9)*, and potential virulence traits were suggested. Nonetheless, the conclusion whether a given gene product or metabolite is relevant for successful infection of a host can only be drawn after *in vivo* infection experiments *(10)*.

Molecular analysis of virulence is based on the comparison of wild-type with mutant strains, which harbor modifications of

Steffen Rupp, Kai Sohn (eds.), *Host-Pathogen Interactions*, DOI: 10.1007/978-1-59745-204-5_12,

genes or pathways that are suspected to have a role during the pathogenesis of the virulent microorganism. Signs of virulence deficits may be disclosed as altered course of infection with respect to the time, the rate of pathogen dissemination, or the organ fungal burden detected in tissue of infected animals. Furthermore, changes in morphologic development of fungi may be detected only *in vivo* where a pathogen interacts with a host under physiologic conditions.

By this approach, we and others have reported that components of the developmental program that regulates hyphal formation of *C. albicans* are required for successful pathogenesis of *C. albicans* in mice.

2. Materials

2.1. Cell Culture and Preparation of the Fungal Inoculum

1. Genetically stable *C. albicans* strains without extrachromosomal DNA elements were grown on YPD agar plates consisting of 2% (w/v) glucose, 1% (w/v) yeast extract, and 2% (w/v) peptone.
2. Maintenance of extrachromosomal or intrachromosomal DNA constructs with selectable markers in strains exhibiting the respective auxotrophy was achieved by streaking of *C. albicans* on appropriate drop-out media. For example, the *URA3* gene was selected by complete supplemented media without uridine (CSM-URA) consisting of 2% (w/v) glucose, 0.5% (w/v) $(NH_4)_2SO_4$, 0.17% (w/v) yeast nitrogen base without amino acids and ammonium sulfate (Difco), and 0.077% (w/v) CSM-URA supplement (Bio-101) (*see* **Note 1**).
3. For *in vitro* culture of *C. albicans* on solid media, humidified incubators are used at 28°C or 37°C, respectively, without supplementation of carbon dioxide CO_2.
4. For *in vitro* culture of *C. albicans* in liquid media, flasks or tubes were incubated with 120 to 250 rpm on a rotary shaker at 28°C or 37°C, respectively. Cultured cells were not allowed to sediment during incubation.
5. The number of cells in suspension after *in vitro* culture was determined by use of a 100-μm Neubauer chamber (Fisher Scientific).
6. Enumeration of colony forming units (CFU) in organ homogenates was achieved by plating of 1:10 serial dilutions of the homogenate on agar plates and incubation at 28°C or 37°C overnight. CFU were calculated as average number of colonies per plate divided by the plating volume and multiplication with the dilution factor.

2.2. Infection of Host Animals

1. BALB/c mice were from Charles Rivers Breeding Laboratories (Sulzfeld, Germany).
2. 1.0-mL syringes and 0.3-mm needles were from Becton Dickinson (Heidelberg, Germany).
3. Forceps and scalpels were from Aesculap (Medika, Hof, Germany).
4. Cages were placed under an infrared light bulb for mildly elevated temperatures if required prior to injection.

2.3. Assessment of Virulence

1. For monitoring of physiologic data, ordinary scales, measures, and graded water dispensers were used:
2. Turax electric blender and a blending rotor was from IKA (Multimed, Kirchheim, Germany).
3. Tissue lysis solution was 20% (w/v) KOH containing 0.1% (w/v) of fluorescent brightener (Calcofluor white; Sigma, München, Germany). Stained fungal cells were inspected on a Zeiss Axiophot epifluorescent microscope (Carl Zeiss, Jena, Germany).

3. Methods

C. albicans or *A. fumigatus*, which are frequently used for virulence testing, are opportunistic pathogens in humans. In laboratory animals, establishing an experimental mycoses can be facilitated by pharmacologic conditioning of the animals, by injection of sufficient numbers of infectious organisms directly into the site of infection, or by both. Care should be taken to select the appropriate route of infection and the most adequate host for the type of disease that is supposed to be modeled.

Here, instructions are given to set up and perform a systemic *C. albicans* bloodstream infection of mice in order to measure the virulence of a *C. albicans* mutant strain in comparison with a genetically related parental or wild-type strain.

The mouse is a very good choice for the host species because its maintenance is well established, and many facilities are experienced in the housing of genetically defined strains of mice. The tail blood vessels of mice are conveniently accessible for infection, and the organs and soft tissues can be analyzed without the need for specialized surgical instruments. We assume that profound knowledge of animal anatomy is available.

Previous publications vary in the number of fungal cells used for infection *(11)*. This reflects the need for a number of initial experiments to establish the laboratory-specific number of fungal cells used for infection. This number is dependent on the host and its housing conditions, as well as on the virulence of the *C. albicans* wild-type or common ancestor strain used as a standard inoculum *(12)*.

Because the phenotype of genetically engineered *C. albicans* strains is to be compared during an infection of mice, selection of the appropriate pair or group of strains itself becomes a very important task. Care should be taken that the clones under study are primary cloning products derived from a single common ancestor. Ideally, this common ancestor wild-type strain is identical to the standard virulent strain used to perform the initial experiments mentioned above. A control with a complemented mutation is to be included (so-called reintegrant or add-back strain). Additionally, the strains that will be compared should be identical with respect to rate of growth at 37°C *in vitro* and lack or expression of an identical set of auxotrophies for the growth *in vitro (12)*. Experiments may be designed to test for virulence traits under *in vivo* conditions in the mouse but also for the identification of fungal survival factors or biochemical pathways required for replication in a mammalian tissue environment.

Mice are allowed to adjust to the animal facility after their arrival for at least 5 to 7 days. Strain, sex, and age of the animals should be kept constant between repeats of the same experiment. Many researchers infect female BALB/c mice because they are not easily hackled up, and the lateral tail veins used for inoculation of *C. albicans* can easily be located due to the lack of pigment in the skin of albino animals.

3.1. Initial Experiments to Establish the Animal Model

1. Experimental animal infections depend on many factors, which can hardly be predicted for each institution. Therefore, performing the methods outlined in this paragraph, the basic procedure must be established with a virulent *C. albicans* wild-type strain.
2. The number of fungal cells used for infection of a mouse must be titrated.
3. Depending on your local regulations, a severe drop of body weight, the lack of water uptake, or death of the animals will serve as evaluation of the experiment.
4. The course of a wild-type candidiasis should last for 12 to 16 days. This is a good agreement between a short distress for the animals and a sufficiently long period during which the *C. albicans* cells must express their virulence repertoire in order to survive against host defense mechanisms. Considerably shorter durations bear the risk of artifacts due to the large numbers of cells inoculated, which might cause obstruction of arteries or simply cause an anaphylactic shock.
5. Avirulent strains will hardly cause death of the animals at an inoculum density that, like a wild-type strain, would kill the animals after 12 to 16 days.

3.2. Preparation of the Fungal Inoculum

1. Clonal *C. albicans* strains for virulence testing are stored at −70°C. It is assumed that the growth characteristics of the strains have already been established and that phenotypic analyses of the strains have already been performed (*see* **Note 2**).
2. Prior to infection, strains are propagated for at least two subcultures under appropriate conditions (e.g., to select for genetic markers).
3. Two days prior to infection, a liquid culture is inoculated and incubated at 28°C on an orbitary shaker at 120 to 240 rpm. Cells must be kept in suspension and should not be allowed to sediment. Development of hyphae must be avoided (*see* **Note 3**).
4. One day prior to infection, stationary phase blastoconidia are harvested from the overnight culture and washed 3 times in PBS at 4°C. The concentration of cells is enumerated, and the density of cells is adjusted in the inoculum suspension (using PBS) according to the results of the initial set-up experiments, which was 2.5×10^6 mL^{-1} to 5×10^6 mL^{-1} blastoconidia in our hands *(13–15)*.
5. For control of vitality of the cells, the inoculum suspension is diluted to 2000, 1000, and 500 cells mL, respectively, and 100 μL of each suspension is spread on three agar plates each and incubated for 20 to 24 h at 37°C to permit maximum growth. Inoculum suspensions are kept at 4°C for the next day (*see* **Note 4**).
6. On the day of infection, the number of vital cells in each inoculum suspension of the strains to be compared is calculated from the average number of colonies grown from the diluted cell suspensions. If required, the concentration of cells in all inoculum suspensions is adjusted to each other by dilution with PBS (4°C) in order to meet the concentration of the sample with the lowest density of vital cells (*see* **Note 5**).

3.3. Infection of Host Animals

1. Two or 3 days before the infection, five to eight 10-week-old *(16)* female BALB/c mice are combined to form a single test group and are maintained in one or two cages with water and food *ad libitum* (*see* **Note 6**).
2. Animals are individually encoded (*see* **Note 7**), and the initial weight is noted for each individual.
3. Hygiene requirements and specifications to prevent the release of genetically modified organisms must be observed according to your local regulations.
4. The particular conditions of animal testing may vary between different institutions. The animals must not suffer pain or any other ailments.

5. On the day of infection, inoculum suspensions are kept on ice during the transfer to the animal facility.
6. Prior to infection, mice may have to be treated with an analgesic or even be sedated, especially when the person who carries out the experiment is not yet completely familiar with the procedure (*see* **Note 8**).
7. Keeping the cages at elevated temperature for 30 to 60 min will relax the animals and help to locate the tail veins.
8. The inoculum suspension is drawn into a narrow, long, 1.0-mL syringe, which is then equipped with an 0.3-mm needle. A mouse is placed into a glass tube with its tail lolling out of an aperture in the lid. A mesh ensures free access of air on the other end of the tube.
9. The tail is manipulated with your fingers to ensure unhindered access to the blood vessels. On both sides of the tail, an 0.5- to 0.8-mm-wide vein can be located. The needle is probing from the distal end to the base of the tail. After entering the lumen of the vein, 200 μL of the inoculum suspension are injected (*see* **Note 9**).
10. After retraction of the needle, the puncture site is carefully compressed between two fingers for a few seconds to prevent bleeding.
11. The animal is released back to its cage and observed while the next mouse is infected.

3.4. Assessment of Virulence

1. Collection of data includes graphing of the infected animals' body weight, their behavioral changes, appearance of the fur as well as posture (*see* **Note 10**), and survival time after infection. For example, end-point of the infection may be defined by 20% loss of body weight. Ailment of the animals had to be as short as possible: moribund mice were sacrificed immediately, and their final body weight was measured again. Kaplan-Meier (*see* **Note 11**) survival curves were generated *(17)*, for example, with a statistics software package, and compared using the log-rank test *(18)*. A p value <0.05 was considered significant.
2. For the detection of tissue fungal burden, internal organs (e.g., kidneys, liver, and lungs) of dead dissected animals were transferred into sterile tubes and kept in PBS (4°C) (*see* **Note 12**). Tissue was homogenized with a Turax electric blender (*see* **Note 13**) at 15,000 to 20,000 rpm for 20 to 60 s. The number of CFU in the homogenate was enumerated as described above.
3. Alternatively, for direct examination of *C. albicans* cellular morphology in host tissue, organ samples of 10 to 30 mm^3 were incubated for up to 12 h at 60°C in tissue lysis solution.

4. Liquified specimens were centrifuged (1500 × g for 10 min), and the sediment was resuspended in 20 μL of PBS with 0.1% (w/v) of fluorescent brighteners. Samples were mounted onto glass slides for epifluorescence microscopy.

4. Notes

1. We have found CSM drop-out reagents to be very helpful in supporting growth of slow-growing mutants; however, it can be omitted if necessary.
2. Phenotypic analysis should exclude the presence of crippled strains in order to eliminate the possibility of futile animal testing as much as possible.
3. Hyphal filaments cannot be counted well, they may start to aggregate and block the syringe, and therefore are not easily applied for controlled infection via the bloodstream.
4. It should be tested that a strain will not start to die significantly during the overnight storage at 4°C. If this not possible, infect the mice and control vitality in parallel on the same day. It may require the use of more animals or the application of two or more cell densities in the inoculum suspension.
5. Reject your inoculum suspension if the rate of vital cells is unexpectedly low (e.g., under 90%). Care should be taken during the initial set-up experiments to accurately correlate the result of the Neubauer chamber with the enumeration of CFU.
6. The average number of animals depends on the penetration of the phenotype and the stability of the entire system. More than 10 animals is rarely required.
7. Use a marker and paint colored rings on the root of the tail; different numbers of rings encode individual animals. A streak on the top of the tail counts "five." Alternatively, earpunch or earclip methods will work as well, respectively.
8. For example, ketamine; ask your local animal supervisor for instructions.
9. Initially after penetration of the skin, the plunger of the syringe cannot be pressed down. When entering the vein, a marked drop of resistance can be felt. The vein will appear colorless during the time of injection. A few air-bubbles are not a serious problem. If the puncture does not result in a successful injection of inoculum suspension, try it again either on the contralateral side of the tail, or try to inject into the same vein, but chose a spot more proximal to the base of the tail.

10. Seriously ill animals will take up a typical round body posture, which we refer to as a “spherical mouse.”
11. For data handling, estimation of the group size, appropriate graphing, observation frequency, and statistical analysis in general, it might be helpful to ask your local office for the design of clinical trials.
12. Analyses may require that the organs are either shock frozen in liquid nitrogen, fixed in formaldehyde (for histology), or be kept in PBS on ice before further analyses.
13. Alternatively, a glass homogenizer or passage through a 200-μm steel mesh can be used. Most tissues can easily be homogenized due to the reticular structure of the organs.

References

1. Lengeler, K.B., Davidson, R.C., D'Souza, C., Harashima, T., Shen, W.C., Wang, P., Pan, X. et al. (2000) Signal transduction cascades regulating fungal development and virulence. *Microbiol Mol Biol Rev.* **64**, 746–785.
2. Liu, H. (2001) Transcriptional control of dimorphism in *Candida albicans. Curr Opin Microbiol.* **4**, 728–735.
3. Ruiz-Herrera, J., Elorza, M.V., Valentin, E., Sentandreu, R. (2006) Molecular organization of the cell wall of *Candida albicans* and its relation to pathogenicity. *FEMS Yeast Res.* **6**, 14–29.
4. Hoyer, L.L. (2001) The ALS gene family of *Candida albicans. Trends Microbiol.* **9**, 176–180.
5. Naglik, J.R., Challacombe, S.J., Hube, B. (2003) *Candida albicans* secreted aspartyl proteinases in virulence and pathogenesis. *Microbiol Mol Biol Rev.* **67**, 400–428.
6. Ghannoum, M.A. (2000) Potential role of phospholipases in virulence and fungal pathogenesis. *Clin Microbiol Rev.* **13**, 122–143.
7. Gow, N.A., Brown, A.J., Odds, F.C. (2002) Fungal morphogenesis and host invasion. *Curr Opin Microbiol.* **5**, 366–371.
8. Jones, T., Federspiel, N.A., Chibana, H., Dungan, J., Kalman, S., Magee, B.B., Newport, G. et al. (2004) The diploid genome sequence of *Candida albicans. Proc Natl Acad Sci U S A.* **101**, 7329–7334.
9. Nierman, W.C., Pain, A., Anderson, M.J., Wortman, J.R., Kim, H.S., Arroyo, J., Berriman, M. et al. (2005) Genomic sequence of the pathogenic and allergenic filamentous fungus *Aspergillus fumigatus. Nature.* **438**, 1151–1156.
10. Odds, F.C., Gow, N.A., Brown, A.J. (2001) Fungal virulence studies come of age. *Genome Biol.* **2**, Reviews1009.
11. Lo, H.J., Köhler, J.R., DiDomenico, B., Loebenberg, D., Cacciapuoti, A., Fink, G.R. (1997) Nonfilamentous *C. albicans* mutants are avirulent. *Cell.* **90**, 939–949.
12. Odds, F.C., Van Nuffel, L., Gow, N.A. (2000) Survival in experimental *Candida albicans* infections depends on inoculum growth conditions as well as animal host. *Microbiology.* **146** (Pt 8), 1881–1889.
13. Leberer, E., Harcus, D., Dignard, D., Johnson, L., Ushinsky, S., Thomas, D.Y., Schröppel, K. (2001) Ras links cellular morphogenesis to virulence by regulation of the MAP kinase and cAMP signalling pathways in the pathogenic fungus *Candida albicans. Mol Microbiol.* **42**, 673–687.
14. Schweizer, A., Rupp, S., Taylor, B.N., Röllinghoff, M., Schröppel, K. (2000) The TEA/ATTS transcription factor CaTec1p regulates hyphal development and virulence in *Candida albicans. Mol Microbiol.* **38**, 435–445.
15. Taylor, B.N., Hannemann, H., Sehnal, M., Biesemeier, A., Schweizer, A., Röllinghoff, M., Schröppel, K. (2005) Induction of *SAP7* correlates with virulence in an intravenous infection model of candidiasis but not in a vaginal infection model in mice. *Infect Immun.* **73**, 7061–7063.
16. Ashman, R.B., Papadimitriou, J.M., Fulurija, A. (1999) Acute susceptibility of

aged mice to infection with *Candida albicans*. *J Med Microbiol.* **48**, 1095–1102.

17. Kaplan, E.L., Meier, P. (1958) Nonparametric estimation from incomplete observations. *J Am Stat Assoc.* **53**, 457–481.

18. Mantel, N., Haenszel, W. (1959) Statistical aspects of the analysis of data from retrospective studies of disease. *J Natl Cancer Inst.* **22**, 719–748.

Chapter 13

Real-Time and Semiquantitative RT-PCR Methods to Analyze Gene Expression Patterns During *Aspergillus*-Host Interactions

Jin Woo Bok, Nancy P. Keller, and Dimitrios I. Tsitsigiannis

Abstract

Aspergillus species are infamous for causing several plant and animal diseases that directly (e.g., invasive aspergillosis) or indirectly (e.g., consumption of toxic food supplies) can lead to high rates of morbidity in humans and animals worldwide. Despite progress in molecular information and manipulation of *Aspergillus* spp., including genome sequence availability and suitable transformation methodologies, efforts to control *Aspergillus* diseases are still far from satisfactory, due in part to lack of knowledge of fungal virulence attributes. In order to obtain meaningful insights on the disease mechanism(s), it is essential to detect virulence gene expression during host invasion. Here, we describe two PCR-based detection methods of *Aspergillus* gene expression in both plant and mammalian tissues. Moreover, these techniques can be employed for routine screening of large numbers of aspergilli to improve diagnosis, disease monitoring, and therapy of fungal disease.

Key words: *Aspergillus* spp., pathogen, gene expression, cDNA, real-time PCR, reverse transcriptase–PCR.

1. Introduction

Aspergillus species are opportunistic pathogens that generally require wounds or otherwise weakened hosts for colonization *(1)*. *Aspergillus* diseases affect both plants and animals *(1–3)*. For example, *A. flavus*, a common seed-infecting fungus, contaminates such hosts as corn and peanuts with high levels of the carcinogen aflatoxin *(4)*. Extreme levels of toxin production are associated with drought and other environmental stresses on the host crop, and animal diseases caused by ingesting contaminated feeds are known as aflatoxicoses *(5)*. Recent episodes of human and dog deaths

Steffen Rupp, Kai Sohn (eds.), *Host-Pathogen Interactions*, DOI: 10.1007/978-1-59745-204-5_13,

from aflatoxin poisoning have highlighted the need to detect and ameliorate the presence of this toxin and producing fungus in food and feeds worldwide *(6, 7)*. *A. flavus* is also a noted human pathogen, but the most infamous cause of aspergillosis in humans is *A. fumigatus*, a ubiquitous saprophyte. *A. fumigatus* causes a wide range of animal diseases that include fungal sinusitis, asthma, and allergic bronchopulmonary aspergillosis as well as life-threatening invasive aspergillosis (IA) associated with high mortality rates *(8, 9)*. Pathogenicity of *A. fumigatus* depends on the immune status of patients and fungal strain. There is no unique essential virulence factor for development of aspergillosis, and fungal virulence appears to be under polygenetic control *(10)*. In addition to *A. fumigatus* and *A. flavus*, several other *Aspergillus* spp. can cause aspergillosis in the immunocompromised host including the genetic model *A. nidulans*, otherwise known as a GRAS organism *(11)*.

Detection of antibodies, antigens, DNA, or mRNA represents different diagnostic strategies for fungal infections. Antibody detection in animals is limited in value due to cost of the reagents and unpredictable differences in host humoral responses *(12)*. Currently, there is no routine antibody detection method for fungal pathogens of plants. PCR methodologies for the detection of fungal nucleic acids from infected hosts may be the optimal diagnostic approach because it offers the potential of sensitivity, capabilities for multiple sampling, and target specificity *(13–17)*. Fungal mRNA detection is especially powerful as this reveals expression of genes critical for pathogenesis. Here, we detail two PCR-based methods to detect expressed fungal genes in host tissue using real-time PCR for *A. nidulans* and *A. flavus* infected peanut and semiquantitative reverse transcriptase (RT) PCR for *A. fumigatus* infected mouse.

2. Materials

2.1. *Aspergillus* Gene Expression in Peanut

2.1.1. Seed Infections

1. Peanut (*Arachis hypogaea*) seeds, cultivar Florunner (commercial line highly susceptible to *A. flavus* in the field).
2. Potato dextrose agar medium (PDA; Difco, Franklin Lakes, NJ): potato starchose 4 g, glucose 20 g, and 15 g agar in 1 L double distilled water (ddH_2O).
3. 20X salts solution: $NaNO_3$ 120 g, KCl 10.4 g, $MgSO_4{\cdot}7H_2O$ 10.4 g, KH_2PO_4 30.4 g, and then add ddH_2O to 1 L, autoclaved and stored at room temperature.
4. Trace elements: $ZnSO_4{\cdot}7H_2O$ 2.2 g; H_3BO_3 1.1 g, $MnCl_2{\cdot}4H_2O$ 0.5 g, $FeSO_4{\cdot}7H_2O$ 0.5 g, $CoCl_2{\cdot}5H_2O$ 0.16 g, $CuSO_4{\cdot}5H_2O$ 0.16 g, $(NH_4)6Mo_7O_24{\cdot}4H_2O$ 0.11 g, Na_4EDTA 5.0 g, and then add the solids in order to 80 mL of H_2O, dissolving each completely before adding the next.

Heat the solution to boiling, cool to 60°C, adjust the pH to 6.5 to 6.8 with KOH pellets. Cool to room temperature and adjust volume to 100 mL with ddH_2O.

5. *Aspergillus* glucose minimal medium (GMM): 20X salt solution (50 mL/L), trace elements (1 mL/L), D-glucose (10 g/L), agar (for solid media) (15 g/L) in 1 L ddH_2O, adjust pH 6.5 and then autoclave 15 min, at 15 psi (1.05 kg/cm^2), 121°C on liquid cycle.
6. *Aspergillus* species. *A. nidulans* wild-type strain RDIT9.32 and oxylipin mutants Δ*ppoB* (RDIT59.1), Δ*ppoA*;Δ*ppoC* (RDIT54.7), Δ*ppoA*;Δ*ppoB*;Δ*ppoC* (RDIT62.3) *(18)*, and *A. flavus* strain 12S (kindly provided by P. Cotty, USDA-ARS, New Orleans, LA).
7. Sterile 0.01% Tween 80 (MP Biochemical, Solon, OH, cat. no. 103170) in distilled water.

2.1.2. Plant and Fungal RNA Isolation

1. RNeasy Plant Mini kit (Qiagen Ltd., cat. no. 74903).
2. Turbo DNA-free (Ambion, Inc., Austin, TX, cat. no. 1907).

2.1.3. Plant and Fungal cDNA Synthesis

1. iScript cDNA synthesis Kit (Bio-Rad Laboratories, Inc., cat. no. 170-8890).

2.1.4. Real-Time Reverse Transcriptase–PCR (Real-Time RT-PCR)

1. iQ SYBR Green Supermix (Bio-Rad Laboratories, Inc., cat. no. 170-8880).
2. RiboGreen RNA Quantitation Kit (Molecular Probes Inc., Eugene, OR, cat. no. R-11490).
3. AB-0900, Thermo-Fast 96 Semi-Skirted PCR Plate (Abgene, Rockford, IL).
4. Costar plates for RNA quantification.
5. Beacon Designer Software (Premier Biosoft International, Palo Alto, CA).
6. Primers were purchased from Integrated DNA Technologies (IDT) Inc., Coralville, IA.

2.1.5. Data Analysis

1. MyiQ Real-Time PCR Detection System (Bio-Rad Laboratories, Inc.).
2. MyiQ software package (Bio-Rad Laboratories, Inc.).

2.2. *Aspergillus fumigatus* Gene Expression in Mouse

2.2.1. Animals and Organism

1. ICR (Harlan Sprague Dawley, Indianapolis, IN) mice weighing 24 to 27 g.
2. Cyclophosphamide (Sigma, St. Louis, MO, cat. no. C0768).
3. Cortisone acetate (Sigma, cat. no. C3130).
4. *Aspergillus fumigatus* 293.
5. *Aspergillus* GMM.
6. Sterile 0.01% Tween 80 (MP Biochemical, cat. no. 103170) in distilled water.

2.2.2. Fungal RNA Isolation

1. Mouse lung.
2. Tissue grinder (Fisher, Hampton, NH, cat. no. 09-552-28).
3. Miracloth (Calbiochem, cat. no. 475855).
4. RNAse free DNAse (Qiagen Co., Valencia, CA, cat. no. 79254).
5. Triton X-100 (final conc. 1%, EM Science, La Jolla, CA, cat. no. TX1568-1).
6. Ultra pure water (RNAse free, Cayman, Chem. Co., Ann Arbor, MI, cat. no. 400000).
7. TriZol (Invitrogen, Carlsbad, CA, cat. no. 15596-018).
8. Phenol:$CHCl_3$:isoamyl alcohol = 24:23:1 (Ambion, cat. no. 9732), abridged as PCI.
9. Isopropanol (Fisher, cat. no. S77798).
10. 100% ethanol (AAPER Alcohol & Chemical Co., Shelbyville, KY, cat. no. DSP-KY-417).
11. Diethyl pyro-carbonate (DEPC, Sigma, cat. no. D5758).
12. Chloroform (Fisher, cat. no. C607-4).
13. Microfuge 18 (Beckman, Fullerton, CA).

2.2.3. Reverse Transcriptase–PCR (RT-PCR)

1. SuperScript III kit (Invitrogen Co., cat. no.18080-044).
2. Primers (Integrated DNA Technologies [IDT], Inc.).
3. Gene Amp PCR system 9700 (Applied Biosystems, Foster City, CA).
4. TripleMaster PCR system (Eppendorf, Westbury, NY, cat. no. 954140261).

3. Methods

The essential step of real-time and semiquantitative RT-PCR is the extraction of high-quality mRNA from infected hosts (*see* **Note 1**). Current applied methods in plant and animal tissue provide repeatable and reliable results in obtaining high-quality total RNA from fungus-infected tissue. Here, we show two alternative methods for the detection of host or fungal gene expression during host infection. Real-time PCR provides high sensitivity utilizing small samples yet is relatively expensive. An alternative, less costly method is semiquantitative RT-PCR. The success of detection is dependent on host or pathogen specific primers in both methods. If possible, it is important to design primers to allow for size differences between genomic DNA and mRNA in regions where introns are excised, thus allowing for a technical control of DNA contamination. Additionally, the primers for the first round PCR in semiquantitative RT-PCR should have high melting temperature (70°C to 76°C) to increase specificity for the target DNA amplification.

3.1. *Aspergillus nidulans* Gene Expression in Peanut

3.1.1. Seed Infections

1. *Preparation of seeds.* Mature seeds are immersed in distilled water for 5 min (*see* **Note 2**) and then the exterior brown peanut layer (testa) is removed using the fingers. The two cotyledons are separated and the embryo is carefully removed without damaging any of the cotyledon tissue. Then, cotyledons are surface sterilized by placing them in a tea ball infuser and dipping them in a beaker containing 0.05% sodium hypochlorite in sterile water for 3 min (*see* **Note 3**). The tea ball is transferred into a new beaker with sterile distilled water for 30 s (wash step), followed by a 5-s wash with 70% ethanol in a new beaker (additional sterilization step) and one more 30-s wash with sterile distilled water while shaking the tea ball. The cotyledons were drained completely and placed in a Petri dish until the time of infection. All the steps are aseptically performed in a laminar flow hood or a biosafety hood.
2. *Preparation of inoculum.* Streak a PDA Petri plate with *A. flavus* or a GMM plate with *A. nidulans* conidia from a glycerol stock culture. Incubate the plates at 29°C for *A. flavus* or at 37°C for *A. nidulans* for 3 to 5 days under dark conditions (or light if you are using *A. nidulans* strains that have the *veA* gene) until the plates are covered with conidia. Collect the spores in the hood using a spreader and 5 mL of sterile distilled water and transfer the spore suspension to a culture tube. Count the conidia using a hemocytometer to estimate the spore concentration.
3. *Seed infections.* Peanut cotyledons are inoculated with *A. flavus* or *A. nidulans* at a concentration of 10^5 conidia/mL (10 cotyledons/per 10 mL of conidia suspension). Cotyledon treatments include water control (mock inoculation) and wounding or infection with the fungi (*see* **Note 4**). For wounding experiments, cotyledons are scratched 4 times with a razor blade to a depth of ca. 2 to 3 mm. For all treatments, 20 cotyledons are immersed in 20 mL of sterile distilled water (control and wounded) or sterile water with fungal conidia in 50 mL Falcon tubes while shaking for 30 min in a rotary shaker (50 rpm). All cotyledons are incubated in the dark at 29°C for *A. flavus* or 37°C for *A. nidulans* in glass Petri dishes lined with three pieces of moist filter paper and a water reservoir (the lid of a Falcon tube containing 1 mL of sterile water) to maintain high humidity in the artificial chamber (**Fig. 13.1**). Samples for fungal or plant gene expression analysis are collected at desired time points over a time course of 5 days, by which time cotyledons are covered by the fungi. Collected samples are frozen in liquid nitrogen and kept at −80°C until the time of analysis.

Fig. 13.1. Infected peanut cotyledons (cultivar Florunner) with *A. flavus* (strain 12S). Seeds are placed in a glass Petri dish containing filter paper saturated with sterile distilled water and a water reservoir to keep the humidity high and are incubated in the dark. Picture was taken after 6 days incubation at 29°C.

3.1.2. Plant RNA Isolation

1. 100 to 150 mg of frozen cotyledons is macerated using a pestle and mortar in liquid nitrogen. Decant tissue powder and liquid nitrogen into an RNase-free, liquid nitrogen cooled, 2-mL microcentrifuge tube. Allow the liquid nitrogen to evaporate, but do not allow the tissue to thaw. The 2-mL tubes are kept in liquid nitrogen until all the samples have been processed.
2. Total RNA is extracted using the RNeasy Plant Mini kit (*see* **Note 5**). Remove samples from the liquid nitrogen and immediately add 600 μL of RLT buffer (supplied in the RNeasy kit) including β-mercaptoethanol (β-ME; add 10 μL β-ME per 1 mL Buffer RLT just before use). Vortex the samples vigorously to dissolve the homogenized tissue and to disrupt the cells.
3. Transfer the lysate to a QIAshredder spin column (lilac column) of the kit placed in a 2-mL collection tube, and centrifuge for 2 min at 11,300 × *g*. Carefully transfer the supernatant of the flow-through from the collection tube to a new microcentrifuge tube. Be careful not to disturb the cell-debris pellet and avoid collecting the lipid floating on top layer that is formed.
4. Add 0.5 volume of ice-cold ethanol (96% to 100%) to the cleared lysate, and mix immediately by pipetting 5 to 6 times. Transfer 700 μL of the sample to an RNeasy spin column (pink column) placed in a 2-mL collection tube (supplied in the kit). Centrifuge for 30 s at 6700 × *g*. Discard the flow-through and add the remaining solution of each sample into the same RNeasy spin column. Centrifuge for another 30 s at 6700 × *g* and discard the flow-through. Total RNA remains in the spin column membrane.
5. Add 700 μL Buffer RW1 to the RNeasy spin column, close the lid and let the samples stand for 5 min at room temperature.

Then, centrifuge for 1 min at 6700 × *g* to wash the spin column membrane. Discard the flow-through and the collection tube.

6. Place the RNeasy spin columns (pink column) into new supplied collection tubes and add 500 μL Buffer RPE (Buffer RPE is supplied as a concentrate, and ensure that 4 volumes of 96% to 100% ethanol is added to Buffer RPE before use; that is, 44 mL of ethanol in the supplied 11 mL of Buffer RPE) to the RNeasy spin column, close the lid, and centrifuge for 30 s at 6700 × *g* to wash the spin column membrane. Discard the flow-through.
7. Add another 500 μL Buffer RPE to the RNeasy spin column and centrifuge for 2 min at 6700 × *g* to further wash the spin column membrane. Discard the flow-through.
8. Place the RNeasy spin column in a new 1.5 mL collection tube (supplied). Be careful not to carry any residual wash buffer to the final collection tube. Add 40 μL RNase-free water (supplied) directly to the spin column membrane and incubate the samples for 3 min on the benchtop. Centrifuge for 1 min at 11,300 × *g* to elute the RNA. Repeat the elution step using the first eluate to obtain a more concentrated RNA. Incubate the samples for 2 min on the benchtop and centrifuge for 1 min at 11,300 × *g*. Store the RNA (a mixture of plant and fungal RNA) at −80°C until the time you will proceed to the next steps.

3.1.3. Plant and Fungal cDNA Synthesis

1. Samples are treated with Turbo DNAfree (Ambion, Inc.) prior to reverse transcription to remove residual amounts of DNA. Use the rigorous protocol according to the manufacturer's guidelines. The absence of DNA from the RNA samples is confirmed by performing real-time PCR on 50 ng of total RNA using a control primer set (e.g., actin primers). RNA samples that yield a threshold cycle (C_t) value greater than 32 can be considered free of contaminating DNA.
2. Total RNA is quantified using the RiboGreen kit (Molecular Probes). RiboGreen is an ultrasensitive fluorescent nucleic acid stain for quantifying RNA in solution using a fluorescence microplate reader. RNA samples obtained after **step 8** of **Section 3.2** are diluted 20 times, 200 times, and 2000 times and are mixed with the RiboGreen solution following the manufacturer's protocol.
3. cDNA is generated from 1 μg of total RNA using the iScript cDNA synthesis kit (Bio-Rad Laboratories, Inc., Hercules, CA) in a 20-μL reaction that includes 4 μL of 5X Buffer, 1 mL reverse transcriptase, and the remaining 15 μL consists of 1 μg of total RNA and water. The

cDNA synthesis conditions are 5 min at 25°C, 30 min at 42°C, 5 min at 85°C, hold-step at 4°C. Final cDNA synthesis products are diluted to 200 μL.

3.1.4. PCR

3.1.4.1. Real-Time PCR Primer Design

Proper primer design ensures that the selected primers are specific for the desired target sequence, bind at positions that avoid secondary structure, and minimize the occurrence of primer-dimer formation. The primers are designed based on the desired host or fungal gene using the Beacon Designer software. For instance, we have used the peanut *PnLOX2-3* cDNA sequence and the housekeeping gene "actin depolymerizing factor" (actin-DF) as a plant calibrator *(19)* and the *A. nidulans* genes α-glycosidase (AB057788), α-amylase (AF208225), and *amyR* (AF208225) and the housekeeping gene "tubulin" as a fungal calibrator. Beacon Designer software is designed to generate primer pairs suitable for real-time RT-PCR. Use housekeeping genes as reference/control genes that are expressed uniformly throughout the time course of your study. The "Avoid Template Structure" option of the software is selected to limit primer sequences to regions of little or no secondary template structure. PCR efficiency has a significant effect on the accuracy of calculated gene expression levels. Differences in efficiency can be caused by a number of factors, such as the presence of reverse-transcriptase inhibitors or primer-dimer formation. The efficiency of the primer pairs is determined on cDNA derived from peanut seed or *A. nidulans* using fivefold serial dilutions of cDNA (staring from 100 ng per reaction) in two independent experiments. Master mixes, using 2X SYBR Green reaction mix (Bio-Rad Laboratories, Inc.), from each dilution series are split into three reactions in a volume of 20 μL/well. Reactions are subjected to real-time PCR using the MyiQ Real-Time PCR Detection System and analyzed using the MyiQ software package (Bio-Rad Laboratories, Inc.) (**Fig. 13.2A**). Primer efficiencies, which are unique to each primer pair and has a major impact on the accuracy of the calculated expression ratios, are determined using the formula efficiency (E) = $10^{(-1/\text{slope})}$ with the slope determined by the MyiQ Cycler software (**Fig. 13.2B**) *(20)*. For example, in a study to quantify expression of the peanut lipoxygenases *PnLOX2-3 (19)* after *A. flavus* infection, primers RT-PnLOX2-3-F1 (5′-CCTCATCCTCCTCCTTCTTC-3′) and RT-PnLOX2-3-R1 (5′-AAGGTGTCAACGTCCAGG-3′) were used to amplify the 145-bp *PnLOX2-3* PCR fragment, and the primers RT-actin-F1 (5′-GAGGAGAAGCAGAAG-CAAGTTG-3′) and RT-actin-R1 (5′-AGACAGCATATCG-GCACTCATC-3′) were used to amplify the 106-bp actin-DF PCR fragment. The actin-DF primers had an efficiency of 2.17 and the *LOX* primers had an efficiency of 2.05. Primer efficiency close to 2.00 corresponds with 100% efficiency. For 100% efficiency,

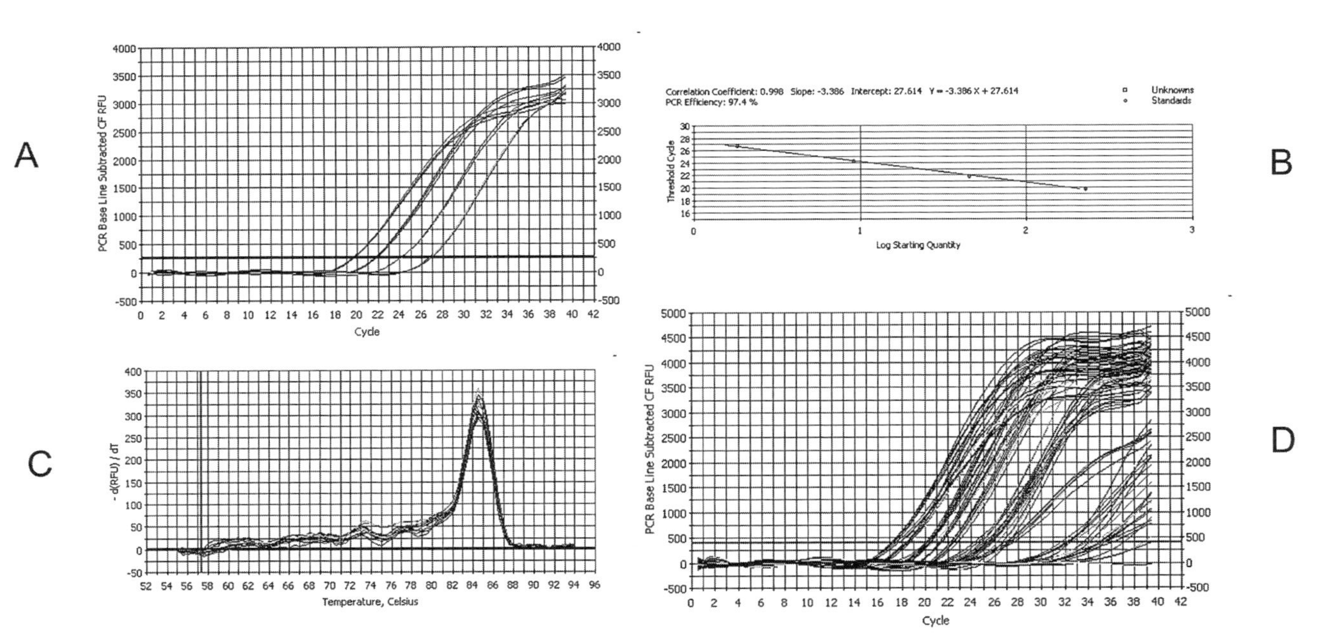

Fig. 13.2. **(A)** An iCycler software screen view showing amplification plots of a target cDNA (peanut actin) 5X dilution series to determine the efficiencies of primer sets. **(B)** Standard curve of threshold values (C_t) against the log of cDNA dilutions, generated by the iCycler software, to calculate the slope and estimate primer efficiency using the formula $(E) = 10^{(-1/\text{slope})}$. **(C)** A melt curve profile depicting the negative first derivative of the fluorescence (−dF/dT) versus temperature plot. The significant change in fluorescence accompanies the melting of the double-stranded PCR products. The plot of −dF/dT versus temperature is displaying these changes in fluorescence as distinct peaks. The software identifies the peaks and assigns melting temperatures from this plot. This analysis confirms the specificity of the chosen primers and reveals the putative presence of primer dimers. **(D)** Fluorescent data of amplification of *PnLOX2-3* and actin products derived from peanut seeds infected with *A. flavus* in real time as temperature increases or decreases. The iCycler iQ system records the total fluorescence generated by SYBR Green binding to double-stranded DNA as temperature changes and plots the fluorescence in real time as a function of temperature. The calculation of the C_t values is generated automatically by the MyiQ software.

there will be a doubling of the amount of DNA at each cycle, for 90% the amount of DNA will increase from 1 to 1.9 at each cycle, so the factor is 1.9 for each cycle, and similarly for 80% and 70% it will be 1.8 and 1.7. Notice that a small difference in efficiency makes a large difference in the amount of final product. One must be sure, however, that the efficiency of PCR is similar for the standards as that of the "unknown" samples.

3.1.4.2. PCR Conditions

Samples are analyzed in a volume of 20 μL using iQ SYBR Green Supermix. Reactions are performed in triplicate using cDNA templates from two independent iScript reactions for each gene (e.g., *LOX* and actin) giving a total of six replicates (20 μL reactions) for each RNA sample. A master mix of SYBR Green Supermix and primers is prepared for each primer pair. Each reaction contains 50 ng of cDNA and 200 nmol of each primer. Primer stock is 5 μM and use 0.8 μL per 20 μL reaction to give a 200 nmol final concentration. Reactions are performed with the MyiQ Real-Time PCR Detection System using the default "two-step amplification plus melting curve" protocol: 95°C for 3 min followed by 40 cycles of 95°C for 1 min (denaturation) and 55°C for 45 s (annealing and elongation). The melt curve analysis is carried out after amplification and consists of 11 min at 55°C followed by 80 to 10 s steps with a 0.5°C increase in temperature at each step (**Fig. 13.2C**). The melt curve analysis is essential to demonstrate the lack of variation and the specificity of PCR products as well as the absence of primer dimers. Amplified products can also be analyzed by agarose gel electrophoresis and sequencing analysis for further confirmation. The calculation of the threshold values (C_t) is generated automatically by the MyiQ software (**Fig. 13.2D**).

3.1.5. Data Analysis

The amount of target RNA (*LOX* or degradative enzyme RNA in this case) is determined at each time point by first normalizing the target RNA to the internal standard RNA (actin for peanut or tubulin for *A. nidulans*) using the $2^{-\Delta C_t}$ formula or $2^{(C_t \text{ internal standard}) - (C_t \text{ target})}$ *(20–22)*. To determine the relative expression ratio of the *LOX* or the fungal gene at each treatment, the normalized *LOX* or fungal gene value is divided by the calibrator value. For instance, the calibrator for gene expression in cotyledons is the average of the normalized values of control (mock-inoculated) samples. The average, standard error, and coefficient of variation (CV) are determined for each set of six replicate reactions. The entire analysis is replicated in three independent experiments for each time point. An example of the results produced is shown in **Fig. 13.3**. **Figure 13.3A** shows real-time PCR analysis of the transcriptional kinetics of *PnLOX2-3* that occurs during *A. flavus* colonization in comparison with mock-inoculation *(19)*. The calibrator for this experiment was the average of normalized transcript from

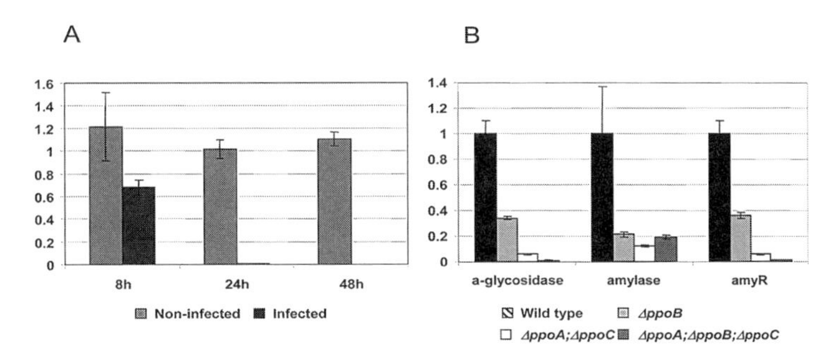

Fig. 13.3. **(A)** Quantification of plant host genes. Suppression of peanut *PnLOX2-3* after *A. flavus* infection as revealed by real-time PCR analysis after 8 h, 24 h, and 48 h. **(B)** Quantification of *Aspergillus* genes. The expression of three *A. nidulans* genes (α-glycosidase, α-amylase, *amyR*) in wild type and the oxylipin mutants Δ*ppoB*, Δ*ppoA*; Δ*ppoC*, Δ*ppoA*; Δ*ppoB*;Δ*ppoC*. The transcript level is represented as the relative quantification of the expression level of each target gene of interest (*PnLOX2-3* in part **A** or α-glycosidase, α-amylase, *amyR* in part **B**) and was determined by first normalizing the target RNA to the internal standard RNA (actin for peanut or tubulin for *A. nidulans*), and then the normalized target RNA value was divided by the calibrator consisting of the average of the normalized values of control (mock or wild-type inoculated samples) at each time point. The average and the standard errors were calculated for each set of six replicate reactions.

mock-inoculated cotyledons. The expression of *PnLOX2-3* is downregulated at 8 h, 24 h, and 48 h postinoculation. Another example (**Fig. 13.3B**) shows the transcriptional kinetics of fungal α-glycosidase, α-amylase, and the transcription factor *amyR* in different *A. nidulans* oxylipin mutants (Δ*ppo*) during the peanut colonization and infection *(18)*. The calibrator for this experiment was the average of normalized transcript from *A. nidulans* wild-type for each gene. The expression of α-glycosidase, amylase, and *amyR* is downregulated in all strains examined during the fungal-host interaction.

3.2. *Aspergillus fumigatus* Gene Expression in Mouse

3.2.1. Animals and Organism

1. Six-week-old, outbreed Swiss ICR mice (Harlan Sprague Dawley) weighing 24 to 27 g are immunosuppressed by intraperitoneal injection of cyclophosphamide (150 mg/kg) on days −4 and −1, and 3 before and after infection, and with a single dose of cortisone acetate (200 mg/kg) on the morning of infection. A clinical isolate, *A. fumigatus* 293, is used for mouse infection.
2. Streak-inoculated fresh conidia from 3-day-old GMM Petri dish cultures grown at 37°C are collected by scraping conidia off the agar top using 0.01% Tween 80 in distilled water. Typically two Petri dishes suffice. Conidia are counted using the hemocytometer and resuspended in 0.85% NaCl to a concentration of 10^8 conidia/mL (*see* **Notes 6** and 7). The mice are anesthetized via halothane inhalation in a bell jar

on day 0. Sedated mice are inoculated by nasal instillation of 50 μL of 10^8 spores/mL (10 mice) or nasal instillation of 50 μL of 0.85% NaCl (10 mice, mock control). All mice are monitored 3 times daily for morbidity. For RNA examination, mouse lung is collected at 3 days after the infection and homogenized for the subsequent experiment described below (we note here that lung biopsy is an important component to human diagnosis, so it would be feasible to perform RT-PCR on human tissue also).

3.2.2. Isolate Pathogen RNA

1. At day 3, all mice are sacrificed in a CO_2 chamber, and all lungs from each group are combined per treatment and homogenized using a tissue grinder (Fisher, cat. no. 09-552-28) with 10 mL of phosphate-buffered saline. Homogenates are passed through 1 ply of Miracloth (Calbiochem, San Diego, CA) to remove large tissue debris (*see* **Note 8**).
2. The flow-through from the 10 lungs is rapidly frozen with liquid nitrogen. The flow-through (ca. 10 mL) is then treated with DNAse (4 units/mL) (RNase Free DNase, Qiagen) and Triton X-100 (final conc. 1%) for 20 min at 37°C. The treated flow-throughs are spun down in 50 mL Oak Ridge tubes at 3000 × *g* and the supernatant discarded.
3. Pellets are resuspended in 4 mL of 1% Triton X-100 and then homogenized for 1 min. The homogenate is spun down in 50 mL Oak Ridge tubes at 3000 × *g* and again the supernatant is discarded.
4. Repeat **step 3**.
5. Pellets are washed twice with sterile distilled water and spun down at 3000 × *g* to discard supernatant.
6. Each pellet (ca. 100 μL) is lyophilized overnight and then mixed with 1 mL TRIzol (Invitrogen Co.). The TRIzol homogenate is incubated for 5 min at 15°C to 30°C to permit the complete dissociation of nucleoprotein complexes.
7. Next, 200 μL chloroform is added to the homogenate, shaken vigorously, and centrifuged at 9600 × *g* for 15 min at 4°C. The supernatant, containing total RNA, is extracted with equal volumes of phenol:chloroform:isoamylalcohol = 24:23:1 (PCI) to get high-quality RNA (*see* **Note 9**).
8. The extracted supernatant is collected and mixed with 500 μL isopropanol (2-propanol) for precipitation of total RNA.
9. After 10 min incubation at room temperature, the mixture is centrifuged at 9600 × *g* for 10 min at 4°C.
10. After removing supernatant, the RNA pellet is washed with 1 mL 75% ethanol, vortexed, and spun at 3800 × *g* for 5 min at 4°C.

11. RNA pellet is air-dried for 10 min (*see* **Note 10**).
12. The dried pellet is dissolved in 40 μL diethyl pyrocarbonate (DEPC, 1 mL/L) treated water by incubating at 60°C for 10 min.
13. A 250-fold dilution is used to measure the concentration of RNA in the sample by spectrophotometer.

3.2.3. Reverse Transcriptase–PCR (RT-PCR)

1. The RNA sample dissolved in DEPC-treated water is used for cDNA synthesis using SuperScript III kit (Invitrogen; *see* **Note 11**).
2. The following 20 μL reaction volume can be used for 10 pg to 5 μg of total RNA.
3. To synthesize the first-strand cDNA, add the following to a nuclease-free microcentrifuge tube:
   (a) 1 μL of oligo(dT) 20 (50 μM) or 50 to 250 ng of random primers or 2 pmol of gene-specific primer.
   (b) 10 pg to 5 μg total RNA.
   (c) 1 μL 10 mM dNTP Mix (10 mM each dATP, dGTP, dCTP, and dTTP at neutral pH).
   (d) Sterile, distilled water up to 13 μL. This mixture is heated to 65°C for 5 min and incubated on ice for at least 1 min. The contents of the tube are collected by brief centrifugation and the following are added:
   (e) 4 μL 5X First-Strand Buffer.
   (f) 1 μL 0.1 M DTT.
   (g) 1 μL RNaseOUT Recombinant RNase Inhibitor (Optional: cat. no. 10777-019, 40 units/μL; *see* **Note 12**).
   (h) 1 μL of SuperScript III RT (200 units/μL; *see* **Note 13**).
4. The contents of the tube are mixed by pipetting gently up and down (*see* **Note 14**).
5. The reaction is inactivated by heating at 70°C for 15 min. The cDNA (the first-strand reaction) can now be used as a template for amplification in PCR (*see* **Note 15**).
6. The following primers are used to amplify an *A. fumigatus* specific gene, *gliI* (a biosynthetic gene in the gliotoxin gene cluster) from the cDNA pool. **Figure 13.4A** shows the relative positions of the primary and secondary primers used in **step 6** and described below.
   (a) Primary amplification. 5′TCGTTGCTGGAGAATCTGCTGTATGA3′ and 5′ACAAAGGGAATCTGGTGGAAGGCGA3′ will amplify a 0.6- to 0.7-kb PCR product using TripleMaster PCR system (Eppendorf) (**Fig. 13.4B**, lanes 2 and 3). (The 0.6- to 0.7-kb is not seen in this gel, this is a common occurrence, hence the requirement of the secondary amplification step. There is a nonspecific

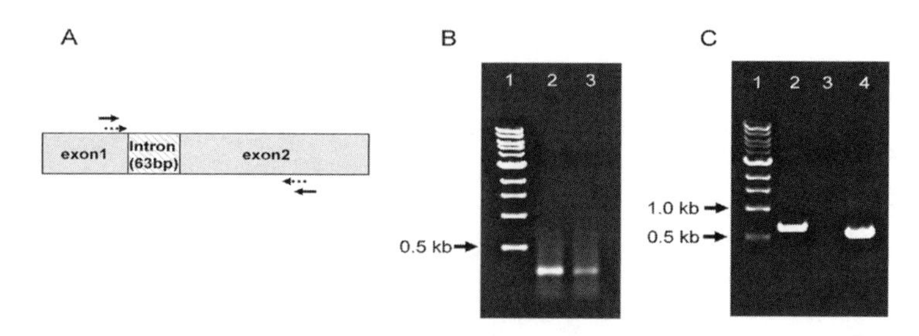

Fig. 13.4. **(A)** Diagram of *gliI* gene (1428 bp) and location of the primary (solid arrows) and secondary (nested) primers (dashed arrows). **(B)** Primary RT-PCR amplification of *gliI* from mouse lung infected with *A. fumigatus* or mock-inoculated. Lane 1, 1 kb size marker; lane 2, 10 μL out of 50 μL of RT-PCR tube of total RNA from mock-inoculated lungs; lane 3, 10 μL out of 50 μL of RT-PCR tube of total RNA from infected lungs. The lower band in both lanes is nonspecific amplification whereas the *gliI* fragment is not seen in this gel. **(C)** Secondary PCR amplification with nested primers of *gliI* from 2 μL out of 50 μL the primary RT-PCR amplifications above. Lane 1, 1 kb size marker; lane 2, PCR from genomic *A. fumigatus* DNA; lane 3, 10 μL out of 50 μL of PCR from RT-PCR product of mock-inoculated lungs; lane 4, 10 μL out of 50 μL of PCR from RT-PCR product of infected lungs. The fragment in lane 4 is smaller than the fragment in lane 2 as an intron was excised in lane 4. (Reprinted with permission from American Society for Microbiology from Bok, J. W., Chung, D., Balajee, S. A., Marr, K. A., Andes, D., Nielsen, K. F., Frisvad, J. C., Kirby, K. A., and Keller, N. P. 2006. GliZ, a transcriptional regulator of gliotoxin biosynthesis, contributes to *Aspergillus fumigatus* virulence. *Infect. Immun.* **74**, 6761–6768.)

band that is not amplified by the secondary primers, **Fig. 13.4C**). Contained in the PCR tube are

(i) 5 μL 10X PCR Buffer
(ii) 1 μL 10 mM dNTP Mix
(iii) 0.2 μL Sense primer (20 pmol)
(iv) 0.2 μL Antisense primer (20 pmol)
(v) 0.2 μL DNA polymerase (5 units/μL)
(vi) 2 μL cDNA (from first-strand reaction)
(vii) Autoclaved, distilled water to 50 μL

Heat the reaction to 95°C for 2 min to denature. Perform 35 cycles of PCR (95°C for 1 min (denaturation), 68°C for 30 s (annealing), and 68°C for 1 min (elongation); primer melting temperature is 76°C based on GC content and expected PCR product is 0.6 to 0.7 kb.

(b) Secondary Amplification. 2 μL of the above PCR reaction is used as a template with nested primers, (5′TGTTGATCGAGACGCCGTTCTG3′ and 5′CAGAGCGGCTCGATTCTGGTG3′, **Fig. 13.4C**, lanes 3 and 4; *see* **Note 16**). The size difference between lane 2 (a genomic DNA control) and lane 4 (cDNA) in **Fig. 13.4C** is due to an intron (63 bp) in *gliI* gene. This result indicates that cDNA was synthesized and detected successfully by this method from infected mouse lung tissue.

4. Conclusions

Aspergillus species are opportunistic pathogens attacking plants, animals, and humans. Pathogenesis is a complex mixture of pathogen, host, and environmental attributes that perhaps can be best revealed during the actual course of pathogen ingress. Here we present two methods that have allowed analysis of both fungal and host gene expression *in situ*. Such studies are key components in elucidating mechanisms of pathogenesis in aspergillosis infections.

5. Notes

1. Intact and high-quality of total RNA isolated from seeds is difficult and the operator needs to be especially vigilant with pipetting technique as plant components can interfere with the quality of extracted RNA.
2. Immersion of peanut seeds in distilled water for 5 min facilitates the removal of testa and prevents the formation of wounds on the cotyledons.
3. All solutions are prepared in water that has a resistivity of 18.3 MΩ-cm and total organic content of less than 5 ppb.
4. Wounding experiments are needed to differentiate the plant host genes that specifically respond to fungal infections and not to wound damage.
5. To avoid RNA degradation problems during the extraction procedure, do not process more than six samples at the same time.
6. We suggest using 0.01% Tween-80 in water to prepare conidial suspensions as conidia are hydrophobic. Immediately make a conidial resuspension in 0.85% NaCl for injection in mouse.
7. Fresh conidial suspensions are recommended for both plant and animal inoculations due to changes in spore physiology over time.
8. Removal of lung tissue by filtration can clog the miracloth. Alternatively, you can use forceps to remove large sections of tissue. The lung can be cut into smaller pieces prior to homogenization, which aids in extraction.
9. After extraction of total RNA from lung tissue with TRIzol, additional extraction with PCI is an important step to increase RNA quality.
10. Do not overdry your RNA pellet. The overdried RNA pellet will be difficult to resuspend in solution.
11. SuperScript III Reverse Transcriptase is a version of M-MLV RT that has been engineered to reduce RNase H activity

and provide increased thermal stability. The enzyme can synthesize cDNA at a temperature range of 45°C to 60°C, providing increased specificity, higher yields of cDNA, and more full-length product than other reverse transcriptases. Because SuperScript III RT is not significantly inhibited by ribosomal and transfer RNA, it can be used to synthesize cDNA from total RNA. For difficult or high-GC-content templates, use a 60°C cDNA synthesis temperature.

12. When using less than 50 ng of starting RNA, the addition of RNaseOUT is essential.
13. If generating cDNA longer than 5 kb at temperatures above 50°C using a gene-specific primer, the amount of SuperScript III RT may be raised to 400 U (2 μL) to increase yield.
14. The reaction temperature should be increased to 55°C for gene-specific primers. Reaction temperature may also be increased to 55°C for difficult templates or templates with high secondary structure.
15. Amplification of some PCR targets (those >1 kb) may require the removal of RNA complementary to the cDNA. To remove RNA complementary to the cDNA, add 1 μL (2 units) of *E. coli* RNase H and incubate at 37°C for 20 min.
16. The second round of PCR with nested primers is recommended to extract low levels of fungal transcripts in host tissues. Ideally, it is good to design primers that encompass an intronic region in the amplified nucleic acid as this allows for an internal control to show that RNA, not DNA, is amplified.

Acknowledgments

This work was supported by NIH 1 R01 Al065728-01 and NSF IOB-0544428 (subcontract S060039) and NSF MCB-0236393 to NPK. We thank David K. Willis (University of Wisconsin-Madison) for introducing us to the real-time RT-PCR methodology and for many fruitful discussions. We thank Marisa K. Trapp for assistance in formatting this article.

References

1. Raper, K. B., and Fennell, D. I. (1965) *The Genus Aspergillus* pp. ix, 686 p., Williams & Wilkins, Baltimore.
2. Campbell, C. K. (1994) Forms of aspergillosis, in (Powell, K. A., Renwick, A., and Peberdy, J. F., Eds.), pp. 313–320, *The Genus Aspergillus*, Plenum, New York.
3. St Leger, R. J., Screen, S. E., and Shams-Pirzadeh, B. (2000) Lack of host specialization in *Aspergillus flavus*. *Appl. Environ. Microbiol.* **66**, 320–324.
4. Payne, G. A. (1998) Process of contamination by aflatoxin-producing fungi and their impact on Crops, in (Sinha, K. K., and Bhatnager, D., Eds.), pp. 279–306,

Mycotoxins in Agriculture and Food Safety, Marcel Dekker, New York.

5. Robens, J. F., and Richard, J. L. (1992) Aflatoxins in animal and human health. *Rev Environ Contam. Toxicol.* **127**, 69–94.
6. Lewis, L., Onsongo, M., Njapau, H., Schurz-Rogers, H., Luber, G., Kieszak, S., Nyamongo, J., Backer, L., Dahiye, A. M., Misore, A., DeCock, K., and Rubin, C. (2005) Aflatoxin contamination of commercial maize products during an outbreak of acute aflatoxicosis in eastern and central Kenya. *Environ. Health Perspect.* **113**, 1763–1767.
7. Leung, M. C., Diaz-Llano, G., and Smith, T. K. (2006) Mycotoxins in pet food: a review on worldwide prevalence and preventative strategies. *J. Agric. Food. Chem.* **54**, 9623–9635.
8. Denning, D. W., and Stevens, D. A. (1990) Antifungal and surgical treatment of invasive aspergillosis: review of 2,121 published cases. *Rev. Infect. Dis.* **12**, 1147–1201.
9. Rementeria, A., Lopez-Molina, N., Ludwig, A., Vivanco, A. B., Bikandi, J., Ponton, J., and Garaizar, J. (2005) Genes and molecules involved in *Aspergillus fumigatus* virulence. *Rev. Iberoam. Micol.* **22**, 1–23.
10. Latge, J. P. (2001) The pathobiology of *Aspergillus fumigatus*. *Trends Microbiol.* **9**, 382–389.
11. Dotis, J., and Roilides, E. (2004) Osteomyelitis due to Aspergillus spp. in patients with chronic granulomatous disease: comparison of *Aspergillus nidulans* and *Aspergillus fumigatus*. *Int. J. Infect. Dis.* **8**, 103–110.
12. Young, R. C., and Bennett, J. E. (1971) Invasive aspergillosis. Absence of detectable antibody response. *Am. Rev. Respir. Dis.* **104**, 710–716.
13. Bretagne, S. (2003) Molecular diagnostics in clinical parasitology and mycology: limits of the current polymerase chain reaction (PCR) assays and interest of the real-time PCR assays. *Clin. Microbiol. Infect.* **9**, 505–511.
14. Scherm, B., Palomba, M., Serra, D., Marcello, A., and Migheli, Q. (2005) Detection of transcripts of the aflatoxin genes *aflD*, *aflO*, and *aflP* by reverse transcription-polymerase chain reaction allows differentiation of aflatoxin-producing and non-producing isolates of *Aspergillus flavus* and *Aspergillus parasiticus*. *Int. J. Food. Microbiol.* **98**, 201–210.
15. Chen, R. S., Tsay, J. G., Huang, Y. F., and Chiou, R. Y. (2002) Polymerase chain reaction-mediated characterization of molds belonging to the *Aspergillus flavus* group and detection of *Aspergillus parasiticus* in peanut kernels by a multiplex polymerase chain reaction. *J. Food. Prot.* **65**, 840–844.
16. Hummel, M., Spiess, B., Kentouche, K., Niggemann, S., Bohm, C., Reuter, S., Kiehl, M., Morz, H., Hehlmann, R., and Buchheidt, D. (2006) Detection of *Aspergillus* DNA in cerebrospinal fluid from patients with cerebral aspergillosis by a nested PCR assay. *J. Clin. Microbiol.* **44**, 3989–3993.
17. Munoz, P., Guinea, J., and Bouza, E. (2006) Update on invasive aspergillosis: clinical and diagnostic aspects. *Clin. Microbiol. Infect.* **12**, 24–39.
18. Tsitsigiannis, D. I., Kowieski, T. M., Zarnowski, R., and Keller, N. P. (2005) Three putative oxylipin biosynthetic genes integrate sexual and asexual development in *Aspergillus nidulans*. *Microbiology.* **151**, 1809–1821.
19. Tsitsigiannis, D. I., Kunze, S., Willis, D. K., Feussner, I., and Keller, N. P. (2005) *Aspergillus* infection inhibits the expression of peanut 13S-HPODE-forming seed lipoxygenases. *Mol. Plant. Microbe. Interact.* **18**, 1081–1089.
20. Rotenberg, D., Thompson, T. S., German, T. L., and Willis, D. K. (2006) Methods for effective real-time RT-PCR analysis of virus-induced gene silencing. *J. Virol. Methods.* **138**, 49–59.
21. Pfaffl, M. W. (2001) A new mathematical model for relative quantification in real-time RT-PCR. *Nucleic Acids Res.* **29**, e45.
22. Livak, K. J., and Schmittgen, T. D. (2001) Analysis of relative gene expression data using real-time quantitative PCR and the 2(-Delta Delta C (T)) Method. *Methods.* **25**, 402–408.

Chapter 14

Antibody-Based Strategy to Identify *Candida albicans* Genes Expressed During Infections

Cornelius J. Clancy, Shaoji Cheng, and M. Hong Nguyen

Abstract

Investigators have long used antibody-based screening strategies to identify *Candida albicans* immunogenic proteins and the genes that encode them during infections. With the recent availability of the *C. albicans* genome sequence and the development of genomic and proteomic technologies, it is now possible to efficiently conduct large-scale screening in standard research labs. *C. albicans* proteins and genes identified with a variety of screening methods have been implicated as important determinants of candidal virulence and exploited as vaccine and therapeutic targets. In this chapter, we describe methods used in our lab, in which sera recovered from patients with candidiasis are used to screen a *C. albicans* genomic DNA expression library. Immunoreactive colonies are detected by reaction with anti-human immunoglobulin, and the corresponding open reading frames are identified using the genome sequence database. The methods are also suitable for use with cDNA expression libraries, and they are complementary to proteomic screening strategies described elsewhere in this volume.

Key words: *Candida, Candida albicans*, antibodies, antigens, proteins.

1. Introduction

There is a long history of studies using antibody-based screening strategies to identify *Candida albicans* immunogenic proteins and the genes that encode them during infections. Beginning in the 1950s, investigators used whole *C. albicans* cells or complex antigenic extracts to detect antibodies in human sera as potential diagnostic strategies *(1)*. From these studies, it was ultimately demonstrated that cell wall mannan is the immunodominant candidal antigen. In the 1980s, investigators found that enolase, a glycolytic enzyme recovered from cytoplasmic extracts, is also strongly antigenic *(2)*. At roughly the same time, immunoblotting

Steffen Rupp, Kai Sohn (eds.), *Host-Pathogen Interactions*, DOI: 10.1007/978-1-59745-204-5_14,

experiments performed on whole cell extracts using sera from patients with systemic candidiasis detected a 47-kDa subcomponent of heat shock protein Hsp90p *(3, 4)*. Of note, patients who recovered from their infections manifested significant anti-47-kDa antibody titers, whereas those who died had low or falling titers. Supportive evidence for a protective role of human recombinant antibody against Hsp90p was obtained in murine models *(5, 6)*. In a recent human clinical trial, the combination of anti-Hsp90p monoclonal antibody and amphotericin B was superior to amphotericin B alone in the treatment of invasive candidiasis *(7)*. Taken together, these studies demonstrated that anticandidal antibodies in human sera can be used to detect cell wall and intracellular antigens involved in diverse biologic functions, including potential vaccine or therapeutic targets.

Investigators interested in identifying novel *C. albicans* proteins expressed during the infectious process began to exploit powerful genomic and proteomic tools in the 1990s. These studies involved probing *C. albicans* cDNA expression libraries or two-dimensional protein gels with sera recovered from patients (including those who were HIV-infected) with invasive forms of candidiasis or laboratory animals *(8–23)*. They demonstrated that sera from patients with candidiasis are rich in antibodies that interact with a large number of *C. albicans* proteins such as heat shock proteins and glycolytic and other metabolic enzymes. In our lab, we adapted these antibody-based strategies to screen a *C. albicans* genomic DNA expression library, with the goal of identifying previously unrecognized virulence factors *(24)*. We identified 65 genes expressed during human candidiasis, including those encoding immunogenic proteins involved in transcription and regulation, stress response and adaptation, metabolism, cytoskeleton and cell wall structure, adherence and flocculation, and hydrolytic enzymes *(25)*. To date, we have implicated at least three of the *in vivo* expressed genes in the pathogenesis of candidiasis *(24, 26–29)*.

In this chapter, we present the screening methods that we use for identification of *C. albicans* genes preferentially expressed within the human host during infection, compared with genes expressed in *C. albicans* during *in vitro* growth in the laboratory. As shown in **Fig. 14.1**, sera recovered from patients with active candidiasis are adsorbed repeatedly against whole *C. albicans* cells, French-pressed cell extracts, and heat-denatured cell lysates in order to remove the antibodies that are reactive with proteins made by *C. albicans in vitro*. The antibodies remaining in the adsorbed sera are used for identification of virulence factors specifically expressed during host infection. At the same time, a *C. albicans* genomic DNA expression library is created in

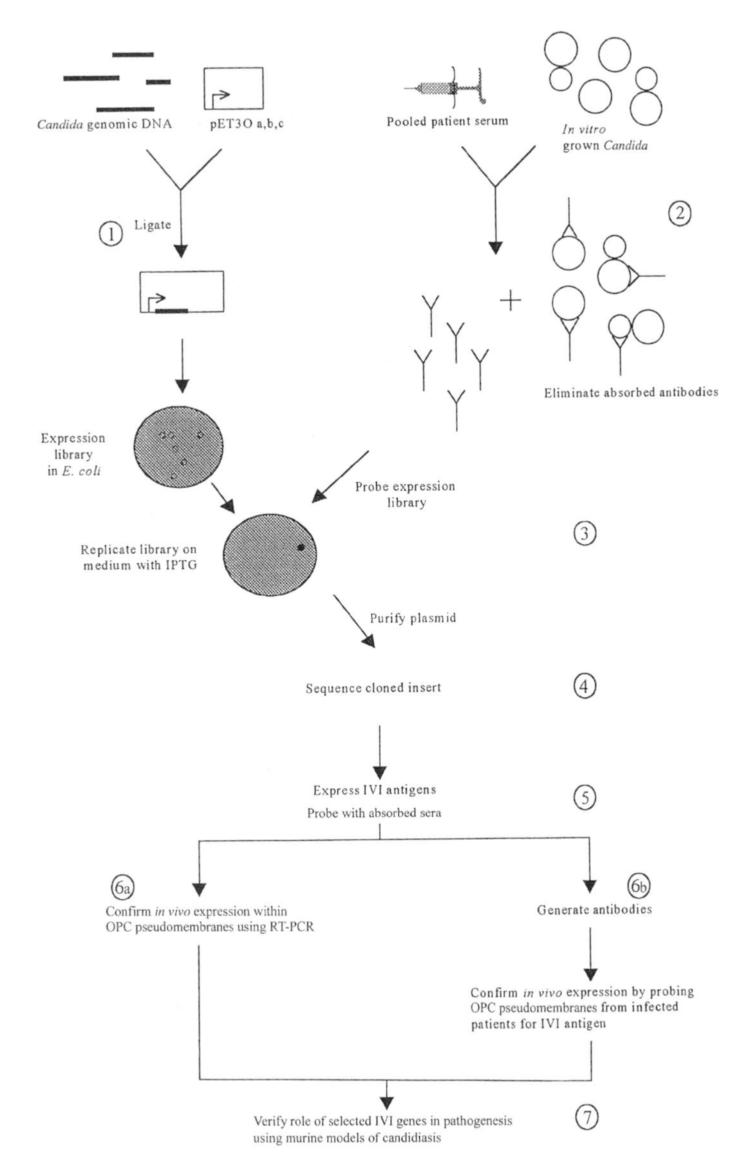

Fig. 14.1. A schematic overview of our screening method (**steps 1** to **5**). In follow-up experiments to those described in this chapter, we have implicated several genes and proteins identified by screening in candidal virulence. As discussed in the Introduction, immunogenic proteins identified by antibody-based methods have also been exploited as vaccine and therapeutic targets.

Escherichia coli. The adsorbed sera are used to screen the expression library, and immunoreactive colonies are detected by reaction with anti-human immunoglobulin. Immunoreactivity is confirmed by rescreening with adsorbed sera, and the corresponding open reading frames are identified using the *C. albicans* genome sequence database.

Of course, our screening methods can also be employed against cDNA expression libraries *(8)*. Elsewhere in this volume, methods for proteomic screening strategies are described, which are complementary to the approach we present. As tools to identify immunogenic proteins and study the host-pathogen interaction, the various antibody-based methods have relative strengths and weaknesses, as reviewed in **Table 14.1**.

Table 14.1
Strengths and weaknesses of genomic and proteomic methods used to identify immunogenic proteins and genes expressed during infection

	Strengths	Weaknesses
Genomic DNA expression library	• Potentially covers the entire genome • Not dependent upon relative level of protein expression by *C. albicans* (i.e., the library expresses both abundant and less abundant proteins)	• Expression in heterologous host does not account for noncanonical codon usage by *C. albicans* • Posttranslational modifications may be absent or differ, depending on species hosting the library • Expression of some proteins may be limited by the vector or toxicity to host species • Might not express full-length proteins • Analysis of gene sequence might be complicated by introns
cDNA expression library	• By altering conditions under which mRNA is harvested, library can focus on specific phenotypes of interest (e.g., hyphal formation) • Sequence analysis is simplified by lack of introns	• Only contains cDNA made from mRNA present under given experimental conditions • High frequency of redundant clones corresponding with abundant and intermediate mRNAs • Shares limitations of genomic library (with exception of introns)
Proteomic (2D PAGE)	• As with cDNA libraries, proteins can be extracted under particular experimental conditions of interest • Proteins have been expressed by *C. albicans* cells rather than recombinant proteins	• Only contains proteins present at time of extraction • Favors most abundant proteins • Potentially limited by ability to isolate and separate proteins

2. Materials

2.1. Construction of a C. albicans Genomic DNA Expression Library

2.1.1. Equipment

1. Microcentrifuge (suitable for Eppendorf tubes).
2. Heating block (suitable for incubating Eppendorf tubes).
3. Gel electrophoresis apparatus.
4. Long-wave UV light.
5. Electroporation system (such as the Gene Pulser Xcell Electroporation System, Bio-Rad. Hercules, CA).
6. 37°C incubator.
7. Centrifuges (suitable for 15- and 50-mL tubes).
8. −20°C and −70°C freezers.

2.1.2. Agar, Media, Reagents, and so Forth

1. TE buffer:(10 mM Tris-HCl, 1 mM EDTA pH 8.0).
2. *Sau*3A and *Bam*HI, and buffers.
3. 10x BSA (1 mg/mL).
4. ddH_2O.
5. 50 mM EDTA (pH 8.0).
6. DNA clean-up kit, for isolating DNA from agarose gels (e.g., Wizard DNA clean-up kit (Promega, Madison, WI) or similar product).
7. Phenol/Chloroform/Isoamyl alcohol (25:24:1; v:v:v).
8. Chloroform/Isoamyl alcohol (24:1; v:v).
9. pET 30a,b,c DNA (Novagen, Gibbstown, NJ), transformed into *E. coli* DH5α.
10. Qiagen Midi-prep (Valencia, CA) (or similar plasmid DNA purification kit).
11. Microcon-100 (Millipore, Bedford, MA).
12. Calf intestinal phosphatase and buffer.
13. T4 DNA ligase and buffer.
14. ElectroTen-Blue electrocompetent cells (Stratagene, La Jolla, CA).
15. SOC medium (Invitrogen Carlsbad, CA).
16. Brain heart infusion agar (BHI).
17. Kanamycin (1000X): dissolved in distilled water at 50 mg/mL. Store at −20°C.
18. LB agar.
19. 15% glycerol.

2.1.3. Supplies

1. Eppendorf tubes (1.5 and 2 mL).
2. Disposable 50 mL, 20 mL plastic centrifuge tubes.

3. Disposable plastic pipettes (10 μL, 200 μL, 1000 μL), inoculating loops (10 μL).
4. Petri dish (100 × 15 mm).
5. Cryovials (2 mL) (for cell storage in 15% glycerol).
6. 0.2-cm electroporation cuvette.
7. Agarose.

2.2. Absorption of Sera Against In Vitro Expressed C. albicans Proteins

2.2.1. Equipment and Supplies

Equipment and supplies are as listed above for library construction, plus the following:

1. Yeast peptone dextrose (YPD) medium (1% (w/v) yeast extract, 1%. (w/v) peptone and 2% (w/v) glucose).
2. Shaking and nonshaking 30°C and 37°C incubators.
3. Flasks (250 mL, Pyrex Fisher, Pittsburgh, PA).
4. ELISA coating buffer (Carb-Bicarb buffer, pH 9.6): Prepare stock solution 1:1 M $NaHCO_3$ and stock solution 2:1 M Na_2CO_3; Filter sterilize and store at 4°C. To prepare working solution mix 4.53 mL from solution 1, 1.82 mL of solution 2 and 0.2 mL 10% NaN_3. Add ddH_2O to 100 mL. Store at 4°C up to 2 weeks.
5. Phenylmethylsulfonyl fluoride (PMSF) (200 mM in DMSO).
6. French press (SLM Instruments, Inc. (Urbana, IL)).
7. Protein concentration assay kit (e.g., BioRad Protein Assay Reagent Package).
8. Phosphate-buffered saline (PBS) buffer: 0.02 M sodium phosphate buffer with 0.15 M sodium chloride, pH adjusted to 7.4.
9. PBS buffer with 0.05% Tween 20 (PBST).
10. PBS + 0.02% NaA_3.
11. Nitrocellulose filter (0.45 μm, 82 mm, Millipore).
12. 1 N HCl.
13. EIA/RIA plate.
14. Anti-human secondary antibody (we use peroxidase-conjugated goat affinity purified antibody to human IgG, IgA, and IgM, obtained from MP Biomedicals, Solon, OH).
15. *O*-phenylenediamine dihydrochloride (OPD) (Sigma, St. Louis, MO).
16. OPD substrate development solution for peroxidase: Prepare stock solution A: 0.1 M Citric acid (4.8 g/250 mL) and stock solution B: 0.2 M Na_2HPO_4 (7.1 g/250 mL); Filter sterilize, store at 4°C. To prepare 12.5 mL working solution (prepare fresh, enough for one 96-well plate) mix 6.25 mL of DDW, 3.0 mL of solution A, 3.25 mL of solution B, 5 μL 30% H_2O_2, and 5 mg (1 tablet) of OPD.

17. Microtiter plate reader.
18. Blocking solution: 1 × PBS, 0.25% gelatin, 0.25% Tween 20, 0.02% NaN_3.

2.3. Screening the C. albicans Genomic DNA Expression Library

2.3.1. Equipment and Supplies

Equipment and supplies are as listed above for library construction, plus the following:

1. Isopropyl-β-d-thiogalactopyranoside (IPTG) (1 M in water; store at −20°C).
2. Replica-Plating Tool (Fisher).
3. Sterile velvet.
4. Chloroform.
5. Air-tight container (capable of holding culture plates).
6. Nonfat dry milk.
7. ECL chemiluminescence kit (Amersham Ltd. Piscataway, NJ).
8. ECL hyper film (Amersham Ltd).
9. pET30 Ek/LIC (Novagen).
10. Nova Blue *E. coli* (Novagen).

3. Methods

3.1. Construction of a C. albicans Genomic DNA Expression Library (see Note 1)

3.1.1. *Sau*3A Partial Digestion of Genomic DNA (see Notes 2 and 3)

1. Place cold Eppendorf tubes (labeled no. 1 to 9) into ice. The tubes should be cold before enzyme is placed into them and kept on ice through out the experiments.
2. Suspend 10 μg of DNA in a final volume of 120 μL TE. Add 15 μL of 10x *Sau*3A buffer and 15 μL of 10x BSA, and mix thoroughly.
3. Dispense 30 μL into tube 1, 15 μL into tubes 2 to 8, and the remainder into tube 9.
4. Add 1 μL (~ 4 units) of *Sau*3A to tube 1 and mix the contents by repeated pipetting.
5. Transfer 15 μL into tube 2. Mix well and continue the dilution scheme through tube 9.
6. Centrifuge the Eppendorf tubes briefly at 4°C to get all liquid to the bottom, and then incubate at 37°C for 30 min (*see* **Note 4**).
7. Place the tubes on ice, and add 10 μL of 50 mM EDTA to a final concentration of 20 mM.
8. Determine the conditions that result in optimal digestions by electrophoresis through 0.8% agarose gel (*see* **Note 5**).
9. Scaling up the optimized conditions, digest 500 μg of DNA.

10. Analyze an aliquot of the digested DNA (~1 μg) by electrophoresis through 0.8% agarose gel to determine if the digestion is satisfactory.
11. Load the complete digestion onto 0.8% agarose gel and run at 2 volts/cm. Examine the stained gel under long-wave UV light (*see* **Note 6**).
12. Cut 1- to 2-kb sized DNA fragments from the gel and purify (e.g., Promega Wizard DNA clean-up kit or comparable product) (*see* **Note 7**).
13. Extract the DNA with Phenol/Chloroform/Isoamyl alcohol (25:24:1), followed by two chloroform extractions.
14. Estimate DNA size and concentration by 0.8% agarose gel electrophoresis.

3.1.2. Ligation of Genomic DNA Fragments into PET Vectors

1. Isolate pET30a, b, and c plasmid DNA from *E. coli* DH5α using Qiagen Midi prep or comparable methods (*see* **Note 8**).
2. Estimate DNA concentration by agarose gel electrophoresis.
3. Digest DNA with *Bam*HI at 37°C for 2 h and verify digestion by agarose gel electrophoresis. A typical digestion consists of: 10 μg vector DNA, 20 μL 10 × buffer, 20 μL 10 × BSA, 10 μL *Bam*HI (10 U/μL), in ddH_2O to 200 μL.
4. Purify/concentrate digested DNA using microcon-100 (Millipore). Wash with 500 μL ddH_2O and resuspend in 100 μL ddH_2O. Save 10 μL for control during ligation.
5. Dephosphorylate the vector DNA. A typical reaction consists of: 90 μL digested vector DNA, 10 μL 10 × calf intestinal phosphatase (CIP) buffer, 1.5 μL CIP (Promega, 1 U/μL):
   (a) Incubate at 37°C for 15 min and then at 56°C for 15 min.
   (b) Add 1.5 μL CIP, incubate at 37°C for 15 min and then at 56°C for 15 min.
   (c) Heat inactivate the enzyme at 75°C for 10 min.
6. Use microcon-100 to purify/concentrate dephosphorylated DNA. Wash with 500 μL ddH_2O and resuspend in 100 μL ddH_2O.
7. Estimate DNA concentration by agarose gel electrophoresis.
8. Perform ligations with varying ratios of vector to insert (1:1, 1:3, 3:1, etc), including digested vector and CIP-treated digested vectors without insert DNA as self-ligation controls. A typical ligation consists of: X μL (200 ng) vector, Y μL insert DNA, 1.5 μL 10X ligase buffer, 0.5 μL T4 DNA ligase (3 U/μL), in ddH_2O to 15 μL:
   (a) Incubate overnight at 14°C (approximately 16 h).
   (b) Heat inactivate the enzyme at 75°C for 10 min.
9. Use microcon-100 to purify/concentrate ligated DNA. Wash with 500 μL ddH_2O and resuspend in 10 μL ddH_2O.

3.1.3. Transformation of E. coli

1. Add 10 μL of ligated DNA prepared above to electrocompetent *E. coli* cells that have high efficiency of transformation (e.g., ElectroTen-Blue electrocompetent cells, Stratagene).
2. Transfer this mixture to a prechilled 0.1 cm electroporation cuvette (Bio-Rad).
3. Electroporate (Gene Pulser Xcell Electroporation System, Bio-Rad) at the following settings: 2.5 kV, 200 Ω, 25 μF (*see* **Note 9**).
4. Quickly add 1 mL of ice-cold SOC medium to the cuvette, mix well, and transfer to a 1.5-mL centrifuge tube.
5. Incubate the transformation mixture at 37°C for 1 h with shaking at 200 rpm.
6. Spread 100 μL of the transformation mixture onto a BHI + kanamycin plate and keep the remaining mixture at 4°C (to be plated if the transformation is good).
7. Incubate the plates at 37°C overnight (~16 h).
8. Count the colonies and determine the ligation and transformation efficiencies (*see* **Note 10**).
9. Collect and pool transformed cells; aliquot to vials containing BHI media with 15% glycerol for storage at −70°C (*see* **Note 11**).
10. Extract plasmid from one of the aliquots.
11. Finally, transform the plasmid into expression host, BL21(DE3).
12. Collect transformed cells and aliquot to vials containing LB media with 15% glycerol for storage at −70°C.

3.2. Absorption of Sera Against In Vitro Expressed C. albicans Proteins (see Note 12)

3.2.1. Protein Isolation from French-Pressed C. albicans Cells

1. Grow *C. albicans* SC5314 in 100 mL YPD at 30°C for at least 18 h (*see* **Note 13**).
2. Collect cells by centrifugation at 1000 × *g* for 10 min at 4°C and discard the supernatant.
3. Wash the cell pellet with 25 mL of carb-bicarb buffer.
4. Resuspend the cell pellet with 20 mL of carb-bicarb buffer containing 1 mM PMSF.
5. Lyse the cells by passing through a French Press (SLM Instruments, Inc.) five times at 1100 p.s.i.
6. Centrifuge the cell lysate at 16,000 × *g* for 30 min at 4°C.
7. Remove the extract from the pellet with a Pasteur pipette.
8. Quantitate the protein in the extract (Bio-Rad Protein Assay Reagent Package or similar method).
9. Aliquot the protein and store at −20°C.

3.2.2. Adsorption of Sera

1. Grow *C. albicans* overnight in YPD at 30°C.
2. The next day, collect the cells by centrifugation, and wash them 3 times with PBS.
3. Resuspend the washed cells in PBS + 0.02% NaA_3 at concentration of 1 × 10^9/mL. Save at 4°C as 100-μL aliquots in 1.5-mL Eppendorf tubes.
4. Add 0.5 to 1.0 mL of the sera to a 2-mL Eppendorf tube containing *C. albicans* cells and mix by inversion at 4°C overnight.
5. After overnight incubation, centrifuge at 16,000 × *g* for 10 min at 4°C.
6. Transfer the sera into a new tube. Save 10 μL as "first direct absorption."
7. Repeat **steps 3** to **5** six times. Save 10 μL after each round of absorption (label as "second" through "seventh direct absorption").
8. After the seventh adsorption with whole cells, the adsorbed sera are further adsorbed using French-press lysed Candida cells. Spread 5 mL of French-pressed *C. albicans* cell extracts onto a nitrocellulose membrane (82 mm, 0.45 μm) pre-wet with PBS in a clean Petri dish; incubate at room temperature for 1 h with agitation.
9. Wash the membrane 10 times with PBST, and transfer to a clean Petri dish.
10. Spread the sera available after the seventh round of direct adsorption onto the membrane. Incubate overnight at 4°C with agitation.
11. Collect the sera. Wash the membrane with 300 μL of PBS. Combine the washed fluid with the sera. Save 10 μL as "eighth direct adsorption."
12. Repeat **steps 8** to **10** using cell extracts that were boiled for 10 min to denature proteins. Save 10 μL as "final adsorption."
13. Aliquot the final adsorbed sera in individual 2 mL cryotubes, and store at −70°C.

3.2.3. Testing the Efficiency of Adsorption by ELISA

1. Dilute French-press isolated protein with carb-bicarb buffer to100 μg/mL.
2. Add 100 μL per well to 96-well EIA/RIA plate (Costar 3361) and incubate at 4°C overnight (*see* **Note 14**).
3. Wash the plate 5 times with PBST.
4. Add 400 μL of blocking solution per well and incubate at 37°C for 1 h.
5. Perform serial twofold dilutions of sera. Dispense 300 μL of a 1:100 dilution of serum in PBS to the first well in each row

of a round bottom 96-well plate, and dispense 150 μL of PBS to the remaining wells. Dispense 150 μL of diluted sera to adjacent wells.

6. After the 1 h incubation, wash the EIA/RIA plate 5 times with PBST.
7. Transfer 100 μL of diluted sera from **step 5** to the corresponding well of the EIA/RIA plate, and incubate at 37°C for 1 h.
8. Wash the plate 5 times with PBST.
9. Dilute anti-human secondary antibody (ICN 55256) 1:1000, and add 100 μL to the wells; incubate at 37°C for 1 h.
10. Wash the plate 3 times with PBST, and then twice with PBS (*see* **Note 15**).
11. Add 100 μL of OPD substrate development solution to the wells, and incubate at 37°C for 10 to 15 min in the dark.
12. Add 100 μL of 1 N HCl to stop the reaction.
13. Read in a microtiter plate reader at OD_{450} (*see* **Note 16**).

3.3. Screening the C. albicans Genomic DNA Expression Library

3.3.1. Screening the Library

1. Serially dilute the genomic expression library and plate on brain-heart infusion (BHI) medium containing kanamycin (KM; 50 μg/mL) for 12 to 14 h at 37°C to generate plates containing 300 to 500 colonies ("parental plates").
2. Keep the parental plates at 4°C for 1 h, along with BHI plates containing KM and IPTG (1 mM) to be used for replica plating. Number the parental and corresponding replica plates for identification; place similar orientation marks on the edges of the bottoms of the paired plates.
3. Replicate the colonies from parental plates using sterile velvet onto the corresponding BHI + KM + IPTG plates.
4. Incubate for 5 h at 37°C to induce expression of the cloned genes. Store the parental plates at 4°C for future use (*see* **Note 17**).
5. Place the replica plates without lids into an air-tight container containing chloroform for 20 min to partially lyse the bacteria and expose cytoplasmic proteins.
6. Overlay the chloroform-lysed replica plates with nitrocellulose membranes for 15 min at room temperature. Mark each membrane in pencil with the plate number and the orientation markings.
7. Remove the membranes and saturate them face-up with PBST with 5% nonfat milk. Incubate for 1 h at room temperature with agitation.
8. Wash the membrane 3 times with PBST at room temperature for 5 min.

9. Incubate the membranes with diluted adsorbed sera in PBST at room temperature for 1 h with agitation (the optimized dilution of adsorbed sera are determined in preliminary experiments).
10. Again, wash the membrane three times with PBST at room temperature for 5 min.
11. Incubate the membranes with peroxidase-conjugated goat anti-human immunoglobulin (reactive with all classes of human immunoglobulins) (Cappel/ICN) in PBST at room temperature for 1 h, with agitation (the optimized dilution of the secondary antibody is again determined in preliminary experiments).
12. Wash the membrane 3 times with PBST and twice with PBS at room temperature (5 min each wash).
13. Cover the membrane with ECL chemiluminescence solution and expose to ECL Hyperfilm, as per product instructions.
14. Merge the film images with parental plates to identify reactive clones.

3.3.2. Confirming Reactive Clones and Analyzing C. albicans DNA on the Inserts

1. Streak the reactive clones on BHI + KM agar plates. Incubate at 37°C overnight.
2. Pick at least three single colonies from each possible reactive clone and grow them overnight at 37°C in BHI broth containing KM (50 μg/mL).
3. Centrifuge and resuspend pellets in an equal volume of fresh medium.
4. Spot 5 μL of each suspension onto BHI + KM and BHI + KM + IPTG plates. Colonies can be spotted in identical grid patterns to each plate (**Fig. 14.2**).
5. Two negative controls are included as controls on each plate: pET30a/BL21(DE3) with no cloned insert, and a random clone that contained an insert but was nonreactive with sera during primary screening. Once reactive clones are identified, they can be included as positive controls.
6. Incubate the plates at 37°C for 3 to 5 h.
7. Keep the BHI + KM plate at 4°C as a master plate.
8. Treat the BHI + KM + IPTG plate with chloroform vapors and immobilize the lysed proteins on nitrocellulose membranes; membranes are probed with adsorbed sera (as in **Section 3.3.1, steps 5** to **14**).
9. Pick reactive clones from the master plate, restreak for isolation, and rescreen several times to ensure the reproducibility of results.

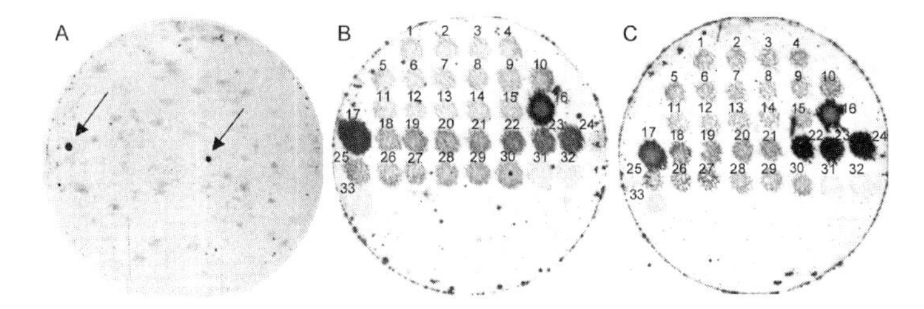

Fig. 14.2. Representative membranes from successive rounds of screening with sera. Clones expressing *C. albicans* proteins that are potentially reactive with antibodies in sera are identified after incubation with peroxidase-conjugated goat anti-human antibody (**A**, colonies indicated by arrows). The reactive clones are streaked to isolation; single colonies from each original clone are resuspended in medium and spotted onto BHI + kanamycin + IPTG plates for rescreening. Representative membranes from second and third rounds of screening are presented (**B, C**), which show both reactive (e.g., no. 16 and no. 17) and nonreactive colonies. Each plate can include positive (nos. 22 to 24) and negative (nos. 31 to 33) controls.

10. Grow the reactive clone in 5 mL of BHI broth containing kanamycin (100 μg/mL) at 37°C overnight.
11. Extract the plasmid DNA from reactive colonies with Qiagen mini-prep.
12. Sequence *C. albicans* DNA in the insert by using pET30 primers (Novagen).
13. Nucleotide and protein sequence analyses from each ORF can be performed using the Wisconsin Genetics Computer Group sequence analysis software package (GCG; now Accelrys), version 7.0, or comparable software.
14. Analyze the sequences by BLAST search and by comparison for homology with the *C. albicans* and *S. cerevisiae* genome databases (CGD: www.candidagenome.org; CandidaDB: http://genolist.pasteur.fr/CandidaDB/; SGD: www.yeastgenome.org).

3.3.3. Identifying the Open Reading Frames (ORFs) That Encode the Antigens Responsible for Reactivity with Sera

1. In the event that an insert contains more than one potential ORF, subclone the first ORF in-frame with the ribosomal binding site into the pET30 Ek/LIC system (Novagen).
2. Design forward and reverse PCR primers to amplify the ORF and attach tail sequences that are complementary to the sequence of the vector.
3. Treat with T4 DNA polymerase in the presence of dATP, and ligate to pET30 Ek/LIC (following the instructions in the Ek/LIC kit).
4. Transform into NovaBlue *E. coli*, and then into the high-level expression *E. coli* strain BL21(DE3).

5. Test the clones containing the subcloned ORF for reactivity with adsorbed sera by the colony blotting method described above.
6. If first ORF is not reactive with sera, use adsorbed sera to perform Western analysis on protein extracted from the reactive clones at serial time points after IPTG induction. The size of the reactive protein might suggest the size of the ORF responsible for its expression. In those instances where two or more ORFs encode potentially reactive proteins of similar estimated size, sequentially subclone and probe these ORFs.

4. Notes

1. We constructed our library from DNA pooled from several *C. albicans* isolates, in order to account for potential differences in the genomes of clinical isolates. Investigators might reasonably decide that this extra work is not necessary and extract DNA from a representative strain, such as *C. albicans* SC5314.
2. We chose *Sau*3A for controlled partial digestion of genomic DNA. It recognizes 5′-GATC-3′ and produces overhanging sticky ends that are compatible with *Bam*HI sites. After partial digestion, selection of a specific size range can be facilitated by centrifugation over sucrose gradients and collection of an appropriate fraction, or by isolation of a specific size range from agarose gels. Partially digested DNA can be easily ligated into a vector digested with *Bam*HI.
3. As with any restriction enzyme, *Sau*3A does not result in random digestion of the genome. As such, there might be biases in the representation of open reading frames within the library. As an alternative or complementary approach, one can create a library from randomly sheared genomic DNA. Sheared DNA can be end-repaired, sized on a low-melting-point agarose gel, and subsequently cloned into the vector at a blunt-ended cloning site.
4. Check the digestion by inverting the tube every 5 min. Stop the incubation as the cell suspension begins to become viscous. Overdigestion will break down DNA into smaller sizes and reduce the yield.
5. Optimal digestions will appear as an even smear across all sizes.
6. Quickly punch the 1- and 2-kb marker bands with a sharp pencil point under long-wave UV light. Use the punched markers as orientation when cutting the gel.

7. We found that 1- to 2-kb sized fragments were optimal when screening the library. Larger genomic DNA fragments in the library increase the likelihood of multiple open reading frames existing within the insert. Smaller DNA fragments increase the amount of screening necessary to cover the genome.
8. For our purposes, the pETabc system is ideal for several reasons: (a) target genes are cloned under control of strong bacteriophage T7 transcription and translation signals, and expression is induced by providing a source of T7 RNA polymerase in the host cell; (b) the presence of a lac operator permits tight regulation of expression by IPTG, thereby decreasing the possibility of toxicity stemming from overexpression of cloned DNA; (c) the presence of a histidine$_6$-tag immediately after the translation start codon and just upstream of the multiple cloning site allows simple purification of translated proteins using nickel chromatography columns. *C. albicans* proteins are readily expressed from this vector by *E. coli.*
9. The time constant for electroporation should be around 4.2 ms.
10. Ligation efficiency represents the percentage of vectors in the library that contain inserts. The ideal ratio should be ≥90%. The transformation efficiency with commerical kits is typically ≥10^8 transformants/μg plasmid DNA.
11. The number of clones necessary to cover the whole *C. albicans* genome can be estimated using the formula $N = Ln(1 - P)/Ln(1 - f)$, in which P is a chosen probability of desired success (usually 0.95 to 0.999) and f is the fraction of the genome represented by the average size clone. For our library with an average insert size of 1.5 kb, f is 1.5 kb/17,000 kb (the genome size of *C. albicans*). As such, 339,515 and 521,917 clones are minimal for probabilities of success ranging between 95% and 99.9%.
12. The purpose of adsorbing sera against *in vitro* expressed *C. albicans* proteins is to enrich for antibodies directed against proteins that are expressed exclusively or preferentially during the infectious process. These steps can be omitted if the identification of *in vivo* induced proteins is not an experimental objective. In this event, proceed to the experiments in **Section 3.3**.
13. Growth conditions can be altered, depending on the morphology against which you wish to absorb sera. Under the conditions described here, *C. albicans* will grow primarily in the yeast morphology.
14. This is a high protein binding plate that is appropriate for ELISA. An alternative to overnight incubation is to incubate at 37°C for 1 h.

15. The wash with PBS is important to remove Tween 20, which will inhibit the peroxidase reaction.
16. Each round of absorption should see a decrease in anti-*Candida* immunoglobulin titers until they are maximally absorbed.
17. In our experience, the incubation period that optimized expression of most *C. albicans* proteins was between 3 and 5 h. This can be corroborated in each lab during preliminary screening experiments.

References

1. Ponton J, Omaetxebarria MJ, Elguezabal N, Alvarez M, Moragues MD. (2001) Immunoreactivity of the fungal cell wall. *Med Mycol.*; 39 Suppl 1:101–10.
2. Martinez JP, Gil ML, Lopez-Ribot JL, Chaffin WL. (1998) Serologic response to cell wall mannoproteins and proteins of *Candida albicans. Clin Microbiol Rev.*; 11:121–41.
3. Matthews RC, Burnie JP, Tabaqchali S. (1984) Immunoblot analysis of the serological response in systemic candidosis. *Lancet*; 2(8417–18):1415–8.
4. Matthews R, Burnie J, Smith D, Clark I, Midgley J, Conolly M, Gazzard B. (1988) Candida and AIDS: evidence for protective antibody. *Lancet*; 2(8605):263–6.
5. Matthews R, Hodgetts S, Burnie J. (1995) Preliminary assessment of a human recombinant antibody fragment to hsp90 in murine invasive candidiasis. *J Infect Dis.*; 171:1668–71.
6. Matthews RC, Rigg G, Hodgetts S, Carter T, Chapman C, Gregory C, Illidge C, Burnie J. (2003) Preclinical assessment of the efficacy of mycograb, a human recombinant antibody against fungal HSP90. *Antimicrob Agents Chemother.*; 47:2208–16.
7. Pachl J, Svoboda P, Jacobs F, Vandewoude K, van der Hoven B, Spronk P, Masterson G, Malbrain M, Auon M, Garbino J, Takala J, Drgona L, Burnie J, Matthews R; Mycograb Invasive Candidiasis Study Group. (2006) A randomized, blinded, multicenter trial of lipid-associated amphotericin B alone versus in combination with an antibody-based inhibitor of heat shock protein 90 in patients with invasive candidiasis. *Clin Infect Dis.*; 42:1404–13.
8. Swoboda RK, Bertram G, Hollander H, Greenspan D, Greenspan JS, Gow NA, Gooday GW, Brown AJ. (1993) Glycolytic enzymes of *Candida albicans* are nonubiquitous immunogens during candidiasis. *Infect Immun.*; 61:4263–71.
9. Alloush HM, Lopez-Ribot JL, Chaffin WL. (1996) Dynamic expression of cell wall proteins of *Candida albicans* revealed by probes from cDNA clones. *J Med Vet Mycol.*; 34:91–7.
10. Lopez-Ribot JL, Monteagudo C, Sepulveda P, Casanova M, Martinez JP, Chaffin WL. (1996) Expression of the fibrinogen binding mannoprotein and the laminin receptor of *Candida albicans in vitro* an in infected tissues. *FEMS Microbiol Lett.*; 142:117–22.
11. Alloush HM, Lopez-Ribot JL, Masten BJ, Chaffin WL. 3-phosphoglycerate kinase: a glycolytic enzyme protein present in the cell wall of *Candida albicans.* (1997) *Microbiology*; 143:321–30.
12. Gil-Navarro I, Gil ML, Casanova M, O'Connor JE, Martinez JP, Gozalbo D. (1997) The glycolytic enzyme glyceraldehyde-3-phosphate dehydrogenase of *Candida albicans* is a surface antigen. *J Bacteriol.*; 179:4992–9.
13. Sentandreu M, Elorza MV, Valentin E, Sentandreu R, Gozalbo D. (1995) Cloning of cDNAs coding for *Candida albicans* cell surface proteins. *J Med Vet Mycol.*; 33:105–11.
14. Eroles P, Sentandreu M, Elorza MV, Sentandreu R. The highly immunogenic enolase and Hsp70p are adventitious *Candida albicans* cell wall proteins. (1997) *Microbiology*; 143:313–20.
15. Pitarch A, Pardo M, Jimenez A, Pla J, Gil C, Sanchez M, Nombela C. (1999) Two-dimensional gel electrophoresis as analytical tool for identifying *Candida albicans* immunogenic proteins. *Electrophoresis*; 20:1001–10.

16. Pardo M, Ward M, Pitarch A, Sanchez M, Nombela C, Blackstock W, Gil C. (2000) Cross-species identification of novel *Candida albicans* immunogenic proteins by combination of two-dimensional polyacrylamide gel electrophoresis and mass spectrometry. *Electrophoresis*; 21:2651–9.
17. Pitarch A, Diez-Orejas R, Molero G, Pardo M, Sanchez M, Gil C, Nombela C. (2001) Analysis of the serologic response to systemic *Candida albicans* infection in a murine model. *Proteomics*; 1:550–9.
18. Pendrak ML, Klotz SA. (1995) Adherence of *Candida albicans* to host cells. *FEMS Microbiol Lett.*; 129:103–13.
19. Lopez-Ribot JL, Alloush HM, Masten BJ, Chaffin WL. (1996) Evidence for presence in the cell wall of *Candida albicans* of a protein related to the hsp70 family. *Infect Immun.*; 64:3333–40.
20. Fernandez-Arenas, E., Molero, G., Nobela, C., Diez-Orejas, R., and Gil, C. (2004) Contribution of the antibodies response induced by a low virulent *Candida albicans* strain in protection against systemic candidiasis. *Proteomics*; 4:1204–1215.
21. Fernandez-Arenas E., Molero G., Nombela C, Diez-Orejas R., and Gil, C. (2004) Low virulent strains of *Candida albicans*: unravelling the antigens for a future vaccine. *Proteomics*; 4:3007–3020.
22. Pitarch, A., Abian, J., Carrascal, M., Sanchez, M., Nombela, C., Gil, C. (2004) Proteomics-based identification of novel *Candida albicans* antigens for diagnosis of systemic candidiasis in patients with underlying hematological malignancies. *Proteomics*; 4:3084–3106.
23. Pitarch, A., Jimenez, A., Nombela, C., Gil, C. (2006) Decoding serological response to Candida cell wall immunome into novel diagnostic, prognostic and therapeutic candidates for systemic candidiasis by proteomic and bioinformatics analyses. *Mol Cell Proteomics*; 5:79–96.
24. Cheng S, Clancy CJ, Checkley MA, Handfield M, Hillman JD, Progulske-Fox A, Lewin AS, Fidel PL, Nguyen MH. (2003) Identification of *Candida albicans* genes induced during thrush offers insight into pathogenesis. *Mol Microbiol.*; 48:1275–88.
25. Nguyen MH, Cheng SJ, Clancy CJ. (2004) Assessment of *Candida albicans* genes expressed during infections as a tool to understand pathogenesis. *Med Mycol.*; 42:293–304.
26. Cheng S, Nguyen MH, Zhang Z, Jia H, Handfield M, Clancy CJ. (2003) Evaluation of the roles of four *Candida albicans* genes in virulence by using gene disruption strains that express *URA3* from the native locus. *Infect Immun.*; 71:6101–3.
27. Cheng S, Clancy CJ, Checkley MA, Wozniak KL, Seshan KR, Jia HY, Fidel P, Cole G, Nguyen MH. (2005) The role of *Candida albicans NOT5* in virulence depends upon diverse host factors *in vivo*. *Infect Immun.* 73:7190–7.
28. Badrane H, Cheng S, Nguyen MH, Jia HY, Zhang Z, Weisner N, Clancy CJ. (2005) *Candida albicans IRS4* contributes to hyphal formation and virulence after the initial stages of disseminated candidiasis. *Microbiology*; 151:2923–31.
29. Raman SB, Nguyen MH, Zhong Z, Cheng S, Jia HY, Iczkowski K, Clancy CJ. (2006) *Candida albicans SET1* encodes a histone 3 lysine 4 methyltransferase that regulates morphogenesis and contributes to the pathogenesis of invasive candidiasis. *Mol Microbiol.*; 60:697–709.

Chapter 15

Identification of the *Candida albicans* Immunome During Systemic Infection by Mass Spectrometry

Aida Pitarch, César Nombela, and Concha Gil

Abstract

Over the past two decades, mass spectrometry (MS) has ceased to be a fairly exotic technique banished from the protein scientists' mind to become a seminal tool for deciphering the information encoded in the genomes of many biological species. Clues to this shift in the *modus operandi* for characterizing their proteomes stem from the progressive availability of full genome sequences and well-annotated protein databases of many model (micro)organisms, the development both of soft ionization methods for large biomolecules (peptides and proteins) and of innovative instrumentation designs, and the introduction of sophisticated search algorithms able to correlate MS information with sequence databases, to name but a few. Here we integrate the typical MS-based strategy for identifying proteins of *Candida albicans*, an opportunistic fungal pathogen of humans, which have proved to be present during systemic infection and targeted by the immune system as a consequence of its interaction with the host (i.e., the *C. albicans* immunome).

Key words: *Candida albicans*, immunome, mass spectrometry, tandem mass spectrometry, peptide mass fingerprinting, peptide sequencing, *de novo* sequencing, MALDI-TOF, nanoelectrospray, proteomics.

1. Introduction

The past few years have witnessed how the study and systematic characterization of the *C. albicans* proteome (the protein complement expressed by the genome of this opportunistic pathogen fungus at a given time point) *(1, 2)* has undergone not only a renaissance but also a progressive and exciting expansion *(3, 4)*. This certainly arises from (i) the recent completion of genome sequencing for *C. albicans* and related species *(5, 6)* and the ensuing creation of well-annotated genomic and proteomic databases *(3, 7–11)*, (ii) advances and recent improvements in the mass spectrometry (MS) technology for the protein and peptide

Steffen Rupp, Kai Sohn (eds.), *Host-Pathogen Interactions*, DOI: 10.1007/978-1-59745-204-5_15,

analysis, such as the development of soft ionization methods for converting large, polar and nonvolatile biomolecules to gas-phase ions (*see* **Note 1**) *(12–14)*, and (iii) the introduction of sophisticated search algorithms capable of efficiently correlating MS information with sequence databases *(15, 16)*. All these steps forward have played, without a shred of doubt, a crucial role in the successful use of MS as the technique of choice both for high-throughput identification of the proteomes of *C. albicans* and other biological species (i.e., the gateway to link the genome of an [micro]organism to its proteome) and for the characterization of posttranslationally modified proteins *(17–20)*.

A mass spectrometer measures the mass-to-charge ratio (*m/z*) of gas-phase ions (*see* **Note 2**). This basically comprises three parts that are maintained under vacuum: an ion source, a mass analyzer, and an ion detector (**Fig. 15.1**) *(18, 21–24)*. Overall, the MS technique, which is often preceded by site-specific proteolysis, implies the following steps:

1. The conversion of peptides or small-sized proteins to gas-phase ions by soft ionization techniques (*see* **Note 1**), such as matrix-assisted laser desorption ionization (MALDI) and electrospray ionization (ESI). These generate gas-phase ions from solid-phase and liquid-phase samples, respectively (*see* **Note 3**).

2. The separation of the produced gas-phase ions (singly or multiply protonated ion species; *see* **Note 3**) on the basis of their *m/z* values (taking advantage of some physical property, such as time of flight, or electric or magnetic fields) in the mass analyzer. The most widely used mass analyzers in proteomics are (i) time-of-flight (TOF), (ii) quadrupole (Q) (*see* **Note 4**), (iii) ion trap (IT), and (iv) Fourier-transform ion cyclotron resonance (FT-ICR) analyzers.

3. Optional fragmentation of specifically selected ion species (*see* **Note 3**) by collision-induced dissociation (CID; *see* **Note 5**) within the mass spectrometer in order to obtain further sequence data. Peptide fragmentation is carried out in tandem mass spectrometry (MS/MS) systems, resulting either from the combination of two mass analyzers (tandem-in-space instruments, such as TOF-TOF, triple quadrupole [QqQ], and hybrid quadrupole-TOF [QqTOF]) or from use of the same mass analyzer two or more times (tandem-in-time instruments, such as IT and FT-ICR; *see* **Note 6**).

4. Detection of the selected and/or fragmented ion species at each *m/z* value in the ion detector and its conversion into a readable or graphic display in the data system. This leads to the generation of mass spectra, which are plots of relative ion abundance (*y*-axis), normalized to the most abundant ion in the spectrum, against *m/z* values (*x*-axis).

Although MALDI was originally coupled with a TOF mass analyzer (as this is configured to MALDI, that is, to a pulsed [such as laser] ionization method) and ESI was suited to QqQ or IT analyzers (because the continuous ionization process carried out with ESI is useful to perform scans in these scanning analyzers), several combinations of ionization techniques and mass analyzers, with their own strengths and weaknesses, have recently emerged (**Fig. 15.1**). These have been designed to improve sensitivity, resolving power, mass accuracy, scan rate, dynamic range, cost-effectiveness, throughput, MS/MS capabilities, versatility, and quantification of MS analyses *(18, 21, 23)*.

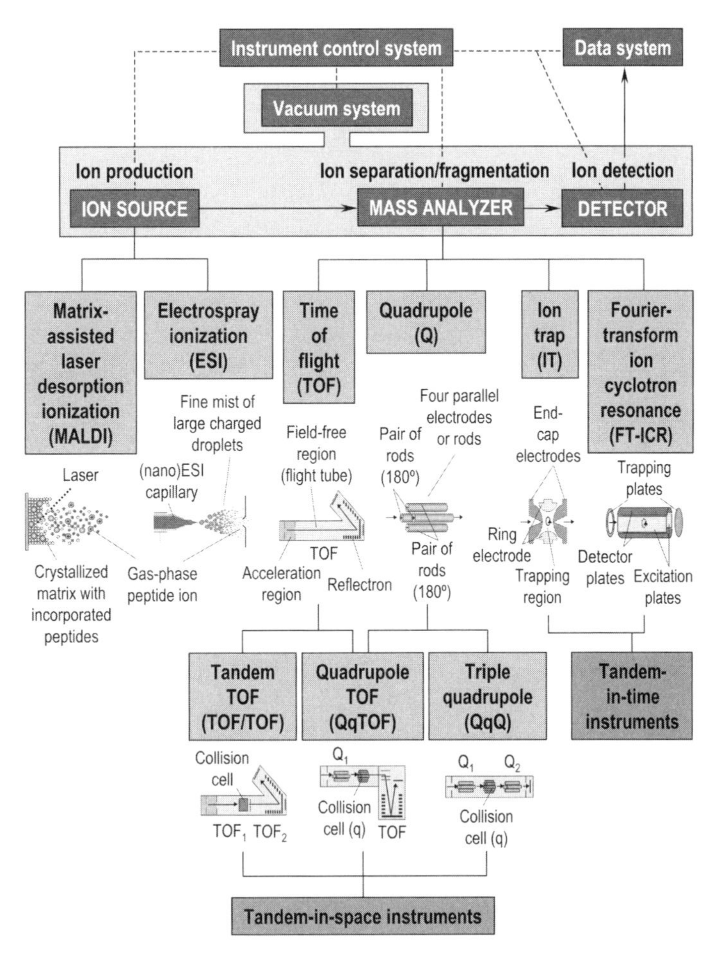

Fig. 15.1. Basic components of a mass spectrometer. All mass spectrometers contain at least three main components: the ion source (the area of ion production), the mass analyzer (the area of ion separation and/or fragmentation), and the ion detector (the area of ion detection). All internal components (the mass analyzer, ion detector, and parts of the ion source) are maintained under high vacuum to avoid ion collisions. All instruments are connected to a vacuum system, an instrument-control system, and a data system. The type of instrument and its capabilities are essentially defined by ion source and mass analyzer *(21)*. ESI and MALDI interfaces can be combined with different mass analyzers. Examples of the main types of ion sources, mass analyzers, and tandem mass spectrometer systems (tandem-in-space and tandem-in-time mass spectrometers) are shown *(22)*.

The general strategy for protein identification using MS is shown in **Fig. 15.2**. The first step is to analyze the peptide sample (*see* later) by peptide mass fingerprinting (also known as peptide mass mapping) using MALDI-TOF MS. This high-throughput

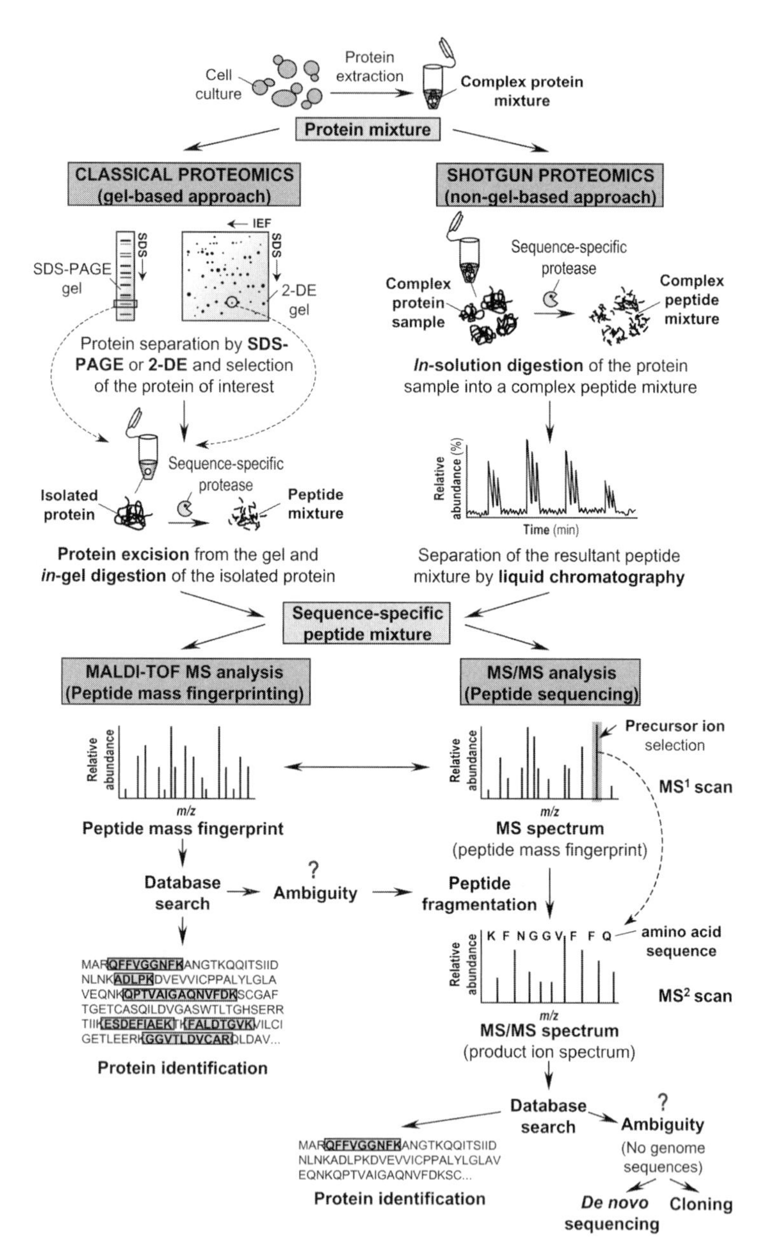

Fig. 15.2. General strategies for protein identification by MS. In a classic proteomic experiment *(17, 18)*, a protein mixture is separated by 2-DE. The protein of interest is then excised from the 2-DE gel and digested with a protease (usually trypsin). The resulting peptides are analyzed using MS. In a shotgun proteomic analysis *(60)*, a protein sample is digested into a complex peptide mixture. This is then directly loaded onto a two-dimensional or multidimensional liquid chromatographic (LC) system, which is coupled to a mass spectrometer (often a MS/MS instrument equipped with a ESI source). In both cases, the peptide samples can be mass fingerprinted or sequenced by MS or MS/MS analyses, respectively.

technique draws on the molecular masses of each sequence-specific peptide obtained in the MS spectrum for a DNA and/or protein database search. If the protein cannot unambiguously be characterized in this way, it is then submitted to MS/MS (peptide fragmentation analysis) for partial amino acid sequencing or peptide fragmentation tagging. This low-throughput approach results in structural information of the fragmented peptide(s), such as the amino acid sequence and the type or site of attachment of post-translational modifications. However, if the protein of interest cannot be confidently identified by the latter method either, it turns out to be an unknown protein, and/or it is not present in the DNA or protein databases, then its amino acid sequences deduced *de novo* from the MS/MS spectra (either by manual interpretation or with computer assistance) can be used to identify the protein by cross-species homology (*see* **Note 6**) *(25, 26)* or, alternatively, to design oligonucleotides for cloning the corresponding novel gene *(27, 28)*.

Both in MS and MS/MS approaches, which provide complementary data for protein identification, peptide sample can be obtained using different methods (*see* **Note** 7 and **Fig. 15.2**). The classic proteomic procedure is to cleave an individual protein (isolated electrophoretically from a complex protein sample; *see* **Chapter 26**) into its constituent peptides by *in situ* (in-gel) digestion with a protease or chemical reagent with specific cleavage site(s), leading to a mixture of sequence-specific peptides. Alternatively, the complex sample of proteins can be converted into peptides by site-specific proteolysis or chemical methods (in-solution digestion), and the resultant sequence-specific peptide mixture can subsequently be separated by liquid chromatography (LC) and on-line analyzed by MS (LC-MS approach; *see* **Notes 3** and **8**) *(21, 29)*.

The goal of the current chapter is to integrate the typical MS-based strategy for identifying *C. albicans* proteins that have been shown to be present during systemic infection and capable of eliciting specific antibody responses (as a consequence of its interaction with the host) by using serologic proteome analysis (SERPA) or classic immunoproteomics (two-dimensional electrophoresis followed by quantitative Western blotting and mass spectrometry) and data mining procedures (*see* **Chapter 26** for further details). Characterization of the full set of *C. albicans* antigens (i.e., the *C. albicans* immunome) targeted by the immune system under a certain physiologic or pathologic condition (commensal colonization or invasive disease) (**Fig. 15.3**) *(30–32)* is one direct way both to investigate how this fungal pathogen interacts with its host to produce a clinical disease (that is, the interface between the host immune system and pathogen proteins) and to uncover potential candidates for diagnosis, prognosis, risk stratification, clinical follow-up, therapeutic monitoring, and/or vaccine design for candidiasis *(4)*. MS and MS/MS analyses have enabled the unambiguous characterization of many *C. albicans* immunogenic proteins present during the host-pathogen interaction (hitherto unidentified by classical

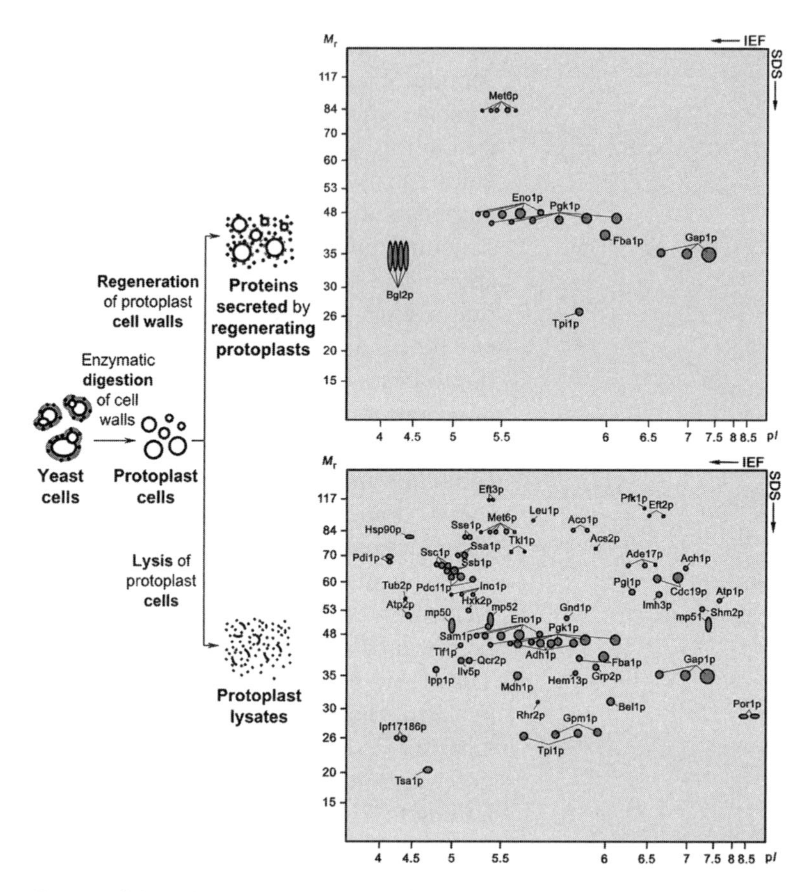

Fig. 15.3. Schematic 2-D representations of the *C. albicans* immunome present during systemic infection and identified by MS. Illustrative 2-D patterns of MS-identified *C. albicans* immunogenic proteins that stimulate the human immune system during systemic candidiasis are depicted (*see* Refs. *25* and *30* for further details). *Top panel*, *C. albicans* cell wall–associated immunogenic proteins *(30)* (secreted from protoplasts in active wall regeneration *(72)*). *Bottom panel*, soluble *C. albicans* cytoplasmic immunogenic proteins *(25)* (from protoplast lysates; *see* **Chapter 26**).

methods) *(4, 25, 30, 33)*. For instance, classic and noncanonical cell wall–associated proteins (such as β-1,3-glucosyltransferase or glucan 1,3-beta-glucosidase and some glycolytic enzymes, respectively) *(30)* and different functional categories of housekeeping proteins (including [i] chaperones and heat shock proteins, [ii] metabolic proteins, [iii] translation apparatus-involved proteins, and [iv] miscellaneous proteins) *(25)* have been identified as *C. albicans* antigens that induce specific human antibodies during systemic candidiasis by MS analyses (**Fig. 15.3**).

The two complementary MS techniques described here (peptide mass fingerprinting and peptide sequencing) focus on the identification of two-dimensional electrophoresis (2-DE)-separated *C. albicans* proteins. Therefore, we will also cover the methods for peptide sample preparation of individual 2-DE–isolated proteins for MS and MS/MS analyses. However, the protocols for protein separation and visualization will not be reiterated, because these are detailed in **Chapter 26**, to which the reader should be directed. Remarkably, these MS strategies have proved to be successful to characterize not only the *C. albicans* immunome (and associated host antibody response) but also the *Candida* proteomes related to the cell wall, response to nitrogen starvation, expression of a gene in an ectopic gene loci, septin cytoskeleton, yeast-to-hypha transition, virulence, macrophage responses, modes of action of antifungal agents, and resistance mechanisms to azole drugs, among others, providing novel models of the host-fungus interaction (for both basic and applied research), many of which are not predictable from genomic studies *(3, 4, 33, 34)*.

2. Materials

All solutions and buffers must be prepared with ultrapure water, as supplied by Nanopure or Milli-Q 18 MΩ/cm resistivity systems (Millipore, Bedford, MA) and best-quality chemical reagents (sequencing or HPLC grade). All reagents should be prepared before use either in prerinsed, low-binding, siliconized microcentrifuge tubes (*see* **Note 9**) or in glass vials with Teflon-lined caps.

2.1. Preparation of Protein Digests for MS and MS/MS Analyses

1. 0.5- or 1.5-mL low-binding, siliconized microcentrifuge tubes (*see* **Note 9**).
2. Gel-loading pipette tips (Bio-Rad, Hercules, CA) (*see* **Note 9**).
3. Vacuum centrifuge (SpeedVac; Thermo Savant, Holbrook, NY).
4. Laminar-flow hood.

2.1.1. Excision and Mincing of C. albicans Immunoreactive Protein Spots from Stained 2-DE Gels

1. Coomassie-, silver-, or SYPRO Ruby-stained, preparative 2-DE gel containing *C. albicans* protein spots of interest (*see* **Chapter 26**).
2. Clean scalpel.

2.1.2. Destaining and Washing of the Protein Gel Particles

1. Wash/dehydration solution: 50% (v/v) acetonitrile (ACN) in 25 mM ammonium bicarbonate, pH 8.5.
2. Dehydration solution: 100% ACN.

2.1.2.1. From Coomassie Blue–Stained Protein Spots

2.1.2.2. From Silver-Stained Protein Spots

1. Silver destaining solution: 100 mM sodium thiosulfate, 30 mM potassium ferricyanide. Prepare a fresh 1:1 (v:v) solution of these reagents.
2. Wash solution: Milli-Q grade water (Millipore).

3. Equilibration solution: 25 mM ammonium bicarbonate, pH 8.5.
4. Dehydrating solution: 100% ACN.

2.1.3. Reduction and S-Alkylation of the Gel Particles

1. Reducing solution: 10 mM dithio-erythritol (DTE) in 25 mM ammonium bicarbonate, pH, 8.5. Prepare a fresh solution.
2. Alkylating solution: 55 mM iodoacetamide (*see* **Note 10**) in 25 mM ammonium bicarbonate, pH 8.5. Prepare a fresh solution.
3. Wash/rehydration solution: 25 mM ammonium bicarbonate, pH 8.5.
4. Dehydrating solution: 100% ACN.

2.1.4. In-Gel Digestion

1. Digestion solution: 12.5 ng/μL trypsin (sequencing grade; Roche, Mannheim, Germany) in 25 mM ammonium bicarbonate, pH 8.5 (*see* **Note 11**). Dissolve trypsin in 0.01% (w/v) trifluoroacetic acid (TFA).
2. Ammonium bicarbonate solution: 25 mM ammonium bicarbonate, pH, 8.5.

2.1.5. Extraction of the Tryptic Peptides

1. Extraction/dehydration solution: 50% (v/v) ACN, 1% (v/v) TFA.
2. Dehydrating solution: 100% ACN.
3. Sonication bath.

2.1.6. Microscale Desalting and Concentration of the Peptide Sample

1. In-tip reverse-phase resins ($\text{ZipTip}_{\text{C18}}$ microcolumn; Millipore).
2. Equilibration solution: 0.1% (v/v) TFA.
3. Wetting solution: 50% (v/v) ACN.
4. Wash solution 1: 0.1% (v/v) TFA.
5. Wash solution 2: Milli-Q grade water.
6. Elution solution: 60% (v/v) methanol, 1% (v/v) acetic acid.

2.2. High-Throughput Protein Identification by Peptide Mass Fingerprinting Using MALDI-TOF MS Analysis

2.2.1. On-Target Sample Preparation for MALDI MS Analysis

1. Wash solution 1: Milli-Q grade water.
2. Wash solution 2: 100% (v/v) methanol.
3. Matrix solution: 3 mg/mL α-cyano-4-hydroxycinnamic acid (Sigma) in 50% (v/v) ACN and 0.01% (v/v) TFA (*see* **Note 12**). Prepare a fresh solution.
4. Peptide mass standard mixture (Sequazyme kit; PerSeptive Biosystems, Framingham, MA).
5. MALDI target plate (96 × 2 spot Teflon-coated plate; PerSeptive Biosystems) (*see* **Note 13**).

2.2.2. Acquisition and Processing of MALDI-TOF Mass Spectra from Tryptic Digests

1. MALDI-TOF mass spectrometer (Voyager-DE STR; PerSeptive Biosystems) (*see* **Note 13**).
2. Data Explorer software (Voyager software; Applied Biosystems, Framingham, MA) (*see* **Note 13**).

2.2.3. Database Searching Based on Peptide Mass Fingerprinting

1. *C. albicans* genomic database (CandidaDB; http://genolist.pasteur.fr/CandidaDB).
2. SWISS-PROT/TrEMBL nonredundant protein database (http://www.expasy.ch/sprot).
3. National Center for Biotechnology Information nonredundant (NCBInr) database (http://www.ncbi.nlm.nih.gov).
4. ProFound search engine (http://prowl.rockefeller.edu).
5. Mascot search engine (http://www.matrixscience.com).
6. MS-Fit search engine (http://prospector.ucsf.edu).

2.3. Low-Throughput Protein Identification by Peptide Sequencing Using NanoESI (nESI) MS/MS and MS^n Analyses

2.3.1. Sample Preparation for nESI MS Analysis

1. nESI source (Protana, Odense, Denmark) (*see* **Note 14**).
2. Gold-coated borosilicate capillaries, preferably with opened needle tips (Protana).
3. High-voltage clip.
4. Tridimensional (*x-y-z*) manipulator (Protana).
5. Video camera equipped with macro-lenses.
6. Vacuum centrifuge (SpeedVac).

2.3.2. Acquisition and Processing of MS/MS and MS^n Spectra by nESI-IT-MS Analysis

1. LCQ IT mass spectrometer (ThermoFinnigan, San Jose, CA) (*see* **Note 13**).
2. Xcalibur software (ThermoFinnigan) (*see* **Note 13**).

2.3.3. Database Searching Using Partial Amino Acid Sequences

1. CandidaDB (http://genolist.pasteur.fr/CandidaDB).
2. SWISS-PROT/TrEMBL database (http://www.expasy.ch/sprot).
3. NCBInr database (http://www.ncbi.nlm.nih.gov).
4. Sequest search engine (http://fields.scripps.edu/sequest/index.html).
5. Mascot search engine (http://www.matrixscience.com).
6. MS-Tag search engine (http://prospector.ucsf.edu).
7. Sequest search engine (http://fields.scripps.edu/sequest).
8. Sonar search engine (http://proteometrics.com).
9. BLAST program (http://www.ncbi.nlm.nih.gov/entrez).
10. FASTS program (http://fasta.bioch.virginia.edu/).

11. LutefiskXP *de novo* sequencing tool (http://www.hairyfatguy.com/Lutefisk).
12. PEAKS *de novo* sequencing tool (http://bioinformaticssolutions.com).

3. Methods

3.1. Preparation of Protein Digests for MS and MS/MS Analyses

Before analyzing a relevant protein by MS, this should be cleaved previously (by enzymatic or chemical procedures with specific cleavage site(s)) into its constituent peptides (*see* **Note 7**), in such a way that these have molecular weights within the mass range of the mass spectrometers and good fragmentation efficiency. Protein cleavage can be carried out by two different techniques: in-gel digestion (*35, 36*) and in-solution digestion *(29, 37)* (*see* **Fig. 15.2** and **Section 1** for further details).

The protocols outlined below are based on the in-gel digestion method (in which the protein sample, separated by 2-DE, is *in situ* digested). This begins after the *C. albicans* proteins of interest (identified as components of the *C. albicans* immunome that elicit specific host immune responses) have been separated from a complex protein sample (a given *C. albicans* [sub]proteome) as spots in a Coomassie-, silver-, or SYPRO Ruby-stained 2-DE gel (*see* **Chapter 26**). These protocols are also compatible with one-dimensional polyacrylamide gels (SDS-PAGE gels). The flowchart in **Fig. 15.4** summarizes the different steps in this procedure. These involve (i) excision and mincing of the protein spot(s) of interest from an appropriately stained, preparative 2-DE gel, (ii) destaining and washing of the excised gel piece(s), (iii) reduction and S-alkylation of the washed gel piece(s), (iv) its/their in-gel digestion using a sequence-specific protease, (v) peptide extraction from the gel piece(s), and (vi) desalting of its/their extracted peptides. Currently, all these steps can also be performed automatically using commercially available workstations for spot picking, in-gel digestion, and spotting peptides on a MALDI target plate (*see* **Note 15**). When working manually, these steps should preferably be carried out in a laminar-flow hood (wearing prerinsed, powder-free gloves) in an endeavor to avoid, or at least reduce, contamination with human keratins (*see* **Note 16**).

3.1.1. Excision and Mincing of C. albicans Immunoreactive Protein Spots from Stained 2-DE Gels

The selection and excision process of protein spots from an appropriately stained, preparative 2-DE gel containing the proteins of interest (*see* **Note 17**) for further MS analysis must be performed on high-quality 2-DE experiments that clearly yield well-resolved, undistorted protein spots and no streaking patterns (especially

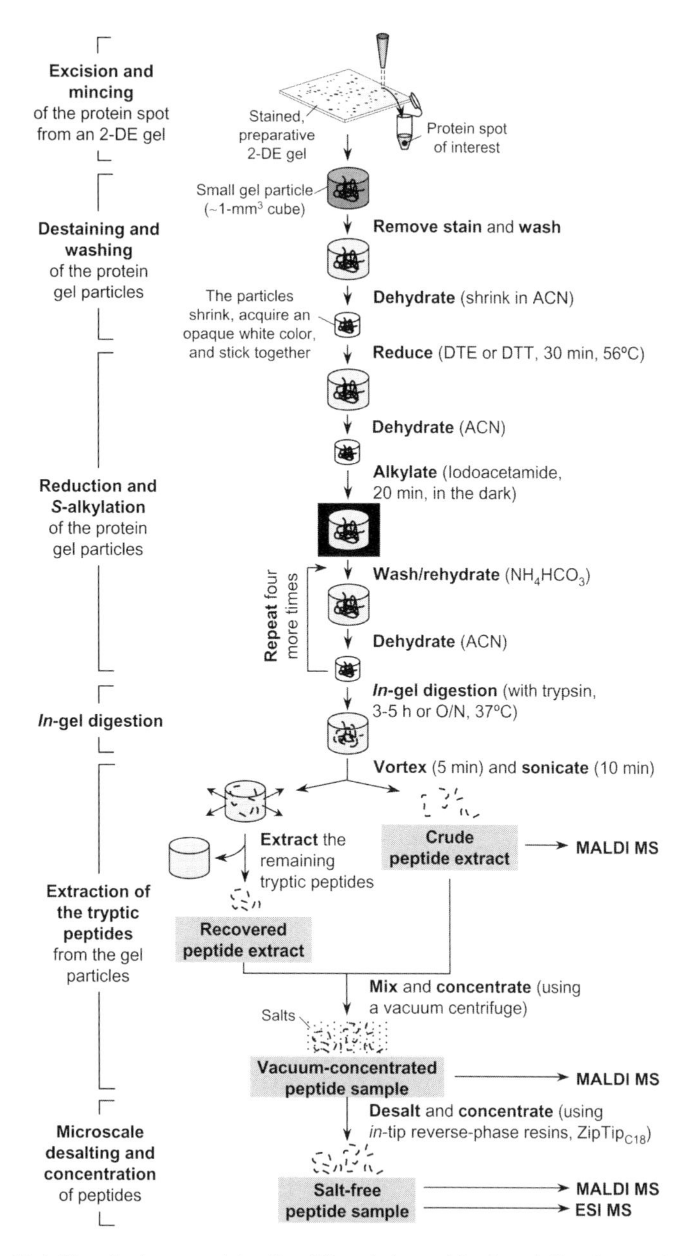

Fig. 15.4. Flowchart summarizing the different steps of the in-gel digestion method.

through the related 2-DE gel region) to make good, easy, and reliable sampling. Furthermore, it is also crucial to cut as close to the protein spot as possible to (i) minimize the amount of background gel and polyacrylamide in the enzymatic reaction, and, therefore, improve the detection limit of subsequent MS analysis, and (ii) avoid including any fragment of other surrounding protein spot(s) in the sample and, thus, prevent wrong MS-identification results. The amount of protein in the spot and, overall, of the extracted peptides is often associated with the density of the protein spot staining.

1. Rinse the Coomassie-, silver-, or SYPRO Ruby-stained, preparative 2-DE gel containing proteins of interest (*see* **Chapter 26**) with two changes of water (10 min each).
2. Manually select and excise the protein spot(s) of interest from the gel as closely as possible with a clean scalpel (*see* **Note 18**), and transfer each spot to a prerinsed, 0.5- or 1.5-mL microcentrifuge tube (*see* **Notes 9** and **18**).
3. Excise a gel piece of similar size from a protein-free region of the gel for use as a control for subsequent MS analysis (*see* **Note 19**).
4. Cut the excised gel piece(s) into small particles (~1-mm^3 cubes) using a scalpel (*see* **Note 20**). Do not pulverize them (*see* **Note 21**).

3.1.2. Destaining and Washing of the Protein Gel Particles

Prior to enzymatic digestion, several in-gel destaining and/or washing steps should be carried out on the gel particles from Coomassie blue–stained and silver-stained protein spots to remove residual detergents (like SDS) and staining agents (such as Coomassie blue or silver) that may interfere with the digestion procedure and subsequent MALDI or ESI (*see* **Note 3**) MS analysis.

3.1.2.1. From Coomassie Blue–Stained Protein Spots

The removal of Coomassie blue from protein gel particles can (i) improve the enzymatic digestion efficiency, by eliminating the ionic interactions between the basic residues of proteins and the sulfonic acid group of the Coomassie blue, and (ii) avoid the suppression of certain mass-to-charge (*m/z*) signals in following mass spectra.

1. Add enough wash/dehydration solution to cover the Coomassie blue–stained gel particles (*see* **Note 22**). Vortex for 5 to 10 min. Quick spin to submerge all particles, remove the solution (pale blue) carefully, and discard it (*see* **Note 23**).
2. Repeat this step until color is entirely removed from the gel particles (~1 to 2 times; *see* **Note 24**).
3. Dehydrate the gel particles with ACN (*see* **Note 25**). Rapid spin and discard the supernatant.
4. Completely dry the gel particles at room temperature in a vacuum centrifuge for 10 to 20 min.

3.1.2.2. From Silver-Stained Protein Spots

The removal of residual silver ions from protein gel particles (stained with a MS-compatible method; *see* **Chapter 26** and Ref. *36*) can (i) enhance the sensitivity and quality of the *m/z* signals in the subsequent MS analysis, and (ii) improve the identification efficiency by peptide mass database analysis (*38*). The protocol presented here is adapted from that described by Gharahdaghi et al. (*38*).

1. Add enough silver destaining solution to immerse the silver-stained gel particles (*see* **Note 22**), and incubate until the excised spot is completely destained (~10 min; *see* **Note 24**). Quick spin to submerge all particles. Carefully remove the solution (pale yellow) and discard it (*see* **Note 23**).
2. Rinse the gel particles in water to remove any trace of destaining solution (*see* **Note 22**). Rapid spin, and carefully discard the supernatant.
3. Rinse them in equilibration solution for 5 to 10 min, quick spin, and discard the supernatant. Repeat this step again.
4. Dehydrate the gel particles with ACN for 15 min (*see* **Note 25**). Spin and discard the supernatant.
5. Vacuum-dry the gel particles at room temperature for 10 to 20 min.

3.1.3. Reduction and S-Alkylation of the Gel Particles

Reduction of the disulfide bonds of cysteine residues from proteins embedded in the gel particles and subsequent S-alkylation of their generated thiol side chains can enhance the sensitivity and coverage of the *m/z* signals from these cysteine-containing peptides in subsequent mass spectra. It is of note that although 2-DE–separated proteins are already reduced and S-alkylated (*see* **Chapter 26** and **Note 10**), the accomplishment of these steps is still recommended. After this reduction/S-alkylation step, several rehydration and dehydration series followed by complete drying are performed to prepare the gel fragments for efficient digestion enzyme uptake.

1. After destaining and washing, add enough reducing solution to cover the shrunk gel particles (*see* **Note 22**), and incubate at 56°C for 30 min. Cool to room temperature (~5 min). Quick spin to submerge all particles, and carefully remove the solution (*see* **Note 23**).
2. Dehydrate the gel particles with ACN for 5 to 10 min (*see* **Note 25**). Spin and discard the supernatant.
3. Add the same volume of alkylating solution (*see* **Note 10**) as used in **step 1** to the gel particles, and incubate at room temperature for 20 min in the dark. Spin and carefully remove the solution.
4. Rinse the gel particles in wash/rehydration solution for 15 min while gently vortexing. Spin and discard the supernatant.
5. Repeat **step 2** (*see* **Note 25**).
6. Repeat **steps 4** and **5** (rehydration and dehydration, respectively) alternately four more times.
7. Completely dry the gel particles in a vacuum centrifuge for 15 to 20 min.

3.1.4. In-Gel Digestion

Although chemical methods of protein fragmentation are a good alternative to proteolysis for cleavage of insoluble or membrane proteins (*39*), the latter is still the most commonly used procedure in proteomics. Trypsin is the protease of choice (*see* **Note 11**), because this yields tryptic peptides with a suitable size (~800 to 2500 Da) for optimal MALDI and ESI MS analysis and a sequence-specific characteristic (i.e., peptides with C-terminal basic [arginine and lysine] residues if these are not followed by a proline) that facilitates ionization and following MS analysis (**Fig. 15.5**).

1. Rehydrate the gel particles with enough (~15 μL) digestion solution (to cover them; *see* **Note 26**) by vortex mixing for 5 min, and quick spin at 4°C to submerge all particles.
2. Incubate at 4°C (on a bucket containing ice) for 40 min with occasional vortexing (*see* **Note 27**). Check that all gel fragments are completely submerged in the solution during this incubation time. Add more digestion solution if all the initially added volume is absorbed by the gel particles.

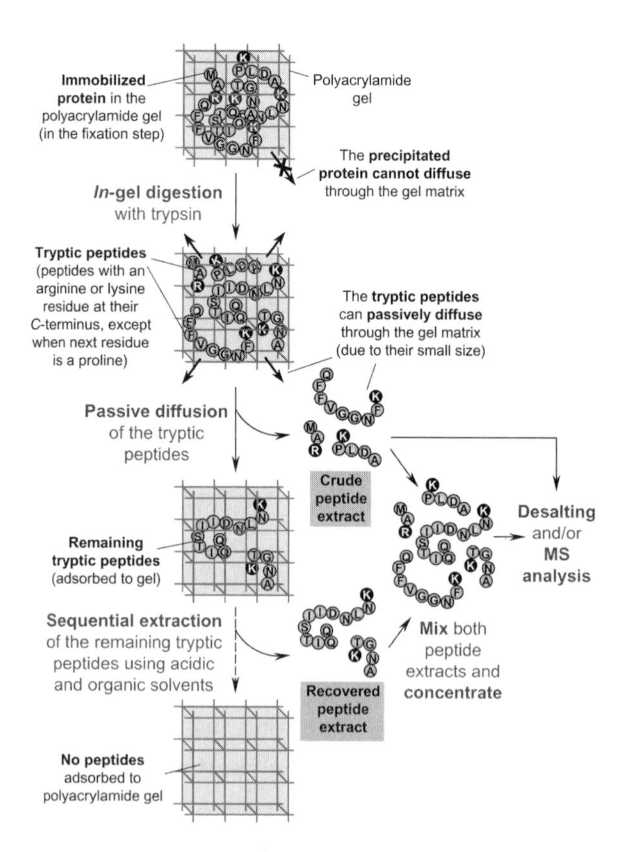

Fig. 15.5. Schematic representation of the procedures for in-gel digestion of a given protein with trypsin and extraction of the resulting tryptic peptides from the polyacrylamide gel matrix.

3. Carefully remove excess trypsin solution from the sample and discard (*see* **Notes 23** and **28**).
4. Top off the gel particles with enough ammonium bicarbonate solution to keep them immersed throughout digestion. Vortex the sample and rapid spin.
5. Incubate at 37°C for 3 to 5 h (or overnight).

3.1.5. Extraction of the Tryptic Peptides

Given their relatively small size, peptides produced as a result of the enzymatic reaction can passively diffuse through the gel matrix and be recovered in solution by sequential extraction using acidic and organic solvents (*see* **Fig. 15.5** and **Note 29**). The final volume of all these peptide extracts, particularly for low-abundance protein samples, should be reduced since this is typically very large for subsequent MALDI or ESI MS analysis.

1. After digestion, vortex the gel particles for 5 min, sonicate for 10 min, and then spin for 2 to 3 min. Transfer the supernatant to a new microcentrifuge tube (*see* **Note 23**) and store the tube (designed as crude extract) on ice (*see* **Note 30**).
2. Add a sufficient volume of extraction/dehydration solution to the gel fragments to cover them. Vortex for 5 min and sonicate for 10 min to extract additional tryptic peptides from the gel particles. Spin for 2 to 3 min, dispense the supernatant into a separate microcentrifuge tube, and keep the tube (named as recovered peptide extracts) on ice.
3. Repeat **step 2** two more times. Combine the resulting supernatants with that of **step 2**.
4. Add enough ACN to the gel particles, vortex for 5 min, and then sonicate for 10 min to extract remaining tryptic peptides from them. Spin for 2 to 3 min, transfer the supernatant to the microcentrifuge tube designed as recovered peptide extracts (*see* **step 2**), and store it on ice.
5. Repeat **step 4** two more times. Pool the resulting supernatants with the earlier extract (that of **step 2**).
6. Combine all supernatant extracts: crude and recovered peptide extracts (**steps 1** and **5**, respectively; *see* **Note 30**). Clarify by centrifugation for 5 to 10 min and carefully transfer supernatant to a new microcentrifuge (*see* **Note 31**).
7. Concentrate extracted peptides by reducing the final volume of the total supernatant (to 10 to 20 μL) using a vacuum centrifuge for ~15 to 20 min.
8. Store the vacuum-concentrated peptide sample at −20°C until further analysis, or alternatively proceed to desalting (*see* **Section 3.1.6**) or MALDI MS analysis (*see* **Section 3.2**).

3.1.6. Microscale Desalting and Concentration of the Peptide Sample

Salts, buffers, and other contaminants present in the extracted peptide sample, which interfere with MALDI and, especially, with ESI (*see* **Note 3**), can be removed by reverse-phase chromatography (RP-HPLC) prior to MS analysis. This clean-up step can also be carried out manually (without any sophisticated equipment) using commercially available reverse-phase microcolumns (in-tip reverse-phase resins). In this method, the pre-equilibrated peptide sample is trapped within the reverse-phase material, desalted by several wash steps, and then eluted. This procedure results in highly purified, salt-free, and concentrated peptide solutions that can be directly applied to MALDI or ESI MS analysis. The current protocol is adapted from the method described by the manufacturer supplier (Millipore) of these in-tip reverse-phase resins ($ZipTip_{C18}$ microcolumns).

1. Completely dry the peptide sample in a vacuum centrifuge for ~20 to 30 min, and reconstitute in 10 μL equilibration solution (*see* **Note 32**).
2. Wet the $ZipTip_{C18}$ microcolumn by slowly aspirating 10 μL wetting solution into the microcolumn. Expel this solution into a waste container. Repeat this process nine more times.
3. Equilibrate the ZipTip by slowly drawing 10 μL equilibration solution into the microcolumn. Dispense this solution to the waste. Repeat this step nine more times.
4. Bind the tryptic peptides to microcolumn by gently drawing the 10 μL reconstituted peptide sample into the ZipTip. Expel it out of the microcolumn and apply the eluate back to the ZipTip. Repeat 10 cycles to enhance peptide binding (*see* **Note 33**).
5. Remove any contaminants from the adsorbed peptides onto the ZipTip by slowly drawing 10 μL wash solution 1 into the microcolumn. Dispense this solution to the waste. Repeat this process nine more times.
6. Repeat **step 5** substituting wash solution 1 by water.
7. Elute the peptides from the microcolumn by slowly up-taking and displacing 5 μL elution solution from the ZipTip into a clean microcentrifuge tube. Repeat this process nine more times.
8. Store this 5-μL salt-free peptide sample at −20°C until further analysis, or alternatively proceed to MALDI and/or ESI MS analysis (*see* **Sections 3.2** and/or **3.3**, respectively).

3.2. High-Throughput Protein Identification by Peptide Mass Fingerprinting Using MALDI-TOF MS Analysis

The resulting peptides, extracted from the polyacrylamide matrix, desalted and concentrated, are then mass fingerprinted by MS (**Fig. 15.2**). This leads to an experimental peptide mass profile (fingerprint or map) that is specific to the MS-analyzed protein. This set of experimental peptide masses (i.e., its peptide mass fingerprint) is subsequently compared with the theoretical peptide masses derived from an *in silico* digestion with trypsin (or with the corresponding enzyme used to obtain the experimental protein digest) of all protein sequences present in the available protein or translated genomic database(s). This approach, referred to as peptide mass fingerprinting *(39–41)*, is currently the method of choice for high-throughput identification of proteins and peptides from (micro)organisms with completely sequenced genomes (like *C. albicans* at present) *(5)*.

Although protein identification by peptide mass fingerprinting can be performed using any mass spectrometer, MALDI-TOF MS analysis (**Fig. 15.1**) remains the preferred technique for this type of screening because (i) MALDI is easy-to-automate (allowing high-throughput analysis), predominately produces singly charged ions (yielding intrinsically easy-to-interpret mass spectra), and tolerates moderate buffer and salt concentrations in the protein digest (enabling its direct analysis without extra desalting steps) (*see* **Note 3**), and (ii) TOF instrument is a pulsed analyzer (suited to MALDI), and one of the simplest MS analyzers with high sensitivity and wide mass range.

In MALDI-TOF MS (**Fig. 15.6**) (*42*), the peptide or protein sample is co-crystallized with a low-molecular-weight organic compound (called matrix) that absorbs light at the wavelength of the laser. The co-crystalline sample and matrix are systematically irradiated with the light from a laser pulse, under high vacuum (*see* **Note 34**), to promote their desorption and ionization. The gas-phase molecular ions formed are then accelerated to the same kinetic energy at a fixed point and initial time by the application of a high voltage in an electric field. Subsequently, these enter a field-free region (flight tube) where they drift through it at a velocity inversely proportional to their *m/z*, allowing their separation. The drift time (also known as time of flight, that is, the time that these require to traverse the length of the field-free drift tube and strike the ion detector) is recorded to calculate their *m/z* values.

3.2.1. On-Target Sample Preparation for MALDI MS Analysis

Proper on-target sample preparation is absolutely essential for efficient peptide or protein ionization or, in other words, for successful MALDI MS results. This is because the sample homogeneity and consistency on the MALDI target plate may drastically influence the acquisition time and resolution of the generated mass spectra *(43)* (*see* **Note 35**). Consequently, the co-crystallization process of the sample and matrix (in which peptide sample is incorporated into the crystal structure of the matrix to allow its subsequent desorption and ionization by a laser pulse; *see* **Fig. 15.6**) should carefully be monitored using the sample visualization systems of the instrument (a video camera). The dried MALDI sample should hopefully appear as an even crystalline layer on the surface of each target well.

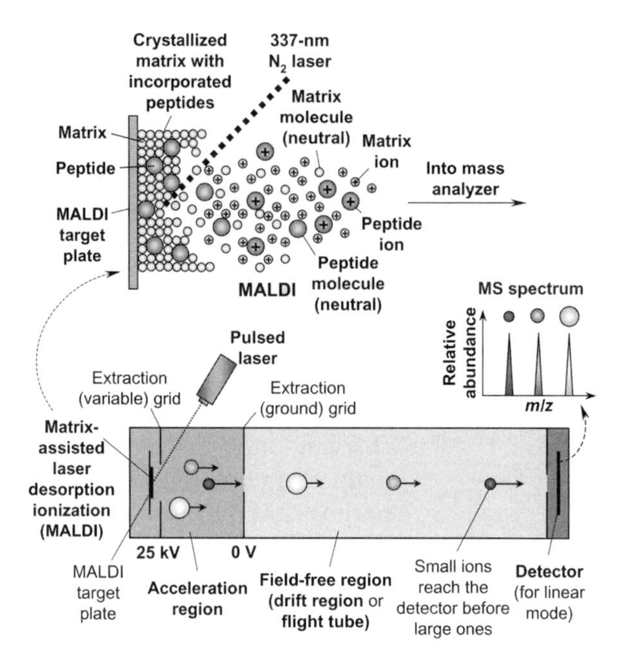

Fig. 15.6. Biochemical bases of MALDI-TOF MS. The peptide or protein sample of interest is incorporated into the crystal structure of a saturated matrix solution (*see* **Note 36**) to enable its subsequent desorption and ionization by systematic irradiation with the light from a laser pulse. The gas-phase ions produced in the ion source are then accelerated to a fixed kinetic energy, at a fixed point and initial time, by applying a fixed voltage. These are then guided into a high-vacuum field-free flight tube, where they travel at a velocity that is inversely proportional to their *m/z* values. The time required for each ion to traverse this region is measured after striking the ion detector and used to calculate its *m/z*.

Although several specific procedures of on-target sample preparation have been reported for MALDI MS analysis (*see* **Note 36**) *(43)*, the protocol described below is based on the classical dried droplet method *(13)*. In this procedure, commonly used in proteomics, the sample and matrix solutions (mixed together or added one after the other; *see* **Note 36**) are deposited on the target plate and allowed to co-crystallize through solvent evaporation. All these steps can be automated by using commercially available robotic pipettors or sample preparation workstations.

1. Rinse the MALDI target plate with water, and wipe dry with a lint-free tissue (*see* **Note 37**). Repeat this rinse step with wash solution 2. Perform two or more clean cycles.
2. Deposit 0.5 to 1 μL peptide sample as a droplet onto one position (well) on the MALDI target plate (*see* **Notes 38** and **39**).
3. Immediately add an equal amount (0.5 to 1 μL droplet) of matrix solution to the sample droplet on the target (*see* **Note 39**). Mix thoroughly by slowly up-taking and displacing the total volume from the micropipette several times (*see* **Notes 36** and **40**).

4. Leave to air dry at room temperature to permit co-crystallization (*see* **Notes 36** and **41**).
5. Apply 0.5 to 1 μL peptide standard mixture to the target plate in an adjacent position to the samples, followed by an equal amount of matrix (as described in **step 3**), for external calibration (*see* **Section 3.2.2**). Allow the solvent to evaporate at room temperature.
6. Proceed to insert the MALDI target plate (*see* **Note 42**) into the mass spectrometer through a vacuum lock and record mass spectra (*see* **Section 3.2.2**).

3.2.2. Acquisition and Processing of MALDI-TOF Mass Spectra from Tryptic Digests

The mass spectra generated by MALDI-TOF MS analyses contain data on the molecular weights for all sequence-specific (tryptic) peptides that can be detectable in the protein digests (i.e., their peptide mass fingerprints). Data acquisition can be automated by programming laser firing positions, laser power (*see* **Note 43**), extraction, focusing, and detection voltages, number of accumulated mass spectra (*see* **Note 44**), and length of the ion extraction delay (*see* **Note 45**), among other parameters. It is of prime importance that measured peptide masses are as accurate as possible in order to minimize further false-positive protein matches. This can be attained by external and/or internal *m/z* calibration of the acquired MALDI-TOF mass spectra (*see* **Notes 46** and **47**). Mass calibration and peak annotation can also be performed automatically, allowing high-throughput analysis of many samples.

1. Perform MS analyses on a MALDI-TOF mass spectrometer, equipped with delayed extraction and operated in positive ion reflector mode (*see* **Notes 45** and **48**). Average ~100 laser shots per spectrum (from a 337-nm N_2 laser) at a laser power just above threshold to ionize the calibration or peptide sample (*see* **Notes 43** and **44**). Set the acceleration voltage to 15 to 25 kV, and the reflector voltage to 20 kV.
2. Optimize the instrument parameters by adjusting the signal from a standard sample to best possible sensitivity and mass resolution.
3. Acquire an averaged mass spectrum (*see* **Note 44**) of each sample over the *m/z* range of 800 to 3500 Da.
4. Externally calibrate the acquired mass spectrum of the sample of interest using the calibration file generated from the mass spectrum of a peptide standard mixture (calibration sample) placed in an adjacent position to the sample (*see* **Note 46** and **Section 3.2.1**).
5. Internally recalibrate it using its own matrix ions and/or trypsin autolysis peptide ions (*see* **Note 47**).
6. Detect and label peaks corresponding with monoisotopic peptide ions in the sample mass spectrum (*see* **Note 49**).

Define mass peaks using a correct isotopic distribution, signal-to-noise threshold >20 and resolution >10,000, and excluding all ions of matrix, trypsin autolysis products, peptide standards added for internal calibration (*see* **Note 47**), and/or known contaminants (e.g., tryptic peptides of keratins from humans or other species). Process raw data using an appropriate MS analysis program (like the instrument-dependent Data Explorer software).

7. Annotate the peak list (experimental, tryptic, monoisotopic peptide masses; *see* **Note 49**) and use it for database searching (*see* **Section 3.2.3**).

3.2.3. Database Searching Based on Peptide Mass Fingerprinting

Once the list of accurately measured peptide masses is obtained from the protein digest, this is exploited to query translated genomic or protein databases using search engines freely available on the Internet, such as ProFound, Mascot, and MS-Fit, among others (*see* **Note 50**) *(15, 16)*, in order to identify the protein from which the peptides are derived (**Fig. 15.7**). Unambiguous protein identification can be reached if there are (i) at least five matching peptide masses with a mass accuracy ≤30 ppm (*see* **Notes 45** and **51**), (ii) ~15% sequence coverage (*see* **Note 51**), and (iii) a significant gap between the first and the next best (false-positive) database hits *(44)*. These criteria and other parameters are condensed into an appropriate scoring function (such as a relative score, probabilistic score, Bayesian algorithm, or genetic algorithm) *(16)*, which enables the estimation of confidence levels for random matching and automation of the analysis.

1. Input the list of experimental peptide masses obtained by MALDI-TOF MS analysis into an appropriate peptide mass fingerprint search program (e.g., Profound, Mascot, or MS-Fit) *(15, 16)* by its copy/paste or as data file into query (space for *m/z* values).
2. Set search parameters.
   (a) Select the NCBInr, SWISS-PROT/TrEMBL or CandidaDB databases to use for the search (*see* **Note 52**).
   (b) Choose trypsin as the enzyme to generate the peptides. Specify up to one missed cleavage site (*see* **Note 53**).
   (c) Select monoisotopic mass values (*see* **Note 49**), and set peptide mass tolerance at ± 50 ppm (*see* **Note 54**).
   (d) Carry out the search (at least the first one) without species, molecular weight or isoelectric point restriction (*see* **Note 52**).
   (e) Specify type of (i) fixed modifications (like S-carbamidomethylation of cysteine residues), and (ii) variable modifications (such as oxidation of methionine residues,

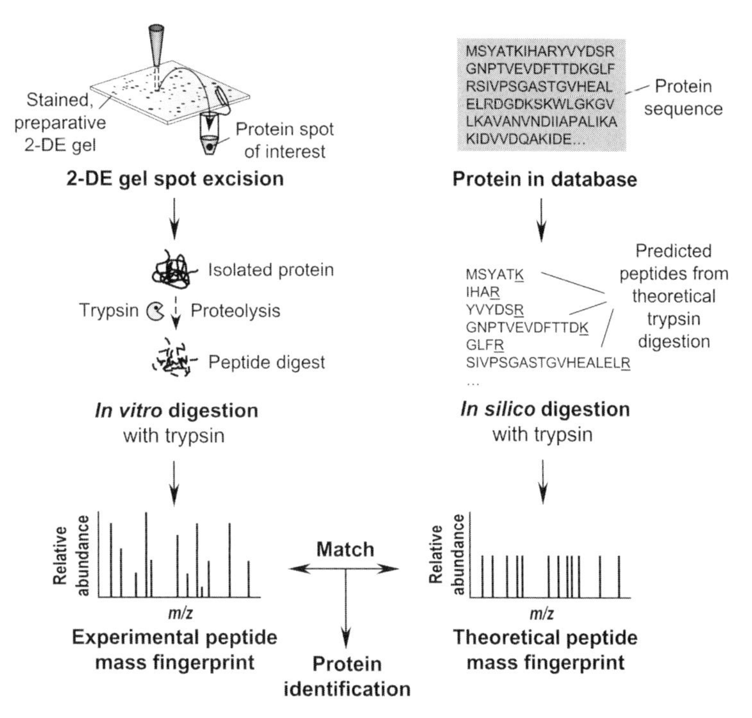

Fig. 15.7. Strategy for protein identification by peptide mass fingerprinting. The protein of interest is excised from the 2-DE gel and digested with a sequence-specific protease (commonly trypsin). The masses of the resultant sequence-specific peptides are measured by using MS (usually MALDI-TOF MS), resulting in a characteristic experimental mass spectrum (known as peptide mass fingerprint). This set of experimental peptide masses, measured with high accuracy, is subsequently compared with the theoretical peptide masses obtained from an *in silico* digestion of all protein sequences present in the available protein or translated genomic database(s), taking into account the specificity of the protease used. If there is a high score of peptide mass matches and protein sequence coverage with a protein in the database, then the protein is unambiguously identified.

acrylamide modification of cysteine residues, and acetylation of protein *N*-terminus, among others) that should be considered for the search (*see* **Notes 17** and **55**).

3. Perform the database search. Check the gap between random and significant protein scores, coverage of the matched peptides, and the number of matching peptide masses to determine if unambiguous protein identity has been attained (*see* **Note 51**). Confirm if species, molecular weight, and isoelectric point are in line with the experimental 2-DE data. Click on the accession number of the first hit to display further protein information.

4. Proceed to peptide sequencing by MS/MS if ambiguous protein identification is achieved (*see* **Section 3.3**).

3.3. Low-Throughput Protein Identification by Peptide Sequencing Using nESI MS/MS and MS^n Analyses

Unfortunately, high-throughput protein identification by peptide mass fingerprinting may prove unsuccessful as a consequence of (i) the detection of insufficient peptides in the peptide mass fingerprint, (ii) the presence of a protein mixture in the sample, (iii) excessive posttranslational modifications in the protein, and/or (iv) nonexistence of the given protein in the sequence databases (owing to [micro]organisms with non–fully sequenced genomes and/or incomplete or unannotated sequence databases) *(17, 18)*. In these cases, other more efficient methods, such as amino acid sequencing by peptide fragmentation analysis *(45, 46)*, are certainly required to obtain protein structural information that allows unambiguous protein identification. Although peptide fragmentation can be attained using different techniques (*see* **Notes 5** and **56**), that generated by collision-induced dissociation (CID) is the most commonly used method for peptide sequencing in proteomics.

Peptide fragmentation through CID, a process whereby a gas-phase peptide ion is selected, isolated, and activated to dissociate by inducing multiple collisions with a neutral, inert gas within a tandem mass spectrometer (*see* **Note 57**), can currently be achieved with several tandem-in-space and tandem-in-time instrument designs (**Fig. 15.1**) coupled either with MALDI or with ESI *(18, 21, 23)*. The protocols outlined below are based on the use of an IT mass spectrometer equipped with a nanospray interface because of its prominent role in MS/MS experiments (*see* **Note 13**). In fact, nESI displays key features for peptide sequencing by MS/MS, such as high ionization efficiency, high sensitivity and low flow rate that enables low sample consumption and extended measurement time (*see* **Note 14**) *(47)*. On the other hand, IT instruments have become popular mass analyzers not only because of their relatively low cost, small space requirements, good sensitivity, and versatility, but also because of their distinctive ability to perform multiple fragmentation reactions (MS^n) (*see* **Note 6**) *(48)*.

3.3.1. Sample Preparation for nESI MS Analysis

In nESI, an acidic, aqueous solution of the peptide sample is sprayed into the inlet orifice of the mass spectrometer through a narrow capillary tube (a metal-coated glass capillary with a spraying orifice diameter of 1 to 2 μm; *see* **Note 14**). The application of a high, positive voltage to the capillary at atmospheric pressure leads to the formation of a cone (known as the Taylor cone) from which charged droplets are sputtered. As the solvent evaporates, which can be assisted by heat and a gas flow (usually N_2 gas), these droplets are reduced in size, yielding desolvated ions (*see* **Note 58**). This desolvation process takes place before entrance into the high-vacuum region of the mass spectrometer (**Fig. 15.8**).

The sample purity is the key to good MS results, because nESI is extremely sensitive to impurities, especially salts and nonvolatile buffers, which suppress sample ionization (*see* **Note 59**). Consequently, peptide samples must be desalted prior to nESI analysis (*see* **Section 3.1.6 and Fig. 15.4**).

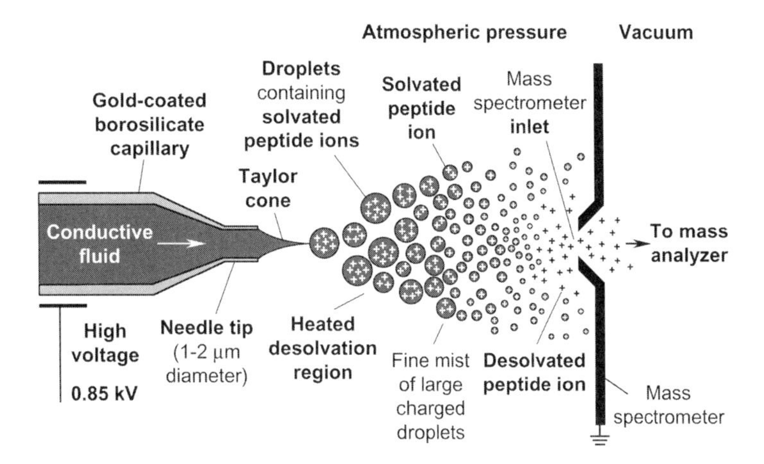

Fig. 15.8. Biochemical principle of the nESI process. The nESI is a miniaturized ESI source that consists of a metal-coated glass capillary with an internal tip diameter of 1 to 2 μm (*see* **Note 14**). The liquid sample is forced through this narrow capillary (on which a high voltage is applied at atmospheric pressure) to produce a fine mist of large charged droplets (all at the same polarity). The solvent evaporation reduces the droplet size, resulting in desolvated ions. These are then guided into the high-vacuum region of the mass spectrometer.

1. Load 1 to 2 μL salt-free peptide sample (prepared as described in **Section 3.1.6**) into a gold-coated borosilicate capillary (*see* **Note 60**) using a micropipette with a gel loader tip. Expel potential air bubbles by carefully shaking the capillary like a fever thermometer (*see* **Note 61**). Use a new capillary for each analysis to avoid the contamination risk with other samples.
2. Mount the loaded capillary in the nanospray needle holder (*see* **Note 61**).
3. Mount the capillary/holder assembly on the tridimensional manipulator in front of the mass spectrometer, and attach the high-voltage clip.
4. Connect the needle holder to a syringe that provides the air pressure (*see* **Note 60**).
5. Center the needle tip of the capillary on axis and 1 to 2 mm from the inlet orifice of the mass spectrometer. Use the manipulator to align them. Monitor this position under video-microscopy control.
6. Gently pressurize the capillary by air using the syringe. Confirm that a tiny sample droplet appears at the needle tip under video-microscopy (*see* **Note 62**).
7. Turn on the electrospray voltage (at 0.85 kV; *see* **Note 63**), and set the temperature capillary to 120°C. Check the nESI phenomena under video-microscopy.
8. Proceed to the MS/MS spectrum acquisition (*see* **Section 3.3.2**). Optimize the voltage and position of the needle tip by monitoring a sample ion.

3.3.2. Acquisition and Processing of MS/MS and MS^n Spectra by nESI-IT-MS Analysis

Distinct stages of mass analysis are required for the acquisition of MS/MS or MS^n spectra (**Fig. 15.9**). In the first phase, the ionized peptide sample is analyzed by operating in MS scan mode to obtain its peptide mass fingerprint and allow the precursor ion selection. In the second stage (MS/MS scan mode), the selected peptide ion (precursor ion) from the first MS scan is fragmented through CID (*see* **Note 57**), and an MS/MS spectrum (also known as product ion spectrum) is generated by sequentially ejecting product ions from low *m/z* to high *m/z* (*see* **Note 64**). Intriguingly, ion trap mass spectrometers have the intrinsic ability to perform a further fragmentation stage (MS^3 scan mode), in which a selected product ion from the MS^2 (MS/MS) scan is subfragmented, resulting in a subfragmentation (MS^3 product ion) spectrum (*see* **Note 6**) *(48)*. Multiple isolation and fragmentation stages (MS^n) can be achieved in the same way if there is sufficient sample. MS^3, or overall MS^n, spectra can be particularly useful (i) for glycosylation or phosphorylation analysis and/or (ii) when these yield data that were ambiguous or nonexistent in their corresponding original MS^2 or MS^{n-1} product ion spectra, respectively.

1. Perform MS/MS analyses on a quadrupole ion trap mass spectrometer, fitted with a nESI source and operated in the positive ion mode (*see* **Notes 48** and **65**).
2. Optimize the instrument settings with a standard sample solution (*see* **Note 66**).
3. Acquire MS/MS spectra using data-dependent scanning in "triple-play" mode (*see* **Note 67**), which comprises three sequential scans:
   (a) A full-range MS scan in which ions are collected under a total of 100 to 200 microscans with a maximum ion injection time of 400 ms, covering the mass range from *m/z* 175 to 2000 Da.
   (b) A narrow-range, high-resolution zoom scan on a selected ion from the first MS scan (*see* **Note 68**) to resolve its isotopic distribution and determine its charge state (*see* **Note 58**).
   (c) An MS/MS scan on this ion, using an isolation width of 3.0 *m/z* units and a normalized collision energy ranging from 20% to 45%, depending on the charge of the precursor ion (*see* **Notes 58** and **69**).
4. Repeat the two last scan events again with other further precursor ions (*see* **Note 68**).
5. Carry out MS^3, or higher-order MS, experiments (if there is sufficient sample) to obtain more accurate sequence information (*see* **Note 6** and **Fig. 15.9**).
6. Process all mass spectra using an appropriate MS analysis program (like the instrument-dependent Xcalibur software).

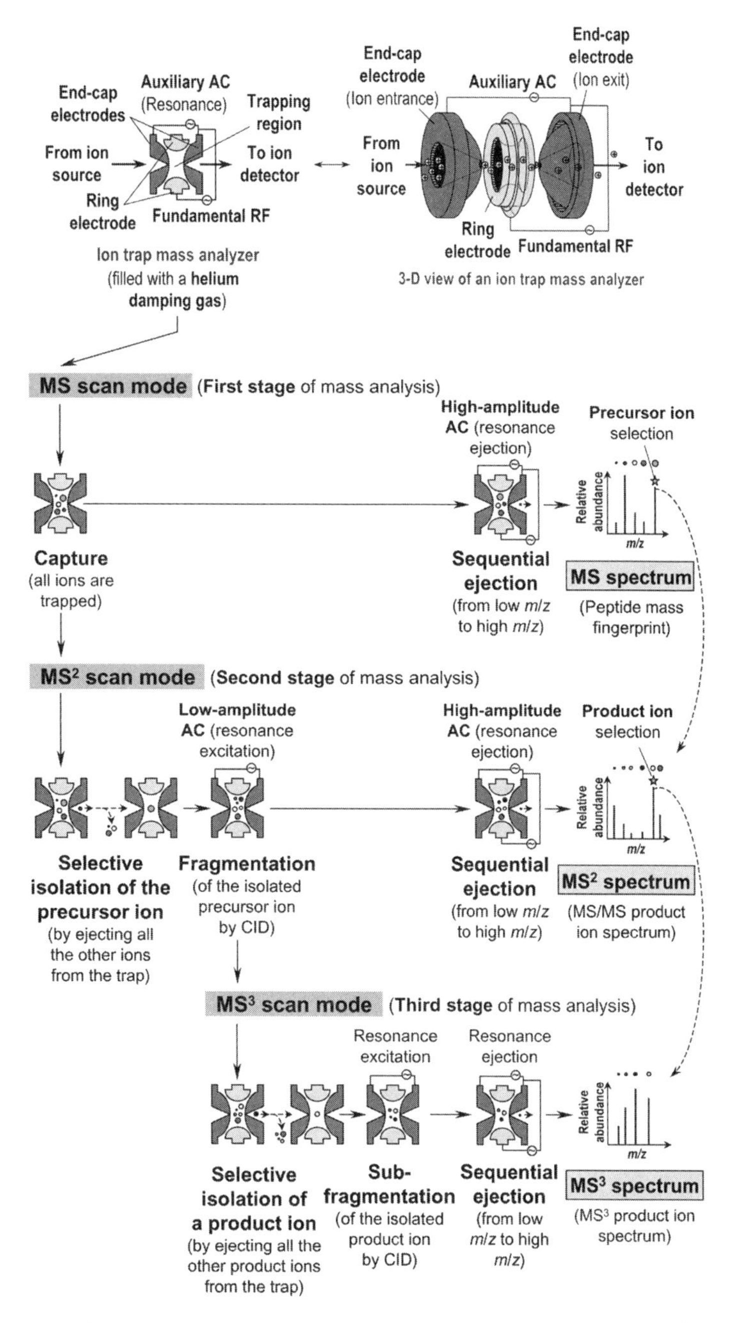

Fig. 15.9. Schematic representation of the workflow in the acquisition of an MS^3 spectrum by IT MS. Multiple stages of mass spectrometry (MS^n, where n is the number of mass analysis stages) can only be achieved in the IT and FT-ICR mass spectrometers (tandem-in-time instruments; *see* **Note 6**). This type of analysis is valuable for phosphorylation and glycosylation studies (*see* **Section 3.3.2** for further details). An ion trap mass analyzer is composed of two end-cap electrodes and a ring electrode and is filled with a helium damping gas. The application of a large fundamental potential (radio-frequency [RF] voltage) to the ring electrode results in the creation of a tridimensional electric field. If the supplementary or auxiliary potential (alternating current [AC] voltage) applied to the end-cap electrodes is of high amplitude, it will lead to resonance ejection; and if this potential is of low amplitude, then it will cause resonance excitation.

3.3.3. Database Searching Using Partial Amino Acid Sequences

The typical database searching strategy for protein identification using MS/MS or MS^n data is similar to that based on peptide mass fingerprinting (**Fig. 15.10**) *(17, 18)*. The experimental peptide masses from the acquired CID product ion mass spectrum are compared with the theoretical peptide masses obtained from *in silico* fragmentation of the corresponding theoretical

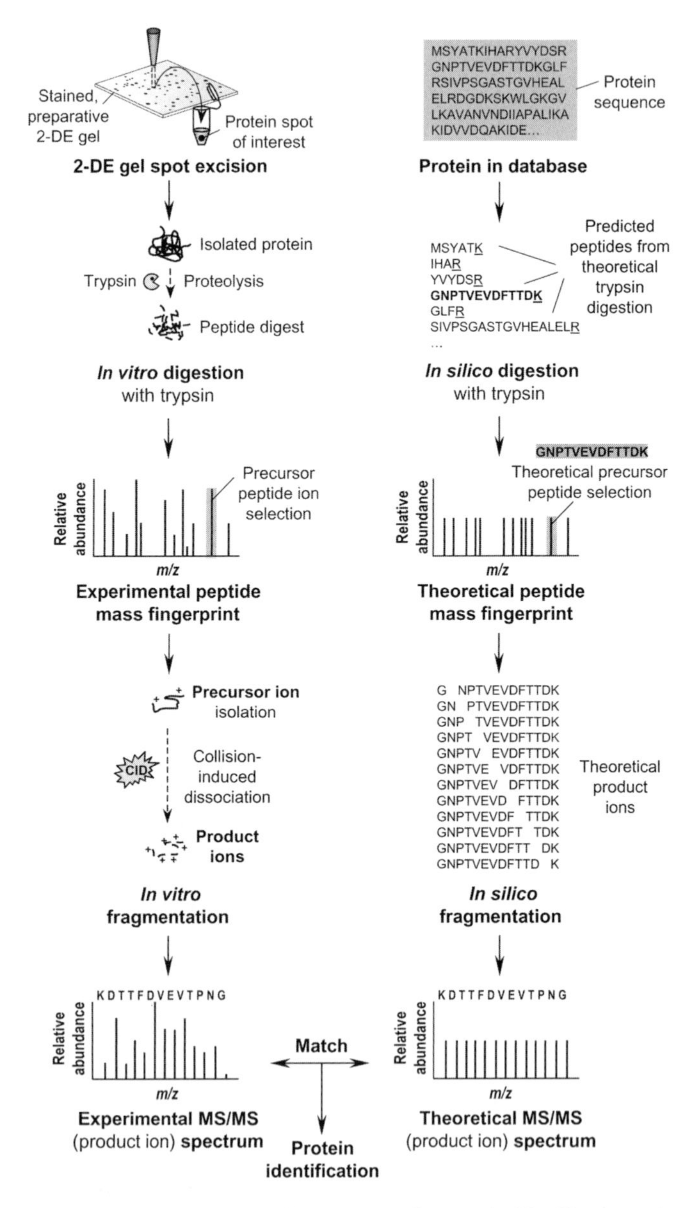

Fig. 15.10. Typical database searching strategy for protein identification using the measured MS/MS data. This approach does not involve the interpretation of MS/MS spectra. Similar to database searching based on peptide mass fingerprinting, two mass spectra (experimental and theoretical CID product ion mass spectra) are compared (*see* **Fig. 15.7** and **Section 3.3.3** for further details).

peptides derived from *in silico* digestion of all protein sequences present in protein or translated genomic database(s), taking into account the specificity of the protease used (*see* **Note 70**).

1. Input the list of experimental peptide masses from their CID product ion spectra into an appropriate peptide sequence search program (e.g., MS-Tag, Mascot, Sequest, or Sonar) *(15, 16)* by its copy/paste or as data file into query (space for *m/z* values) (*see* **Note 71**).
2. Set search parameters.
   (a) Select the NCBInr, SWISS-PROT/TrEMBL or CandidaDB databases to use for the search (*see* **Note 52**).
   (b) Define the mass and charge state of the precursor ion.
   (c) Set precursor ion mass tolerance at ± 50 ppm (*see* **Note 72**) and product (fragment) ion mass tolerance at ± 0.3 to 0.5 Da.
   (d) Select monoisotopic mass values (*see* **Note 49**).
   (e) Choose trypsin as the enzyme used to generate the peptides. Specify up to one missed cleavage site (*see* **Note 53**).
   (f) Carry out the search (at least the first one) without species, molecular weight or isoelectric point restriction (*see* **Note 52**).
   (g) Specify type of amino acid modifications to consider: (i) fixed modifications (like S-carbamidomethylation of cysteine residues) and (ii) variable modifications (such as oxidation of methionine residues, acrylamide modification of cysteine residues, and acetylation of protein N-terminus, among others) (*see* **Notes 17** and **55**).
   (h) Specify the maximum number of unmatched product (fragment) ions.
   (i) Specify ion series expected, and parameters influencing the output format.
3. Perform the database search. Confirm if unambiguous protein identification has been achieved (*see* **Note 73**).
4. Manually interpret all the mass spectra obtained by MS/MS, following the fragmentation rules for peptides *(49)*, to confirm the sequence results.
5. Proceed to *de novo* peptide sequencing if (i) ambiguous protein identification is attained in **steps 3** and **4**, (ii) the protein is not present in the DNA or protein databases, and/or (iii) it proves to be an unknown protein (*see* **Note 70**).
6. *De novo* deduce the amino acid sequences from the product ion spectra, either by manual interpretation (*see* **Note 74**) or with available *de novo* interpretation programs (such as LutefiskXP and PEAKS algorithms) *(15, 16)*.

7. Use the *de novo*–deduced peptide sequences to identify the protein by cross-species homology using BLAST and/or FASTS programs (*see* **Note 70**).
8. Annotate the *de novo* amino acid sequences of the novel protein in appropriate sequence databases, such as SWISS-PROT/TrEMBL database *(25)*.

4. Notes

1. This milestone (the discovery and development of soft peptide/protein ionization techniques) was recognized by the 2002 Nobel Prize in Chemistry. Both Tanaka and Fenn, who developed MALDI *(12)* and ESI *(14)* methods, respectively, were awarded (and shared) the Nobel Prize in 2002 for these key technologies for proteomic research *(50)*, which have undoubtedly had a significant impact and provided a huge amount of proteomic data in the different fields of biology and medicine in recent years.
2. Although *m/z* is often referred to as a unit-less ratio in many scientific publications, it is worth emphasizing that its unit is, however, the Thomson (Th) *(51)*. Mass spectrometers are only capable of separating either positive or negative charged gas-phase ion species (either $[M + nH]^{n+}$ or $[M - nH]^{n-}$, where M is the peptide molecular mass and *n* the ion charge) at the same time.
3. ESI is the method of choice to ionize peptide or protein samples for MS analysis if the sample is limiting, as it has better sensitivity than MALDI *(20, 52, 53)*. On the contrary, if there is no problem with the availability of sample and/or several samples must be analyzed at a time, then MALDI should be used because it is quicker and allows the analysis of many samples in an automated manner. Remarkably, ESI typically generates multiply protonated molecular ions ($[M + nH]^{n+}$; *see* **Note 2** for details on the nomenclature), whereas MALDI generally yields singly protonated molecular ions ($[M + H]^{+}$) (**Fig. 15.11**). Consequently, the interpretation of MALDI spectra to determine the charge state and molecular weight of their peptide ions is clearly easier than that of the ESI spectra. Furthermore, MALDI is more tolerant towards buffers and contaminants than ESI. Conversely, ESI is the preferred interface for combining liquid chromatography (LC) with MS.
4. A single quadrupole analyzer has proved to be inappropriate for proteomic applications, because of its low mass accuracy, low resolution, and limited mass range. However, the combination of three quadrupoles in sequence (triple quadrupole analyzer, QqQ) or the recent configuration of a hybrid quadrupole

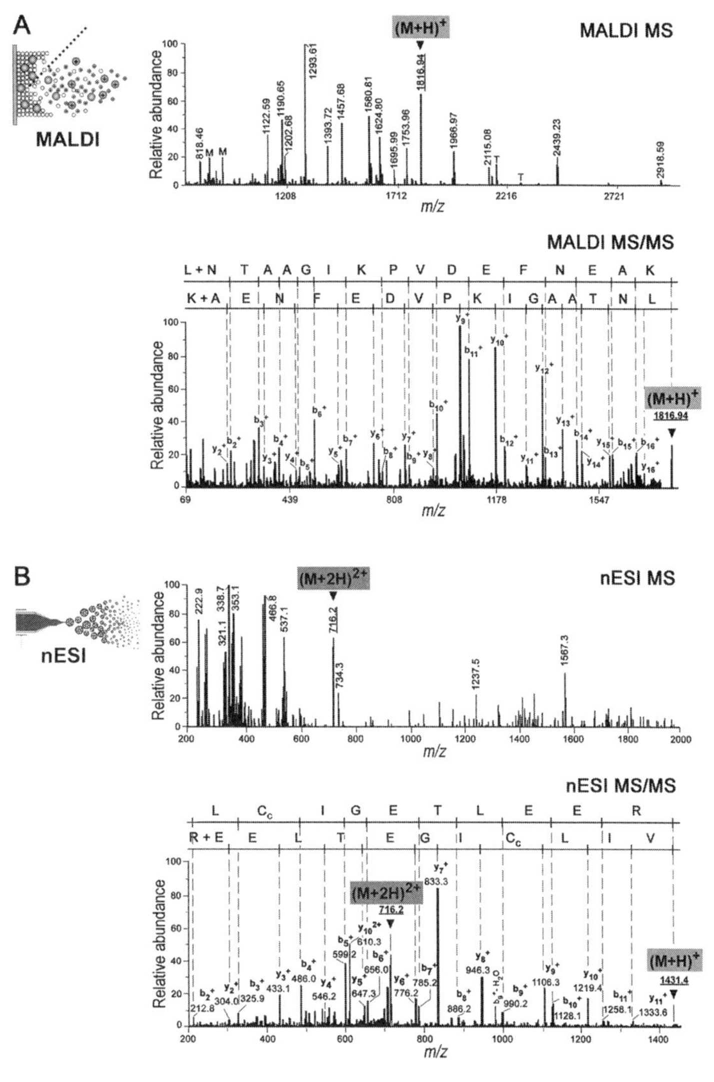

Fig. 15.11. Comparison of representative MALDI and nESI mass spectra from *C. albicans* immunogenic proteins present during systemic infection (*25*). (**A**) Typical MALDI mass spectra. Overall, the *m/z* values of peptide ions from MALDI MS mass spectra correspond with $[M + H]^{1+}$. Therefore, MS/MS spectra will be derived from singly charged MALDI precursor ions (*see* **Fig. 15.6** for details on biochemical principle of the MALDI process). *Top panel*, MALDI-TOF MS mass spectrum (peptide mass fingerprint) of the tryptic digest of *C. albicans* methionine synthase (Met6p). *Bottom panel*, MALDI-TOF/TOF MS/MS mass spectrum acquired from a singly charged precursor ion at *m/z* 1816.9 Th for the tryptic peptide of Met6p with a calculated molecular mass of 1815.9 Da. The series of *b*- and *y*-ions (which correspond with N- and C-terminal fragments of the peptide generated by breakage at its peptide bonds, respectively) are shown. I/L and Q/K as the CandidaDB. (**B**) Representative nESI mass spectra. nESI MS spectra often tend to generate doubly charged ions $[M + 2H]^{2+}$ that appear at half the peptide mass. Remembering that the *C*-terminal residue (Lys or Arg) and the N-terminal amino group are basic, whereby doubly charged peptide ions will be mainly yielded in nESI MS (*see* **Fig. 15.8** for further details on biochemical bases of the nESI process). *Top panel*, MS spectrum (peptide mass fingerprint) of the tryptic digest of *C. albicans* triose phosphate isomerase (Tpi1p) obtained by nESI-IT MS. *Bottom panel*, nESI-IT MS/MS product ion spectrum acquired from a doubly charged precursor ion at *m/z* 716.5 Th for the tryptic peptide of Tpi1p with a calculated molecular mass of 1431.0 Da. The *y* series of ions and those from the *b* series are given. I/L and Q/K as the CandidaDB.

TOF instrument (QqTOF) have proved to be very useful for proteomic analysis (**Fig. 15.1**) *(18, 21)*.

5. Alternative and complementary fragmentation techniques to CID have recently been developed, such as (i) electron transfer dissociation (ETD), performed on linear IT mass analyzers *(54)*, and (ii) electron capture dissociation (ECD), carried out on FT-ICR analyzers *(55)*. In contrast to classical CID, these two new techniques (based on electron transfer of the ions located in the collision cell) are fundamentally useful both in analyzing large peptides and proteins and in localizing posttranslational modifications *(45)*.
6. Unlike other mass analyzers, IT and FT-ICR (tandem-in-time instruments; *see* **Fig. 15.1**) have the distinctive capacity to carry out multiple fragmentation reactions, that is, multiple stages of MS (MS^n). For this reason, these are the ideal mass analyzers for *de novo* peptide sequencing. Remarkably, the seminal contribution of the IT technique to biological research has been highlighted by the 1989 Nobel Prize in Physics, which was awarded to Wolfgang Paul.
7. It is of note that it is currently possible to directly fragment and analyze intact proteins using sophisticated mass spectrometers (such as FT-ICR instruments) through an innovative approach called "top-down" *(45, 56–59)*. This new strategy yields not only sequence information (full amino acid coverage) but also accurate localization and identification of posttranslational modifications in the protein of interest (*see* **Note 5**).
8. The main Achilles' heel of this non-gel-based approach (often referred to as shotgun proteomics) *(60, 61)* (**Fig. 15.2**) is clearly the unfeasibility of visually selecting the protein of interest from a complex protein mixture. This consequently leads to the generation of an enormous amount of irrelevant data to be examined by MS, because every sequence-specific peptide of a given protein must be analyzed. Conversely, this "gel-free" method is currently the best choice to analyze protein samples that cannot be efficiently resolved by 2-DE gels, such as membrane proteins.
9. Before use, all microcentrifuge tubes and pipette tips should be rinsed several times (either with ethanol or with a solution containing 45% (v/v) ACN and 10% (v/v) glacial acetic acid) and air-dried prior to use to eliminate any residual dust (including keratins) from them.
10. It is of prime importance to use the same alkylation reagent as was used for the equilibration of IPG strips (prior to second-dimension separation of 2-DE procedure; *see* **Chapter 26**) in order to avoid generating a second cysteine modification in the protein sample, and therefore, multiple species of the

same peptide. The latter may lead to a significant reduction in its detectability.

11. Although trypsin is the enzyme of choice for protein digestion, other proteolytic agents can also be used in its place. However, the optimum buffer conditions for each protease should be established appropriately for the digestion reaction.
12. The pH of the matrix solution must be below pH 4. This acidic environment enables the sample ionization process by protonation (**Fig. 15.6**). Do not use nonvolatile solvents since these can interfere with the co-crystallization of the sample and matrix. The most commonly used matrices in MALDI MS are (i) α-cyano-4-hydroxycinnamic acid (CHCA), to analyze peptides and protein digests (**Fig. 15.12**), (ii) sinapinic acid (3,5-dimethoxy-4-hydroxycinnamic acid), to analyze proteins and large polypeptides, and (iii) 2,5-dihydroxybenzoic acid (2,5-DHB), to analyze proteins, protein digests, and oligosaccharides released from glycoproteins *(62)*.

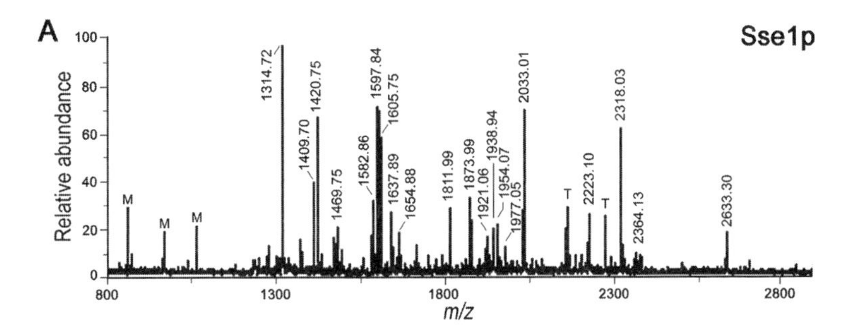

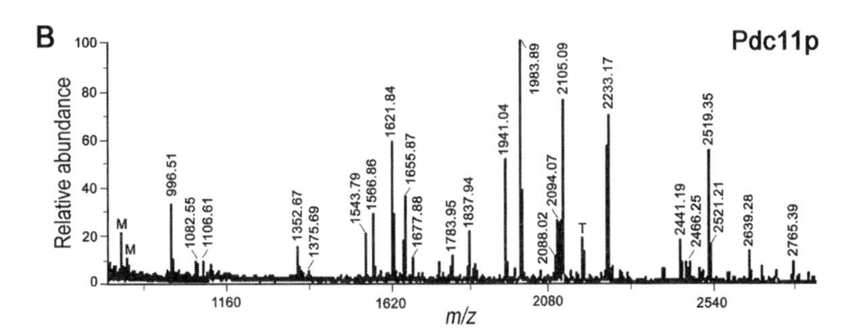

Fig. 15.12. Representative peptide mass fingerprints of the tryptic digests of two identified proteins belonging to the *C. albicans* immunome present during the host-pathogen interaction. Labeled peaks indicate the matched tryptic peptides to two *C. albicans* immunogenic proteins present during systemic infection, that is, heat shock protein of HSP70 family (Sse1p; *top*) and pyruvate decarboxylase (Pdc11p; *bottom*) *(25)*. Peaks of matrix (α-cyano-4-hydroxycinnamic acid) and autolysis products of the trypsin are designated with "M" and "T," respectively. Peptide peaks appear as singly charged ions, $[M + H]^{1+}$ ions. Peptide mass fingerprints were collected on a Voyager DE STR MALDI TOF mass spectrometer. These mass spectra were externally and internally calibrated.

13. MS analyses can be adequately performed on any available equipment (and using its software programs) following the manufacturer's instructions.
14. nESI, a miniaturized ESI source (composed of a gold-coated borosilicate capillary with an internal tip diameter of 1 to 2 μm), is the method of choice for optimal ionization efficiency. nESI operates at very low flow rates (5 to 10 nL/min), resulting in low sample consumption (20–100 nL/min) and extended analysis times for experiment optimization (more than an hour if desired) (*35*).
15. Given that these instruments are enclosed systems, sample contamination (and, therefore, potential wrong identifications and additional noise to the analysis) can be prevented with their use.
16. Contamination of the sample with keratins (derived from skin and hair) or other proteins or substances (e.g., polymers) may occur in any stage of the proteomic experiment as a result of a poor sample handling from the 2-DE procedure to in-gel digestion protein digest. The following measures can be used to minimize sample contamination: wear gloves, rinse sample tubes and pipette tips (*see* **Note 9**), and perform all these steps in clean rooms (or in laminar-flow hoods). It is important to rinse the gloves on their outside surface with ultrapure water (while they are being worn) to reduce interference by powder or other contaminants.
17. Preparative SDS-PAGE gels used for the second-dimension separation should be polymerized overnight to minimize the amount of remaining unreacted acrylamide, and thus prevent alkylation of thiol groups present in proteins (i.e., additional protein modifications) by unpolymerized acrylamide in the SDS-PAGE gel (*see* **Chapter 26**).
18. Alternatively, tips for a 1-mL or 100-μL pipette trimmed to the fitting size of the protein spot can be used to excise it from the stained gel (**Fig. 15.13A**). Transfer of the excised gel piece to a prerinsed tube can be performed using a surgical needle or a pipette tip.
19. A representative mass-to-charge (*m/z*) pattern of trypsin autoproteolysis products and potential contaminant peptides in the gel may be achieved by submitting this control blank to MS analysis.
20. Bent tips for a 100-μL pipette rather than scalpel can be used to mince the excised gel pieces (**Fig. 15.13B**).
21. It is convenient to cut (but not crush or pulverize) each excised gel piece into smaller gel particles (~1 mm^3) in order to improve (i) the removal of residual SDS, Coomassie blue and silver ions, and (ii) the access of the reducing and alkylating reagents as well as trypsin to the gel piece (**Figs. 15.4** and **15.13B**).

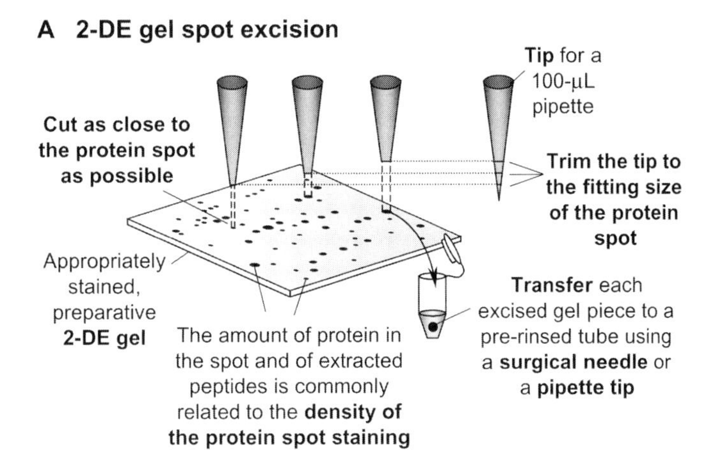

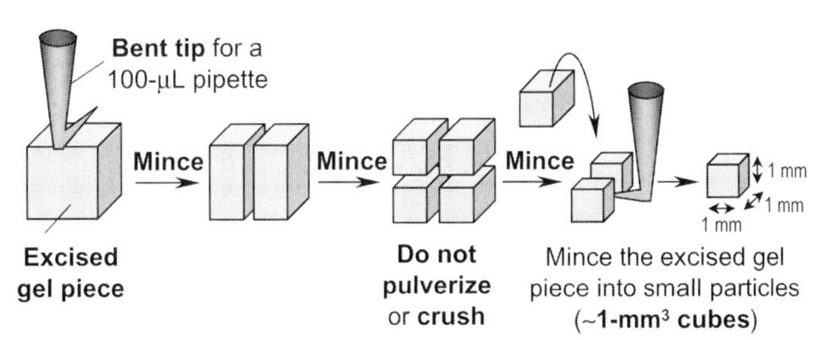

Fig. 15.13. Schematic representation of the proposed procedures for spot excision and mincing. Tips for a 100-μL pipette can be used for (**A**) spot excision by trimming them to the spot size and (**B**) spot mincing after they are bent (*see* **Notes 18** and **20** for further information).

22. Overall, the solvent volumes used in all these steps (from this point on) rely on the size of the excised gel piece. Add enough volume to cover all the gel particles (~5 gel volumes for washing steps).
23. Given the small size of the gel particles, all solutions should be carefully removed using gel-loading pipette tips to prevent any loss of these particles and, therefore, of protein.
24. The gel particles should become completely transparent. For this reason, several wash/dehydration cycles in the Coomassie blue–destaining method should be carried out when >10 pmol of protein is analyzed to remove residual Coomassie staining. However, a single step for silver-destaining procedure is often sufficient.
25. When the gel particles are dehydrated, these shrink, acquire an opaque white color, and stick together (**Fig. 15.4**).
26. More digestion solution volume than can be absorbed by the gel particles should not be added.

27. Potential trypsin autolysis can be bypassed, or at least minimized, during the rehydration step (in which the digestion enzyme passively diffuses through the gel matrix) by performing it at 4°C (on ice) *(27)*.
28. It is important to remove the excess trypsin solution after rehydration is accomplished to reduce the amount of trypsin in the digest.
29. In contrast, the intact protein (before digestion) immobilized in the gel fragments (by precipitation during the fixation process of protein staining; *see* **Chapter 26**) cannot diffuse out of the gel particles (**Fig. 15.5**). This property permits the destaining, washing, reduction, alkylation, dehydration, and rehydration steps to be performed in-gel without any loss of protein material (**Fig. 15.4**).
30. If desired, the crude extracts can be used directly for MALDI-TOF analysis without pooling with all the supernatants of additional peptide extraction steps (**Fig. 15.4**).
31. Cycles of addition of water to total supernatant and subsequent concentration in a vacuum centrifuge can optionally be performed for volatile salt reduction. This may improve the *m/z* signals in the subsequent MS analysis.
32. It is advisable to carry out this step to significantly reduce the ACN concentration in the peptide sample and, thus, enable its proper retention on the ZipTip.
33. It is recommended that the last eluate is dispensed back into the sample microcentrifuge tube (rather than into the waste container) and stored. This should be done exercising extreme caution, as this eluate may be used again if peptides from the sample were not properly bound to the ZipTip.
34. High vacuum is essential to avoid ion collisions.
35. It must be borne in mind that optimal resolution and mass accuracy are the key to unambiguous database searches based on peptide mass fingerprinting.
36. There are several on-target sample preparation methods for MALDI MS analysis (**Fig. 15.14**), each with their own advantages and disadvantages *(43)*:
    (a) The dried droplet method (*13*). Equivolumes of the sample and matrix are mixed together, and a droplet of the mixture is then applied to the target well and allowed to co-crystallize. Alternatively, the matrix can be added to a sample droplet deposited previously on the target plate, and the two solutions are subsequently mixed in the target well using a pipette tip and permitted to dry (*see* **Section 3.2.1**). It is of note that the reproducibility of MALDI MS results is substantially improved by thoroughly mixing the sample and matrix prior to co-crystallization. This method

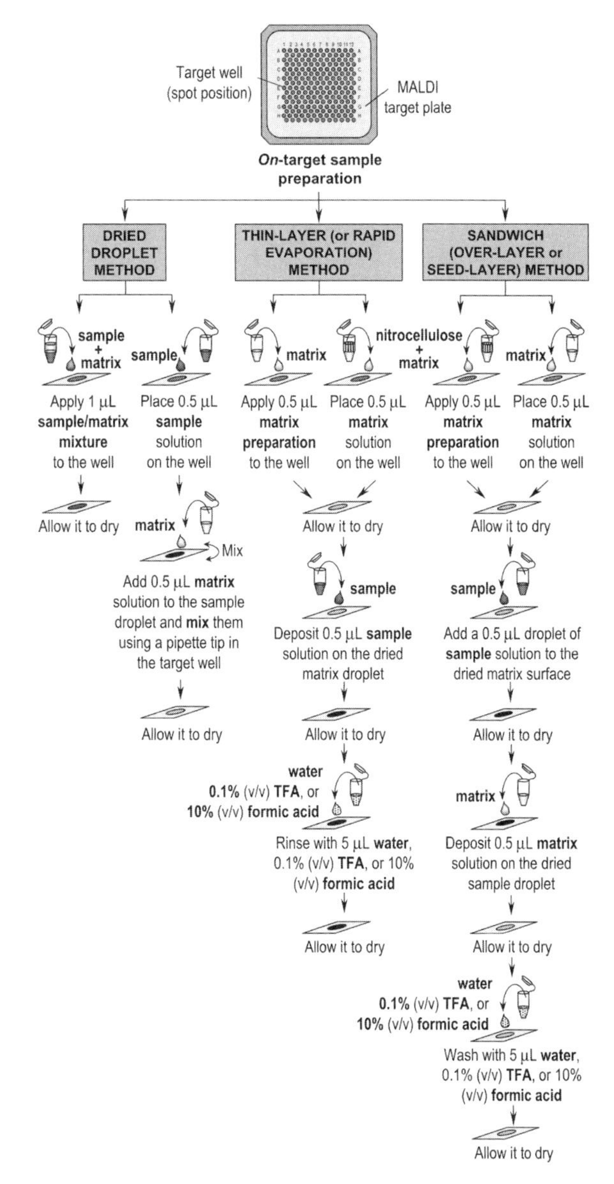

Fig. 15.14. Main methods of on-target sample preparation for MALDI MS analysis of peptides and proteins. Each procedure has its own strengths and weaknesses (*see* **Note 36** for further details). The quality of the on-target sample preparation has a direct impact on the efficiency of desorption and ionization of a given sample in MALDI MS and consequently on the quality of the MS results.

is indisputably useful for high laser repetition rates because of the slow matrix consumption to generate continuous ion packets. However, one limitation of this strategy is the nonhomogeneous distribution of the sample. This can be bypassed either (i) by using more volatile solvents to accelerate drying, or (ii) by vacuum drying the sample/ matrix mixture.

(b) The thin-layer (or rapid evaporation) method *(63)*. The matrix (dissolved in a highly volatile solvent) is spotted or sprayed on the target plate and allowed to dry. The sample is subsequently deposited onto the dried matrix surface on the target plate, permitted to dry and then rinsed with a small amount (a 5- to 10-μL droplet) of water, ice-cold 0.1% (v/v) TFA, or 10% (v/v) formic acid to remove salts exposed on the well surface. This is possible because salts, unlike peptide-matrix crystals, are soluble in ice-cold acidic solution. Alternatively, nitrocellulose (used as a co-matrix) can be incorporated to the matrix solution by mixing in order to enhance peptide binding to the target plate and enable more extensive washing of the sample *(44)*. This method is desirable because (i) potential salts, reagents, and/or buffers present in the peptide mixture and concentrated on the well surface can be washed away, producing a high quality MALDI sample, and (ii) a homogenous matrix layer is obtained, which provides uniform desorption/ionization properties and improves the resolution, sensitivity and reproducibility of the MALDI MS analysis. However, its main disadvantage is that samples with high organic solvent concentrations or high pH can completely dissolve the matrix layer. The destruction of this thin matrix layer can be avoided by applying 0.5 to 1 μL water or 0.5 to 1 μL 10% (v/v) formic acid, respectively, onto the matrix surface prior to sample addition or to the sample solution before its deposition.

(c) The sandwich (over-layer or seed-layer) method *(64)*. A sample droplet followed by another matrix droplet is placed directly over a thin layer of matrix (or matrix in conjunction with nitrocellulose, used as a co-matrix; *see* above). Intriguingly, this method provides excellent reproducibility, and is often used with hydrophobic proteins, high-molecular-mass peptides, whole proteins, or samples with high concentrations of salts, buffers, and/or detergents. However, this procedure is more difficult to automate than those described above.

37. The MALDI target plate should not be wiped with detergents because these are tricky to remove and can lead to background signals in the mass spectra. Furthermore, it is convenient to use Teflon-coated target plates with hydrophilic spot positions (also known as sample anchors) to concentrate the sample and enable rinsing with water or acidic solutions (to eliminate salts and buffers) (*see* **Note 36**) *(43, 65)*.

38. The crude extract (*see* **Section 3.1.5, step 1**), vacuum-concentrated peptide sample (*see* **Section 3.1.5, step 8**), and/or salt-free peptide sample (*see* **Section 3.1.6, step 8**)

can be applied to MALDI MS analysis, because unlike ESI, MALDI can tolerate certain levels of some contaminants and low levels of residual sample compounds (*see* **Note 3 and Fig. 15.4**). In fact, the clean-up (desalting) step (*see* **Section 3.1.6** and **Fig. 15.4**) is not required in many cases for efficient MALDI MS analysis (**Fig. 15.12**).

39. Equal-volume aliquots of the peptide mixture and matrix solution should be used. Given the narrow width of the laser beam, the size of the sample and matrix droplets should be as small as possible. This results in an increase in the analytical sensitivity. It is recommended to use hydrophilic sample anchors on the target plates (*see* **Notes 36 and 37**), as it allows sample size reduction *(65)*.
40. It is crucial to avoid the precipitation of the co-crystalline sample and matrix when these are mixed in order to achieve optimal peptide ionization.
41. Do not heat the co-crystalline sample/matrix mixture on the target well to accelerate the evaporation rate of the solvents, as this alters the matrix crystal formation and peptide incorporation and, consequently, leads to a poor MALDI sample.
42. Because the dried MALDI samples are quite stable, the target plates with matrix/sample preparations can be stored in a cool, dark, and dry environment or in a vacuum for several days or even months, if desired, without significant sample degradation or signal loss.
43. It is important to find the laser firing positions with the best matrix/sample ratio and to use a laser power above the threshold for ion appearance in an attempt to attain high-quality mass spectra. These parameters may have to be empirically determined.
44. Because of the microheterogeneity of the matrix/sample crystals, data should be acquired by accumulation MALDI-TOF mass spectra from laser shots in order to obtain a representative sum spectrum. This, therefore, comes from several individual laser pulses.
45. The incorporation of reflectron and delayed extraction technologies in the MALDI-TOF mass spectrometers has resulted in an increase in mass resolution and, thus, in mass accuracy and sensitivity (*see* **Fig. 15.15**) *(66, 67)*. The reflectron (also referred to as ion reflector) is used as an ion mirror (placed at the end of the flight tube) to reflect ions back through the tube to the ion detector *(66)*. This serves not only to extend the flight length (without increasing the instrument size) but also to correct for small differences in the kinetic energy of peptide/matrix ions with the same m/z (because of uneven energy distribution as these are generated by the laser pulse).

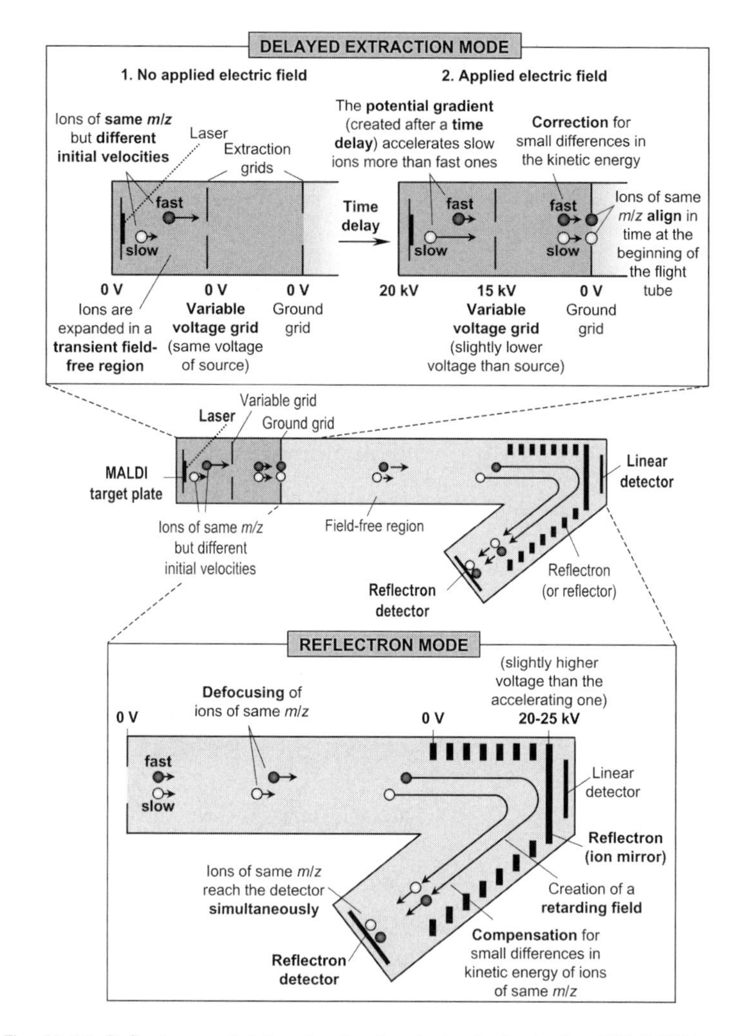

Fig. 15.15. Reflectron and delayed extraction technologies in the MALDI-TOF mass spectrometers. The small differences in the kinetic energy of ions with the same *m/z* can be corrected using these technologies, so that these arrive simultaneously at the plane of the detector. This improves mass resolution and, therefore, mass accuracy and sensitivity (*see* **Note 45** for further details).

As a result, the mass resolution is improved. The timed delay ion extraction is also used to compensate for the initial kinetic energy distribution (with the goal that ions with the same *m/z* reach the detector at the same time) and, consequently, increase the mass resolution *(67)*.

46. External mass calibration is particularly useful when the trypsin autolysis peptide ions cannot be detected in the mass spectra, and these may not, therefore, be calibrated internally (*see* **Note 47**). External *m/z* calibration can be achieved by using the calibration file obtained from:
   (a) A peptide mass standard mixture (calibration sample) deposited on a target well placed in close proximity to

the sample whose spectrum needs to be calibrated. It is advisable to apply a calibration sample next to each sample or every two samples. It must be borne in mind that the calibration sample has to have been calibrated internally (*see* **Note 47**). Acquisition of the calibration spectrum can be done before or after of that of the sample spectrum of interest.

(b) Internal calibration (using trypsin autolysis and/or matrix peaks; *see* **Note 47**) applied to the mass spectrum acquired from another peptide sample located adjacent to the sample of interest. This clearly bypasses the need to spot the calibration sample on the target plate.

47. Internal mass calibration results in better accuracy of the recorded *m/z* measurements. This can also be performed in two different ways:

    (a) By using a combination of trypsin autolysis and/or matrix peaks present in the sample spectrum. The main limitation of this method is that these peaks are infrequent or absent. In these cases, either external calibration (*see* **Note 46**) or internal calibration by addition of a peptide mass standard mixture (*see* later) is required.

    (b) By mixing the standard or calibration sample with the peptide sample of interest prior to depositing a droplet of this mixture on the target plate and acquiring its corresponding mass spectrum. However, this method has several intrinsic disadvantages: (i) it is a time-consuming process, (ii) the calibration peptides may overlap the sample peptides, which will thus be excluded from further database searches, and (iii) an excess of the internal standard may suppress ionization of the sample peptides. Although it is recommended to use equimolar sample/standard amounts, the optimal ratio should be determined by experimentation.

48. In positive ion mode, protonated species are mainly detected (except if there are other cations, like sodium, potassium or ammonium in the sample), whereas in negative ion mode, the detected ions are deprotonated.

49. It is convenient to use the monoisotopic peptide mass (the mass of the first peak, corresponding with the ^{12}C peak, in the peptide isotopic envelope) rather than the average mass (taken at the centroid of the isotopic envelope) of each peptide ion for database searches based on peptide mass fingerprinting, as this results in higher mass accuracy and, consequently, less false-positive results.

50. In theory, the same protein identity will be achieved with all available database search engines if the same parameters are used.

51. Protein identification can be considered as unambiguous when nearly all measured peptide masses match a protein sequence within 50-ppm mass accuracy *(68)*. Overall, the amino acid sequence coverage should be ≥15% to achieve successful protein identification. However, it must be borne in mind that, for example, (i) proteins >100 kDa can unequivocally be identified with a sequence coverage <10% if their 15 to 20 measured peptide masses map to a protein sequence to within 50 ppm, and (ii) proteins <20 kDa can also conclusively be identified with only 3 to 4 detected peptide masses that conversely cover 30% to 50% amino acid sequence *(68)*.
52. It is important to use the NCBInr or SWISS-PROT/TrEMBL databases (the most comprehensive and best-annotated protein databases, respectively) *(10)* rather than CandidaDB database (download in FASTA format from http://genolist.pasteur.fr/CandidaDB *(7)* for MS applications) in the primary database query without constraining species, molecular weight or isoelectric point, in order to allow proper protein identification in case there are protein contaminations (like keratins from human or other species).
53. More than one missed cleavage site should not be considered because it can reduce discrimination and result in many random protein matches.
54. It is important to select an error tolerance window that is, at minimum, the mass accuracy obtained experimentally (*see* **Note 45** and **Section 3.2.2**) to enable inclusion of all the experimental peptide masses in the search. Peptide masses should be measured with a mass accuracy of 30 to 100 ppm for unambiguous database searching applications *(44)*.
55. Cysteine S-carbamidomethylation has to be considered as a fixed modification because cysteine-containing peptides were alkylated with iodoacetamide previously (*see* **Section 3.1.3** and **Note 10**). Artifactual chemical modifications that may have been induced during 2-DE (such as acrylamide modification of cysteine residues [*see* **Note 17**] and oxidation of methionine residues), typical co-translational modifications (removal of the N-terminal methionine and acetylation of the penultimate amino acid) *(25)*, or known posttranslational modifications (such as phosphorylation of serine, threonine and tyrosine, or conversion of N-terminal glutamine to pyroglutamic acid), among others, should be contemplated as common variable modifications.
56. Peptide sequencing can also be performed by post-source decay (PSD) analysis on a MALDI reflector TOF mass

spectrometer *(69)*. However, the generated PSD spectra are often difficult to interpret due to their unpredictable peptide fragmentation patterns.

57. Be aware that the collision energy required to induce peptide fragmentation within an ion trap (IT), triple quadrupole (QqQ), or hybrid quadrupole (QqTOF) analyzer is low, whereas that within a TOF/TOF analyzer is high.
58. One or more protons from the solvent are picked up by the peptide molecules as this evaporates, resulting in the formation of singly or, more commonly, multiple charged ions ($[M + H]^{1+}$ or $[M + nH]^{n+}$, respectively, where n is the ion charge) (*see* **Note 3**). The number of charges that a molecule can acquire relies on the number of its potential charge sites (sites of proton attachment). The charge state of a peptide ion can be determined by sufficient m/z resolution of its corresponding isotopic cluster (**Fig. 15.16**).
59. If there are salts in the peptide sample, the microcapillary orifice can be partially or totally blocked, and the spray can consequently become discontinuous and instable. This leads to a reduction in sensitivity.
60. Metal (gold) coating on the outside of the glass (borosilicate) capillary ensures the transfer of the electrical potential to the fluid at the needle tip of the capillary. Air pressure should, therefore, be applied to the capillary to facilitate the flow of the liquid sample to the tip as well as contact with the metal surface.
61. It is essential to take special care with the needle tip of the capillary (**Fig. 15.8**) when loading liquid sample and mounting the capillary/holder assembly, as this is extremely fragile and can easily be broken.
62. If this does not occur, then apply voltage to the capillary, pressurize it, and carefully press the needle tip of the capillary against the mass spectrometer interface plate under video-microscopy control to open the needle tip.
63. An 0.85-kV potential difference between the needle and the interface plate of the mass spectrometer is sufficient to establish the spray cone and initiate the nanoelectrospray due to the small diameter of the capillary tip (**Fig. 15.8**).
64. The MS and MS/MS scan modes are different in tandem-in-space (e.g., QqQ) or tandem-in-time (like IT) instruments (**Fig. 15.17**) *(18, 20)*.
    (a) Triple quadrupole (QqQ) mass spectrometers. In MS mode, all peptide ions above a certain m/z are transmitted (without switching on the collision cell) to the third quadrupole, which resolves them according to m/z values.

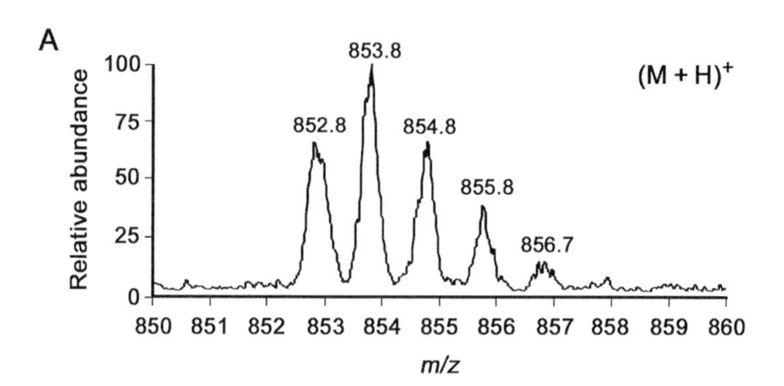

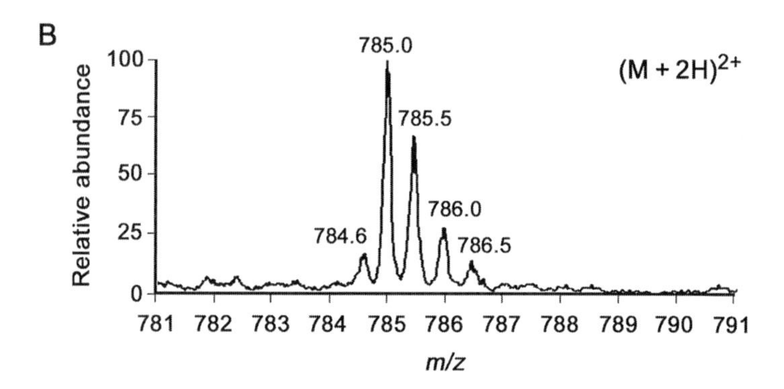

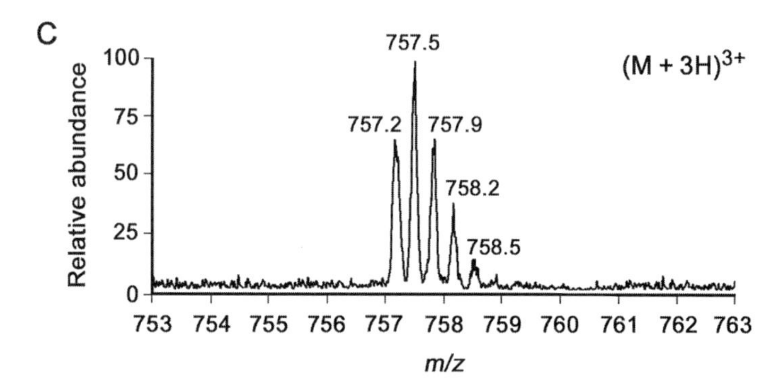

Fig. 15.16. Examples of the isotope distribution of peptide ions at different charge states. The charge state of a peptide ion can be established by adequate *m/z* resolution of its corresponding isotopic cluster. The mass difference between isotopes is 1.0 Da. (**A**) Isotope cluster for a singly charged peptide ion $[M + H]^{+}$. The *m/z* difference between isotopes is 1.0 Th. (**B**) Isotope cluster for a doubly charged peptide ion $[M + 2H]^{2+}$. The *m/z* difference between isotopes is 0.5 Th. (**C**) Isotope cluster for a triply charged peptide ion $[M + 3H]^{3+}$. The *m/z* difference between isotopes is 0.33 Th.

In MS/MS mode, the precursor ion (the ion to be sequenced) is selectively transmitted into the collision cell, where it is fragmented through CID. The resulting product ions are then resolved on the basis of their *m/z* by the third quadrupole.

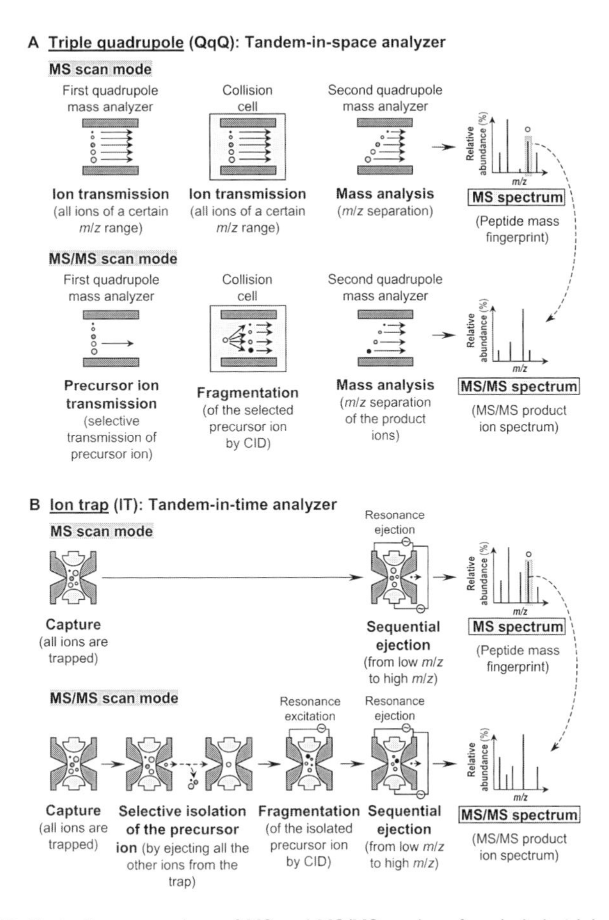

Fig. 15.17. Illustrative comparison of MS and MS/MS modes of analysis in triple quadrupole (QqQ) and ion trap (IT) mass spectrometers. Peptide fragmentation can be performed in tandem-in-space instruments (e.g., QqQ) or tandem-in-time instruments (like IT), resulting from the combination of two mass analyzers or the use of the same mass analyzer two or more times, respectively (*see* **Fig. 15.1**). *See* **Note 64** and **Fig. 15.9** for further details on MS and MS/MS scan modes.

(b) Ion trap (IT) mass spectrometers. In MS mode, the ionized peptide sample is focused into the ion trap (into a small volume with an oscillating electric field), trapped, and sequentially ejected from low *m/z* to high *m/z* (by electronic manipulation of this field). In MS/MS mode, the sample ions are first trapped into the ion trap, and then all except the precursor ion (the ion to be sequenced) are ejected. This isolated ion is resonantly activated to generate fragment ions. The resulting product ions are ejected one after the other according to their *m/z* from the trap.

65. Although these analyses are often carried out manually, they can be automated if desired.

66. IT mass spectrometers usually need little day-to-day tuning for optimal sensitivity and resolution.

67. It is convenient to use a "triple-play" mode data analysis with dynamic exclusion because this avoids reanalysis of precursor ions used previously for MS/MS.
68. It is recommended to select a precursor ion that exceeds a base peak threshold and is among the most intense peptide ions.
69. The collision energy should be adjusted for optimum CID fragmentation.
70. If the protein of interest is present in databases with incomplete gene information (e.g., expressed sequence tag [EST] databases), then the use of "sequence tags" or peptide mass tag searching should be accomplished (*70*). On the contrary, *de novo* sequencing (*de novo* deduction of amino acid sequences from the MS/MS or MS^n spectra by manual interpretation or with computer assistance) is the approach of choice to identify proteins from (micro)organisms with non–fully sequenced genomes and/or unannotated sequence databases, such as, until a short time ago, *C. albicans* (**Fig. 15.18**) *(25, 26)*. *De novo* amino acid sequences are aligned to all protein sequences present in the available protein or translated genomic databases using BLAST or FASTS tools.
71. It is advisable to input the *m/z* values of the product ions with higher intensities if there are a large number of ions in the CID product ion spectrum.
72. Mass accuracy of the precursor ion is crucial to reduce false-positive protein matches (*see* **Notes 46** and **47**).
73. Confident protein identification can be attained if a high number of peptides matchs the same entry in a database. However, the protein should be identified independently by two or three peptides to achieve unambiguous protein identity *(71)*. Therefore, additional confirmation will be required if only a single peptide identifies a protein.
74. Amino acid sequences can be deduced by the mass differences between *y*- or *b*-ion "ladder" series resulting from the CID fragmentation spectra of the selected tryptic peptides and following the fragmentation rules for peptides *(49)* (**Fig. 15.18**).

Acknowledgments

We thank the Merck, Sharp & Dohme (MSD) Special Chair in Genomics and Proteomics, Comunidad de Madrid (S-SAL-0246-2006 DEREMICROBIANA-CM), Comisión Interministerial de Ciencia y Tecnología (CICYT; BIO-2003-00030 and BIO-2006-01989), and Fundación Ramón Areces for financial support of our laboratory.

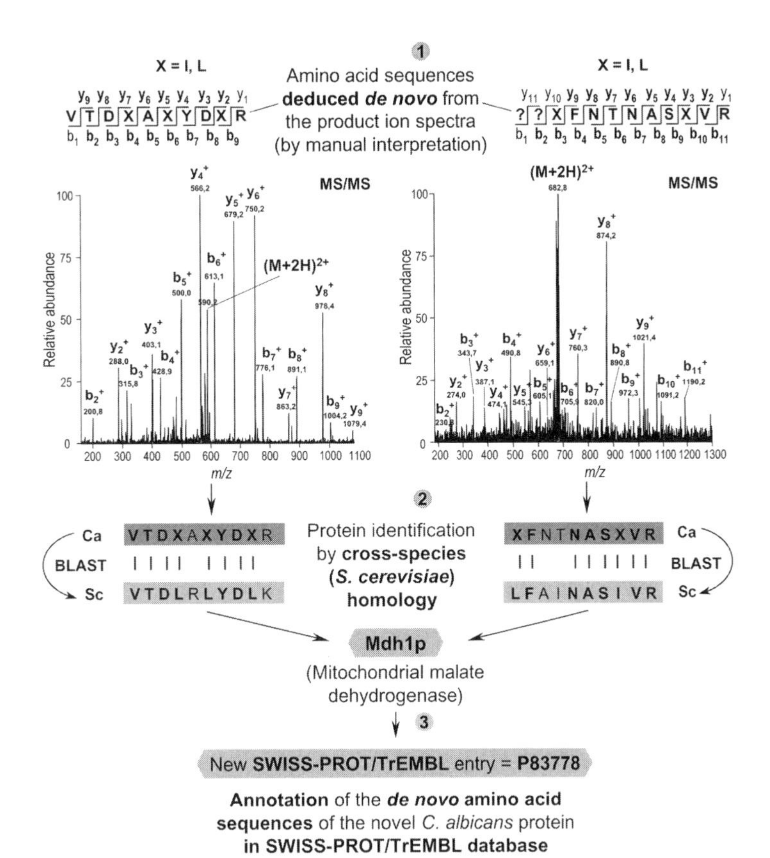

Fig. 15.18. Example of cross-species identification of a *C. albicans* immunogenic protein present during systemic infection by *de novo* sequencing using nESI IT-MS analysis and database searching. This method is useful to identify proteins from (micro)organisms with non–fully sequenced genomes and/or unannotated sequence databases, such as, until a short time ago, *C. albicans*. The protein illustrated in this figure is *C. albicans* mitochondrial malate dehydrogenase (Mdh1p), a protein that had hitherto been unreported as a *C. albicans* antigen and unannotated in sequence databases when it was identified *(25)*. This protein was unambiguously cross-species identified by matching *C. albicans* amino acid sequences (*de novo* deduced from nESI-IT MS/MS data) to proteins from *Saccharomyces cerevisiae* (a yeast organism whose genome was already entirely sequenced) *(6)* databases, taking advantage of the high degree of homology between most sequences from both yeast organisms. This approach can be extended to other (micro)organisms that do not have fully sequenced genomes and/or complete or well-annotated sequence databases *(3, 4, 18)*.

References

1. Wilkins, M. R., Sanchez, J. C., Gooley, A. A., Appel, R. D., Humphery-Smith, I., Hochstrasser, D. F., and Williams, K. L. (1996) Progress with proteome projects: why all proteins expressed by a genome should be identified and how to do it. *Biotechnol. Genet. Eng Rev.* **13**, 19–50.
2. Wasinger, V. C., Cordwell, S. J., Cerpa-Poljak, A., Yan, J. X., Gooley, A. A., Wilkins, M. R., Duncan, M. W., Harris, R., Williams, K. L., and Humphery-Smith, I. (1995) Progress with gene-product mapping of the Mollicutes: *Mycoplasma genitalium*. *Electrophoresis* **16,** 1090– 1094.
3. Pitarch, A., Nombela, C., and Gil, C. (2006) *Candida albicans* biology and pathogenicity: insights from proteomics. *Methods Biochem. Anal.* **49,** 285–330.

4. Pitarch, A., Nombela, C., and Gil, C. (2006) Contributions of proteomics to diagnosis, treatment, and prevention of candidiasis. *Methods Biochem. Anal.* **49,** 331–361.
5. Jones, T., Federspiel, N. A., Chibana, H., Dungan, J., Kalman, S., Magee, B. B., Newport, G., Thorstenson, Y. R., Agabian, N., Magee, P. T., Davis, R. W., and Scherer, S. (2004) The diploid genome sequence of *Candida albicans. Proc. Natl. Acad. Sci. U.S.A.* **101,** 7329–7334.
6. Goffeau, A., Barrell, B. G., Bussey, H., Davis, R. W., Dujon, B., Feldmann, H., Galibert, F., Hoheisel, J. D., Jacq, C., Johnston, M., Louis, E. J., Mewes, H. W., Murakami, Y., Philippsen, P., Tettelin, H., and Oliver, S. G. (1996) Life with 6000 genes. *Science* **274,** 563–546.
7. d'Enfert, C., Goyard, S., Rodriguez-Arnaveilhe, S., Frangeul, L., Jones, L., Tekaia, F., Bader, O., Albrecht, A., Castillo, L., Dominguez, A., Ernst, J. F., Fradin, C., Gaillardin, C., Garcia-Sanchez, S., de Groot, P., Hube, B., Klis, F. M., Krishnamurthy, S., Kunze, D., Lopez, M. C., Mavor, A., Martin, N., Moszer, I., Onesime, D., Perez, M. J., Sentandreu, R., Valentin, E., and Brown, A. J. (2005) CandidaDB: a genome database for *Candida albicans* pathogenomics. *Nucleic Acids Res.* **33,** D353–D357.
8. Arnaud, M. B., Costanzo, M. C., Skrzypek, M. S., Shah, P., Binkley, G., Lane, C., Miyasato, S. R., and Sherlock, G. (2007) Sequence resources at the *Candida* Genome Database. *Nucleic Acids Res.* **35,** D452–D456.
9. Csank, C., Costanzo, M. C., Hirschman, J., Hodges, P., Kranz, J. E., Mangan, M., O'Neill, K., Robertson, L. S., Skrzypek, M. S., Brooks, J., and Garrels, J. I. (2002) Three yeast proteome databases: YPD, PombePD, and CalPD (MycoPathPD). *Methods Enzymol.* **350,** 347–373.
10. Pitarch, A., Sanchez, M., Nombela, C., and Gil, C. (2003) Analysis of the *Candida albicans* proteome. II. Protein information technology on the Net (update 2002). *J. Chromatogr. B Analyt. Technol. Biomed. Life Sci.* **787,** 129–148.
11. Nash, R., Weng, S., Hitz, B., Balakrishnan, R., Christie, K. R., Costanzo, M. C., Dwight, S. S., Engel, S. R., Fisk, D. G., Hirschman, J. E., Hong, E. L., Livstone, M. S., Oughtred, R., Park, J., Skrzypek, M., Theesfeld, C. L., Binkley, G., Dong, Q., Lane, C., Miyasato, S., Sethuraman, A., Schroeder, M., Dolinski, K., Botstein, D., and Cherry, J. M. (2007) Expanded protein information at SGD: new pages and proteome browser. *Nucleic Acids Res.* **35,** D468–D471.
12. Tanaka, K., Waki, H., Ido, Y. (1988). Protein and polymer analyses up to *m/z* 100,000 by laser ionization time-of-flight mass spectrometry. *Rapid Commun. Mass Spectrom.* **2,** 151.
13. Karas, M. and Hillenkamp, F. (1988) Laser desorption ionization of proteins with molecular masses exceeding 10,000 daltons. *Anal. Chem.* **60,** 2299–2301.
14. Fenn, J. B., Mann, M., Meng, C. K., Wong, S. F., and Whitehouse, C. M. (1989) Electrospray ionization for mass spectrometry of large biomolecules. *Science* **246,** 64–71.
15. Palagi, P. M., Hernandez, P., Walther, D., and Appel, R. D. (2006) Proteome informatics I: bioinformatics tools for processing experimental data. *Proteomics* **6,** 5435–5444.
16. Shadforth, I., Crowther, D., and Bessant, C. (2005) Protein and peptide identification algorithms using MS for use in high-throughput, automated pipelines. *Proteomics* **5,** 4082–4095.
17. Pitarch, A., Sanchez, M., Nombela, C., and Gil, C. (2003) Analysis of the *Candida albicans* proteome. I. Strategies and applications. *J. Chromatogr. B Analyt. Technol. Biomed. Life Sci.* **787,** 101–128.
18. Graves, P. R. and Haystead, T. A. (2002) Molecular biologist's guide to proteomics. *Microbiol. Mol. Biol. Rev.* **66,** 39–63.
19. Annan, R. S. and Carr, S. A. (1997) The essential role of mass spectrometry in characterizing protein structure: mapping posttranslational modifications. *J. Protein Chem.* **16,** 391–402.
20. Mann, M., Hendrickson, R. C., and Pandey, A. (2001) Analysis of proteins and proteomes by mass spectrometry. *Annu. Rev. Biochem.* **70,** 437–473.
21. Domon, B. and Aebersold, R. (2006) Mass spectrometry and protein analysis. *Science* **312,** 212–217.
22. Aebersold, R. and Mann, M. (2003) Mass spectrometry-based proteomics. *Nature* **422,** 198–207.
23. Baldwin, M. A. (2005) Mass spectrometers for the analysis of biomolecules. *Methods Enzymol.* **402,** 3–48.

24. Mann, M., Hendrickson, R. C., and Pandey, A. (2001) Analysis of proteins and proteomes by mass spectrometry. *Annu. Rev. Biochem.* **70,** 437–473.
25. Pitarch, A., Abian, J., Carrascal, M., Sanchez, M., Nombela, C., and Gil, C. (2004) Proteomics-based identification of novel *Candida albicans* antigens for diagnosis of systemic candidiasis in patients with underlying hematological malignancies. *Proteomics* **4,** 3084–3106.
26. Pardo, M., Ward, M., Pitarch, A., Sanchez, M., Nombela, C., Blackstock, W., and Gil, C. (2000) Cross-species identification of novel *Candida albicans* immunogenic proteins by combination of two-dimensional polyacrylamide gel electrophoresis and mass spectrometry. *Electrophoresis* **21,** 2651–2659.
27. Shevchenko, A., Jensen, O. N., Podtelejnikov, A. V., Sagliocco, F., Wilm, M., Vorm, O., Mortensen, P., Shevchenko, A., Boucherie, H., and Mann, M. (1996) Linking genome and proteome by mass spectrometry: large-scale identification of yeast proteins from two dimensional gels. *Proc. Natl. Acad. Sci. U.S.A.* **93,** 14440–14445.
28. Masuoka, J., Glee, P. M., and Hazen, K. C. (1998) Preparative isoelectric focusing and preparative electrophoresis of hydrophobic *Candida albicans* cell wall proteins with in-line transfer to polyvinylidene difluoride membranes for sequencing. *Electrophoresis* **19,** 675–678.
29. Washburn, M. P., Wolters, D., and Yates, J. R., III (2001) Large-scale analysis of the yeast proteome by multidimensional protein identification technology. *Nat. Biotechnol.* **19,** 242–247.
30. Pitarch, A., Jimenez, A., Nombela, C., and Gil, C. (2006) Decoding serological response to *Candida* cell wall immunome into novel diagnostic, prognostic, and therapeutic candidates for systemic candidiasis by proteomic and bioinformatic analyses. *Mol. Cell Proteomics* **5,** 79–96.
31. Wilson, R. A., Curwen, R. S., Braschi, S., Hall, S. L., Coulson, P. S., and Ashton, P. D. (2004) From genomes to vaccines via the proteome. *Mem. Inst. Oswaldo Cruz* **99,** 45–50.
32. Sette, A., Fleri, W., Peters, B., Sathiamurthy, M., Bui, H. H., and Wilson, S. (2005) A roadmap for the immunomics of category A-C pathogens. *Immunity* **22,** 155–161.
33. Pitarch, A., Molero, G., Monteoliva, L., Thomas, D. P., López-Ribot, J. L., Nombela, C., and Gil, C. (2007) Proteomics in *Candida* species, in *Candida: Comparative and Functional Genomics.* (d'Enfert, C. and Hube, B., eds), Caister Academic Press, UK, pp. 169–194.
34. Thomas, D. P., Pitarch, A., Monteoliva, L., Gil, C., and Lopez-Ribot, J. L. (2006) Proteomics to study *Candida albicans* biology and pathogenicity. *Infect. Disord. Drug Targets* **6,** 335–341.
35. Wilm, M., Shevchenko, A., Houthaeve, T., Breit, S., Schweigerer, L., Fotsis, T., and Mann, M. (1996) Femtomole sequencing of proteins from polyacrylamide gels by nano-electrospray mass spectrometry. *Nature* **379,** 466–469.
36. Shevchenko, A., Wilm, M., Vorm, O., and Mann, M. (1996) Mass spectrometric sequencing of proteins silver-stained polyacrylamide gels. *Anal. Chem.* **68,** 850–858.
37. McCormack, A. L., Schieltz, D. M., Goode, B., Yang, S., Barnes, G., Drubin, D., and Yates, J. R., III (1997) Direct analysis and identification of proteins in mixtures by LC/MS/MS and database searching at the low-femtomole level. *Anal. Chem.* **69,** 767–776.
38. Gharahdaghi, F., Weinberg, C. R., Meagher, D. A., Imai, B. S., and Mische, S. M. (1999) Mass spectrometric identification of proteins from silver-stained polyacrylamide gel: a method for the removal of silver ions to enhance sensitivity. *Electrophoresis* **20,** 601–605.
39. Yates, J. R., Speicher, S., Griffin, P. R., and Hunkapiller, T. (1993) Peptide mass maps: a highly informative approach to protein identification. *Anal. Biochem.* **214,** 397–408.
40. James, P., Quadroni, M., Carafoli, E., and Gonnet, G. (1993) Protein identification by mass profile fingerprinting. *Biochem. Biophys. Res. Commun.* **195,** 58–64.
41. Thiede, B., Hohenwarter, W., Krah, A., Mattow, J., Schmid, M., Schmidt, F., and Jungblut, P. R. (2005) Peptide mass fingerprinting. *Methods* **35,** 237–247.
42. Marvin, L. F., Roberts, M. A., and Fay, L. B. (2003) Matrix-assisted laser desorption/ionization time-of-flight mass spectrometry in clinical chemistry. *Clin. Chim. Acta* **337,** 11–21.
43. Rappsilber, J., Moniatte, M., Nielsen, M. L., Podtelejnikov, A. V., Mann, M. (2003). Experiencies and perspectives of

MALDI MS and MS/MS in proteomic research. *Int. J. Mass Spectrom.* **226,** 223–237.

44. Jensen, O. N., Podtelejnikov, A., and Mann, M. (1996) Delayed extraction improves specificity in database searches by matrix-assisted laser desorption/ionization peptide maps. *Rapid Commun. Mass Spectrom.* **10,** 1371–1378.
45. Wysocki, V. H., Resing, K. A., Zhang, Q., and Cheng, G. (2005) Mass spectrometry of peptides and proteins. *Methods* **35,** 211–222.
46. Medzihradszky, K. F. (2005) Peptide sequence analysis. *Methods Enzymol.* **402,** 209–244.
47. Wilm, M. and Mann, M. (1996) Analytical properties of the nanoelectrospray ion source. *Anal. Chem.* **68,** 1–8.
48. Jonscher, K. R. and Yates, J. R. (1997) The quadrupole ion trap mass spectrometer–a small solution to a big challenge. *Anal.Biochem.* **244,** 1–15.
49. Biemann, K. (1992) Mass spectrometry of peptides and proteins. *Annu. Rev. Biochem.* **61,** 977–1010.
50. Cook, K. D. (2002) ASMS members John Fenn and Koichi Tanaka share Nobel: the world learns our "secret." American Society for Mass Spectrometry. *J. Am. Soc. Mass Spectrom.* **13,** 1359.
51. Cooks, R. G., Rockwood, A. L. (1991) The "Thomson": A suggested unit for mass spectroscopy. *Rapid Commun. Mass Spectrom.* **5,** 93.
52. Graves, P. R. and Haystead, A. J. (2003) Proteomics and molecular biologist, in *Handbook of Proteomic Methods* (Conn, P. M., ed), The Humana Press Inc., Totowa, NJ, pp. 3–16.
53. Strupat, K. (2005) Molecular weight determination of peptides and proteins by ESI and MALDI. *Methods Enzymol.* **405,** 1–36.
54. Syka, J. E., Coon, J. J., Schroeder, M. J., Shabanowitz, J., and Hunt, D. F. (2004) Peptide and protein sequence analysis by electron transfer dissociation mass spectrometry. *Proc. Natl. Acad. Sci. U.S.A.* **101,** 9528–9533.
55. Zubarev, R. A., Horn, D. M., Fridriksson, E. K., Kelleher, N. L., Kruger, N. A., Lewis, M. A., Carpenter, B. K., and McLafferty, F. W. (2000) Electron capture dissociation for structural characterization of multiply charged protein cations. *Anal. Chem.* **72,** 563–573.
56. Whitelegge, J., Halgand, F., Souda, P., and Zabrouskov, V. (2006) Top-down mass spectrometry of integral membrane proteins. *Expert. Rev. Proteomics* **3,** 585–596.
57. Kelleher, N. L. (2004) Top-down proteomics. *Anal. Chem.* **76,** 197A–203A.
58. Sze, S. K., Ge, Y., Oh, H., and McLafferty, F. W. (2002) Top-down mass spectrometry of a 29-kDa protein for characterization of any posttranslational modification to within one residue. *Proc. Natl. Acad. Sci. U.S.A.* **99,** 1774–1779.
59. Taylor, G. K., Kim, Y. B., Forbes, A. J., Meng, F., McCarthy, R., and Kelleher, N. L. (2003) Web and database software for identification of intact proteins using "top down" mass spectrometry. *Anal. Chem.* **75,** 4081–4086.
60. Swanson, S. K. and Washburn, M. P. (2005) The continuing evolution of shotgun proteomics. *Drug Discov. Today* **10,** 719–725.
61. McDonald, W. H. and Yates, J. R. (2003) Shotgun proteomics: integrating technologies to answer biological questions. *Curr. Opin. Mol. Ther.* **5,** 302–309.
62. Westermeier, R. and Naven, T. (eds.) (2002) *Proteomics in Practice: A Laboratory Manual of Proteome Analysis.* Wiley-VCH, Weinheim, Germany.
63. Vorm, O., Roepstorff, P., and Mann, M. (1994) Improved resolution and very high sensitivity in MALDI-TOF of matrix surfaces made by fast evaporation. *Anal. Chem.* **66,** 3281–3287.
64. Kussmann, M., Lassing, U., Sturmer, C. A., Przybylski, M., and Roepstorff, P. (1997) Matrix-assisted laser desorption/ionization mass spectrometric peptide mapping of the neural cell adhesion protein neurolin purified by sodium dodecyl sulfate polyacrylamide gel electrophoresis or acidic precipitation. *J. Mass Spectrom.* **32,** 483–493.
65. Schuerenberg, M., Luebbert, C., Eickhoff, H., Kalkum, M., Lehrach, H., and Nordhoff, E. (2000) Prestructured MALDI-MS sample supports. *Anal. Chem.* **72,** 3436–3442.
66. Cornish, T. J. and Cotter, R. J. (1994) A curved field reflectron time-of-flight mass spectrometer for the simultaneous focusing of metastable ions. *Rapid Commun. Mass Spectrom.* **8,** 781–785.
67. Vestal, M. L., Juhasz, P., and Martín, S. A. (1995) Delayed extraction matrix-assisted laser desorption time-of-flight mass spectrometry. *Rapid Commun. Mass Spectrom.* **9,** 1044–1050.

68. Jensen, O. N., Wilm, M., Shevchenko, A., Mann, M. (1999) Sample preparation methods for mass spectrometric peptide mapping directly from 2-DE gels. *Methods Mol. Biol.* **112,** 513–530.

69. Kaufmann, R., Spengler, B., and Lutzenkirchen, F. (1993) Mass spectrometric sequencing of linear peptides by product-ion analysis in a reflectron time-of-flight mass spectrometer using matrix-assisted laser desorption ionization. *Rapid Commun. Mass Spectrom.* **7,** 902–910.

70. Mann, M. and Wilm, M. (1994) Error-tolerant identification of peptides in sequence databases by peptide sequence tags. *Anal.Chem.* **66,** 4390–4399.

71. Jensen, O. N., Wilm, M., Shevchenko, A., Mann, M. (1999) Peptide sequencing of 2-DE gel-isolated proteins by nanoelectrospray tandem mass spectrometry. *Methods Mol. Biol.* **112,** 571–588.

72. Pitarch, A., Nombela, C., and Gil, C. (2008) Collection of proteins secreted from yeast protoplasts in active cell wall regeneration. *Methods Mol. Biol.* **425,** 241–263.

Part III
Parasites

Chapter 16

Introduction: Parasites

Klaus Brehm and Carsten G. K. Lüder

Although, by definition, parasites also include pathogenic bacteria, fungi, and viruses, the research field of *medical* (or *veterinary*) *parasitology* exclusively deals with animals that cause disease. This includes a wide variety of organisms from single-celled eukaryotes (protozoa) to multicellular worms (helminths) and even arthropods or annelids. Concerning molecular host-parasite interactions, however, only two of these groups are of major interest and have been investigated to considerable extent. The protozoa, which cause several of the most important infectious diseases such as malaria, sleeping sickness, or toxoplasmosis, and the helminths, which currently infect more than one third of the human world population. One important feature these two groups of pathogens have in common is that the respective diseases are mostly occurring in developing countries (what we call the "Third World") and are usually associated with poverty. Compared with molecular research on "First World" infectious agents such as many bacteria, viruses, or fungi, parasites therefore have received little attention (maybe with the exception of *Plasmodium* and malaria), which has recently even led to the neologism of the "neglected diseases." A further common feature of protozoa and helminths is that both groups are eukaryotes and often display highly complex life cycles that frequently involve transmission between several vertebrate and/or invertebrate hosts. However, concerning their molecular interaction with the host, the type of host response during an infection, their developmental biology, and even their genetics and cell biology, there are numerous striking differences between protozoa and helminths.

The parasitic protozoa constitute a heterogenous group of unicellular eukaryotes that belong to different phylogenetic clades thus indicating that the parasitic lifestyle has been invented

Steffen Rupp, Kai Sohn (eds.), *Host-Pathogen Interactions*, DOI: 10.1007/978-1-59745-204-5_16,

multiple times during evolution. Because the phylogeny of the protozoa is a matter of ongoing debate, it may be most useful (although from a phylogenetic point of view incorrect) to refer to the older classification of Flagellata, Amoebozoa, Apicomplexa (representing the former Sporozoa), Ciliata, and Microspora. Particularly the first three groups contain a wide variety of pathogens that have major impact on health of humans and livestock. For example, an estimated 1.5 million to 2 million people, predominantly children up to an age of 5 years, succumb each year because of infection with *Plasmodium falciparum*, the causative agent of malaria tropica, making this parasite one of the three leading pathogens responsible for human mortality. Furthermore, another apicomplexan parasite, *Toxoplasma gondii*, is considered one of the most commonly and most widely distributed infectious agents with an overall seroprevalence in humans of approximately 30% and with varying prevalences in nearly all mammals and birds throughout the world. Other protozoans (e.g., *Theileria* spp. and *Trypanosoma brucei brucei*) are major obstacles to livestock breeding and meat production in Africa.

Representing a heterogeneous group of organisms, it is not surprising that protozoans also considerably differ in the individual lifestyles within their hosts. A variety of them live—like the vast majority of helminths—exclusively extracellularly in different niches of their host. However, because of their restricted size, a large number of them have explored and employed the possibility to use host cells as a convenient niche to grow, replicate, or differentiate and have thus become obligatory intracellular parasites at least during distinct stages of their life cycles. And even within host cells, protozoans have adapted to different subcellular sites; for instance, cytosol (*Trypanosoma cruzi*), fusion-incompetent parasitophorous vacuoles (e.g., *Toxoplasma gondii, Plasmodium* spp.), or phagolysosomes (*Leishmania* spp.). Importantly, the diverse lifestyles of protozoan parasites also require distinct adaptations and, therefore, constitute a treasury of fascinating pathogen-host interactions. The investigation of the molecular and cellular relationships of protozoans with their host or host cell has indeed become one of the most expanding fields in parasitology. This relates to the facts that (i) a detailed understanding of the parasite-host interaction at the molecular level may open novel avenues to combat parasitic diseases, and (ii) the relative complexity of protozoans—compared with viruses or bacteria—provides microbiologists, molecular biologists, and cell biologists with highly sophisticated and tightly controlled adaptations between two eukaryotic organisms that are of general biological and medical interest.

The complex relationships between protozoans and their hosts imply that it is impossible to provide the audience of a single book with a complete overview on molecular biology methods to study parasite-host interactions. However, one of the common themes that emerged during the past few years is the modulation of apoptosis during infections with protozoans (as with other infectious agents). Intriguingly, different parasites can either induce or inhibit apoptosis with a few protozoans even doing both depending on the parasite stage, the cell type under investigation, and the parasite–host cell relationship. Because apoptosis is—among other functions—crucial for regulating host immune responses, for tissue homeostasis, and as an innate and adaptive effector mechanism particularly against intracellular pathogens, its modulation by protozoans is considered to be of prime importance for the outcome of infection. One of the well-known protozoans that exerts multiple effects on the apoptotic cascades of host cells is *Toxoplasma gondii*, and this is further outlined in the contribution of Hippe and colleagues (**Chapter 19**).

Methods for *in vitro* cultivation of protozoans are certainly a major requisite for the rapidly expanding field of cellular parasitology. It not only allows the production of sufficient amounts of protozoan parasites for further structural, physiologic, molecular, or immunologic analyses but is also crucial for genetic manipulation and hence functional investigations. Although still much more labor-intensive and time-consuming than is expanding bacteria on agar plates or in broth medium, considerable progress has indeed been made in the propagation of protozoans *in vitro*. This is particularly true for several of the apicomplexans including *Plasmodium* spp., which has led to a renaissance of malaria research during the past decade (see **Chapter 18** by Cui and colleagues). An important issue of future work is to cultivate obligatory intracellular parasites axenically. Whereas such culture has already been successfully applied to *Leishmania* parasites, it is not yet applicable to most others (e.g., apicomplexans) and thus restricts parasite analyses without contamination with host cell molecules.

The parasitic helminths belong to two different animal phyla, the Nemathelminthes and the Platyhelminthes, and are, from the phylogenetic point of view, relatively closely related to mammalian hosts (the last common ancestor of parasitic helminths and humans lived just 600 million years ago). This relationship has significant consequences on the infection strategies of parasitic helminths and on the development of anthelmintic drugs. Because of the similar cell biology of helminths and mammals, parasite-specific targets for anthelmintics are relatively hard to identify, and many of the currently used anthelmintic drugs exert heavy side-effects. The need for novel anthelmintics with improved specificity is emphasized by the facts that currently more than 2 billion people worldwide are suffering from helminth diseases and that resistance toward the currently used drugs is increasing. In sharp contrast with this need, both basic science and the pharmaceutical industry in industrialized countries are only reluctantly engaging in molecular helminthologic research due to the fact that the majority of the affected people could simply not afford cost-intensive treatment and expensive drugs.

Apart from their importance as pathogens, parasitic helminths are also highly interesting organisms with respect to immunology research. They provoke strong Th2-dominated immune responses that significantly differ from those responses that are induced by bacteria, fungi, or viruses. Furthermore, helminths typically cause chronic, long-lasting diseases (of years to decades) that are associated with various down-modulatory effects on the host's immune response, including the generation of regulatory T cells and alternatively activated macrophages. Hence, not just since the discovery that helminth infections can have beneficial effects on First World problems such as allergies or autoimmune diseases, the immunology of helminth diseases is one of the most interesting and challenging topics of current research. Another, relatively young topic of helminthology deals with the molecular basis of host-parasite interaction and the influence of host factors on parasite development. During recent years, it has been shown that most of the key developmental molecules that are involved in cell-cell communication and, thus, development of mammals (e.g., cytokines of the insulin, the EGF, or the TGF-β family including corresponding receptors) are also present in helminths leading to the interesting concept of hormonal cross-communication between the paracrine and endocrine systems of host and parasite. According to this theory, phylogenetically conserved cytokines from helminths could activate corresponding receptors of the host, thus manipulating the immune response, and host cytokines could, by stimulating the corresponding parasite receptors, induce parasite development at the appropriate sites within the body. Closely linked to that are questions of parasite development, the mechanisms of which significantly differ between platyhelminths and nemathelminths. Whereas the latter display a rigid and strictly controlled cell lineage, platyhelminths are highly variable in their developmental patterns and are masters of regeneration. Their key to these traits are totipotent stem cells called neoblasts or germinal cells, which are present in all free-living and parasitic platyhelminths and which play a crucial role in planarian regeneration or asexual multiplication of flatworm larvae within the host. The molecular cell biology of this kind of primitive stem cell will surely be one of the most interesting chapters of helminthology in the coming years.

A classic problem of molecular helminthology and particularly of all the research topics mentioned above is the lack of suitable *in vitro* cultivation systems for parasitic helminths and, mainly, for the human parasitic species among them. Because of their highly complex life cycles and their close adaptation to the host environment, it is traditionally very difficult to maintain and study parasitic helminths under *in vitro* conditions. As a consequence of this, cell lines of parasitic helminths or methods to genetically manipulate these organisms are currently not available, which poses a major obstacle to investigations on the molecular basis of host-helminth interactions. However, at least for several model systems such as the nematode *Strongyloides stercoralis* or the flatworm *Schistosoma mansoni*, the respective methodology is emerging and steadily improved. Another suitable system for *in vitro* cultivation of a parasitic helminth that holds great potential for the development of cell lines and methods of genetic manipulation is discussed by Spiliotis and Brehm (**Chapter 17**) and deals with the cestode *Echinococcus multilocularis*.

Chapter 17

Axenic *In Vitro* Cultivation of *Echinococcus multilocularis* Metacestode Vesicles and the Generation of Primary Cell Cultures

Markus Spiliotis and Klaus Brehm

Abstract

Parasitic helminths are a major cause of disease worldwide, yet the molecular mechanisms of host-helminth interaction and parasite development are only rudimentarily studied. A main reasons for this lack of knowledge are the tremendous experimental difficulties in cultivating parasitic helminths under defined laboratory conditions and obtaining sufficient amounts of parasite material for molecular analyses. For one member of this neglected group of pathogens, the fox-tapeworm *Echinococcus multilocularis*, we have established and optimized *in vitro* cultivation systems by which the major part of the parasite's life cycle, leading from early metacestode vesicles to the production of protoscoleces, can be mimicked under laboratory conditions. The methodology comprises co-cultivation systems for host cells and parasite larvae by which large amounts of parasite vesicles can be generated. Furthermore, we have established an axenic (host cell–free) cultivation system that allows studies on the influence of defined host factors on parasite growth and development. On the basis of this system, the isolation and maintenance of primary *Echinococcus* cells that are devoid of overgrowing host cells is now possible. The availability of the primary cell culture system constitutes a first step toward the establishment of genetic manipulation methods for the parasite that will be of great interest for further research on infection strategies and development of *Echinococcus* and other cestodes.

Key words: *Echinococcus*, cestode, helminth, parasite, in vitro cultivation, feeder cells, primary cells, development.

1. Introduction

Of the 2 billion helminth infections in humans worldwide, those that involve the larval stages of cestodes such as the fox-tapeworm *Echinococcus multilocularis*, the dog-tapeworm *Echinococcus granulosus*, or the pork-tapeworm *Taenia solium* are among the most serious and life-threatening *(1)*. In all three cases, infection

Steffen Rupp, Kai Sohn (eds.), *Host-Pathogen Interactions*, DOI: 10.1007/978-1-59745-204-5_17,

of the intermediate host is initiated by oral uptake of infective eggs containing the oncosphere larval stage. Upon hatching in the small intestine of the host, the oncosphere penetrates the intestinal wall and gains access to the host's viscera where the parasite further develops toward the metacestode (also called *cysticercus* or *hydatid cyst* in the case of *T. solium* and *E. granulosus*, respectively). This developmental stage eventually gives rise to the protoscolex, which is the infective form for the definitive host. In *T. solium* infections, muscle and brain are the preferred target organs for cysticercus development, whereas *Echinococcus* metacestodes preferentially target the liver (*E. multilocularis, E. granulosus*) and the lung (*E. granulosus*). In all three cases, the molecular basis of organ-tropism, host-parasite interaction, and immune evasion (cestode larvae reside for years to decades within the host) is not known. Respective research, which is vital for the development of novel cestodicidal drugs, first requires the establishment of suitable *in vitro* cultivation systems.

From the practical point of view, *E. multilocularis* displays several advantages over the other two species that qualify it as a laboratory model system for molecular studies on host-cestode interactions. The most important feature is that *E. multilocularis* metacestode vesicles grow much faster and are easier to handle in experimental systems than *T. solium* cysticerci or *E. granulosus* hydatid cysts. Furthermore, *E. multilocularis* larval material can be routinely kept in the laboratory for several years through serial passages in small rodents, which is not possible in the case of *T. solium* or *E. granulosus (2)*.

The first attempts to establish *in vitro* cultivation systems for *E. multilocularis* larvae were undertaken in the 1990s by Hemphill and Gottstein *(3)* and Jura et al. *(4)* using co-cultivation of host feeder cells and metacestode vesicles. However, for the large-scale production of parasite vesicles with comparable maturity or for studies concerning the influence of defined host factors on parasite development, both of these methods display considerable drawbacks. The most important disadvantage is the continuous presence of host cells in both systems (fibroblasts in the "tissue block" method of Hemphill et al.; hepatocytes in the "sandwich configuration" method of Jura et al.) because, upon addition of defined host factors to the culture, it is not possible to clearly distinguish between direct effects on parasite development and indirect effects that are mediated by host cells.

As a first step to improve existing *E. multilocularis* cultivation systems, we have developed a method in which metacestode vesicles are incubated in liquid culture with host feeder cells (hepatocytes). In this system, it is crucial that the parasite is continuously co-cultivated with actively proliferating feeder cells; that is, freshly trypsinized hepatocytes are added weekly, whereas "old" feeder cells are removed. This co-cultivation system allows the large-scale production of metacestode vesicles that display a comparable degree of maturity *(5)*.

The metacestode vesicles obtained in the co-cultivation system are then used to set up axenic cultures. We found that long-term cultivation in the absence of feeder cells is only possible when reducing, oxygen-free conditions are applied; that is, when reducing substances such as β-mercaptoethanol and L-cysteine are added to the culture medium and nitrogen is used as the gas phase *(6)*. Under these conditions, feeder cells are eliminated within 1 week of cultivation. After this first step of *axenization*, metacestode vesicles can be kept for several weeks in the presence of serum-containing medium. However, vesicle growth and differentiation toward the protoscolex stage can only be obtained in the presence of conditioned medium that has previously been incubated with actively proliferating feeder cells, indicating that host cells secrete growth factors that are necessary for parasite development *(6)*.

The axenic cultivation system displays several experimental advantages over classic co-cultivation methods. First, vesicles can be generated that are devoid of host cells allowing, for example, the isolation of *Echinococcus* genomic DNA and the establishment of cDNA libraries that are essentially free of host contamination *(7, 8)*. Second, defined host factors can be added to the culture medium and all effects that are measured regarding parasite growth and development can be attributed to a direct action of the growth factors on parasite receptors (or of inhibitors on parasite molecules) *(9)*. Third, through trypsinization of metacestode vesicles, "pure" preparations of *Echinococcus* primary cells can be established. In former attempts to establish such cell cultures (from parasite material after growth in laboratory hosts or in co-cultivation systems), the slowly growing parasite cells were routinely overgrown by host cells (e.g., fibroblasts) and degenerated very fast due to the absence of reducing conditions *(6, 10, 11)*. We observed that *Echinococcus* primary cells can be kept for several weeks in culture when reducing conditions are applied and when parasite vesicle fluid is used as the growth medium. On the basis of this culture method, it should, in the near future, be possible to establish genetic manipulation methods for *E. multilocularis*.

In this chapter, we will outline in detail the experimental steps that are necessary to keep *E. multilocularis* in laboratory rodents, the setup of co-cultivation systems with host hepatocytes, the establishment of axenic cultures including experimental design to study the effects of host growth factors on *Echinococcus* development, and the generation of *Echinococcus* primary cell cultures.

2. Materials

2.1. Maintenance and Isolation of In Vivo Cultivated E. multilocularis Larval Tissue

1. 200 mL 1x PBS (phosphate-buffered saline: 15 mM NaH_2PO_4, 100 mM NaCl, 85 mM Na_2HPO_4, pH 7.4; autoclave before use).
2. Antibiotic stock solution: Ciprofloxacin (2 mg/mL; Bayer).
3. Sterile tweezers, scalpels, Petri dishes, and plastic beakers (250 mL).
4. Sterile tea strainer (steel), sterile syringe (10 or 20 mL).

2.2. In Vitro Co-culture of E. multilocularis Metacestode Vesicles with Host Cells

1. Dulbecco's Modified Eagle's Medium, 4.5 g glucose/L (DMEM, Biochrom, cat. FG 0435), supplemented with 10% heat inactivated fetal bovine serum (FBS, Biochrom, cat. S 0115).
2. Antibiotic stock solution (store at −20°C): penicillin/streptomycin solution (PenStrep solution, Biochrom, cat. A 2212, stock 10 mg/mL each, use 10 μL/mL medium).
3. Trypsin/ethylenediamine tetraacetic acid solution (Trypsin/EDTA, Biochrom, cat. L2143).
4. 200 mL 1 x PBS (*see* **Section 2.1**).
5. Sterile Falcon/Greiner tubes (50 mL); sterile plastic tea strainers; sterile plastic beaker (250 mL).
6. Sterile cell culture flasks (75 cm^2).
7. Sterile plastic disposable pipettes (10 mL).

2.3. Axenic In Vitro Cultivation of E. multilocularis Metacestode Vesicles

1. Dulbecco's Modified Eagle's Medium, 4.5 g glucose/L (*see* **Section 2.2**).
2. Antibiotic stock solution: penicillin/streptomycin solution (*see* **Section 2.2**).
3. Trypsin/EDTA solution (*see* **Section 2.2**).
4. Trypan blue solution (0.5% in 1x PBS), sterile filtered (or trypan blue 0.5%, Biochrom, cat. L 6323).
5. β-mercaptoethanol (Sigma-Aldrich, cat. M6250), 1:100 diluted in H_2O, sterile filtered, stored as aliquots at −20°C.
6. Bathocuproine disulfonic acid (Sigma, cat. B-1125), 10 mM in H_2O, sterile filtered, stored as aliquots at −20°C.
7. L-Cysteine (Sigma, cat. C-1276), 100 mM in H_2O, sterile filtered, stored as aliquots at −20°C.
8. Sterile 1xPBS (~1 L; *see* **Section 2.1**).
9. Bottle top filters (500 mL), pore size 0.2 μm, (Nalgene, cat. 291-4520).

10. Sterile cell culture flasks (175 cm^2); sterile cell culture flasks (75 cm^2).
11. Sterile plastic disposable pipettes (10 mL).
12. Plastic tea strainers (autoclaved); sterile 250 mL plastic beakers.
13. Nitrogen (N_2); gas canister with flexible tube.
14. Provisional sieves made of sterile metal tubes and medical gauze (*see* **Note 7**).

2.4. Studying the Influence of Defined Growth Factors on Echinococcus In Vitro Development

1. Dulbecco's Modified Eagle's Medium (4.5 g glucose/L), containing β-mercaptoethanol, bathocuproine disulfonic acid, L-cysteine, antibiotic stock solution, and 1x PBS (250 mL) as described in **Section 2.3**.
2. Sterile,15 mL tubes (scaled; e.g., TPP, cat. 91115).
3. 24-well cell culture trays (sterile); Ziploc freezer bags.
4. Nitrogen (N_2); gas canister with flexible tube.

2.5. Isolation and In Vitro Cultivation of E. multilocularis Primary Cells

1. 30-μm sieve (polyester tissue conically attached in a beaker), polyester filter tissue 30 μm (Hartenstein, cat. PES 5); autoclaved.
2. Trypsin/EDTA solution (*see* **Section 2.2**) prewarmed to 37°C.
3. 30-μm sieve (polyester tissue conically attached in a beaker), polyester filter tissue 30 μm (Hartenstein, cat. PES 5); autoclaved.
4. Sterile plastic tea strainer and beakers (250 mL).
5. Ziploc freezer bags.

3. Methods

3.1. Maintenance and Isolation of In Vitro Cultivated E. multilocularis Larval Tissue

E. multilocularis metacestode tissue, isolated from naturally infected hosts (small rodents) or human patients, can be maintained for years to decades in laboratory rodents through serial intraperitoneal passages. For laboratory maintenance, we strongly recommend use of Mongolian jirds (*Meriones unguiculatus*; gerbils) as they are highly permissible hosts for *Echinococcus* and are easy to handle. Upon intraperitoneal infection with ~1 mL of parasite suspension, parasite development takes about 2 to 3 months until the animal has to be sacrificed. Up to 30 g of *Echinococcus* larval material (containing metacestode tissue and protoscoleces) can be isolated from one infected jird for subsequent use in both *in vitro* and *in vivo* cultivation (*see* **Note 1**).

1. Sacrifice jird using CO_2 (do not use decapitation or drug administration) and extract parasite material immediately. Soak the dorsal hide with 70% ethanol and open peritoneum under sterile conditions with tweezers and scalpel. Extract metacestode tissue with a fresh pair of tweezers and scalpel and place it into a Petri dish. Avoid contamination with gerbil tissue, particularly intestine, stomach, gall bladder, esophagus, or urethra.
2. Cut metacestode tissue into thin slices using a scalpel. Put a metal tea strainer on a plastic beaker (do not use a glass beakers—risk of breakage!). Put the metacestode slices in the tea strainer and strain through the sieve using the back end of a syringe plunger. Rinse every now and then with 1x PBS. Remaining material that is hard to strain is usually host tissue (pink colored) and can be discarded. The isolated parasite material consists of small, intact vesicles, protoscoleces that are still invaginated, debris of larger metacestode vesicles, and different host cells (e.g., fibroblasts).
3. Washing: Pour the strained material into 50-mL Falcon tubes. Top up to 50 mL with 1x PBS and let sediment. After ~10 min, the tissue forms a precipitate at the bottom of the tube. Pour off supernatant by decanting. In case of more than 15 mL of metacestode material, repeat washing once (otherwise one washing step is sufficient).
4. Antibiotic treatment: Add 0.5 vol of 1x PBS (and not more than 0.5 vol!) to each parasite tissue pellet. Add ciprofloxacin (stock 2 mg/mL, use 6 μL/mL), mix well and incubate overnight at 4°C to 8°C. Do not incubate for longer than 12 to 16 h if the material is to be used for *in vitro* cultivation (*see* **Section 3.2**).
5. *In vivo* infection: Wash parasite tissue 3 times with 1x PBS to reduce the amount of antibiotics. Inject into gerbils intraperitoneally (100 μL to 1 mL per infection) using a 0.9-mm-diameter injection needle (20 gauge).

3.2. In Vitro Co-culture

For *in vitro* co-cultivation, we have tested several cell lines as possible feeder cells *(6)*. Although cell lines of nonhepatic origin can generally be used for cultivation (*see* **Note 2**), best results are obtained with hepatocytes, particularly rat Reuber RH- hepatoma cells (ATCC no. CRL-1600). The method described below allows for a large-scale production of metacestode vesicles (up to 200 mL of vesicles from 1 mL of starting material) that are all in the same developmental stage (or maturity) (**Figs. 17.1** and **17.2**). During the first 3 to 4 weeks of co-cultivation, small metacestode vesicles will develop which will reach a diameter of ~2 to 5 mm until protoscolex production starts (*see* **Note 3**).

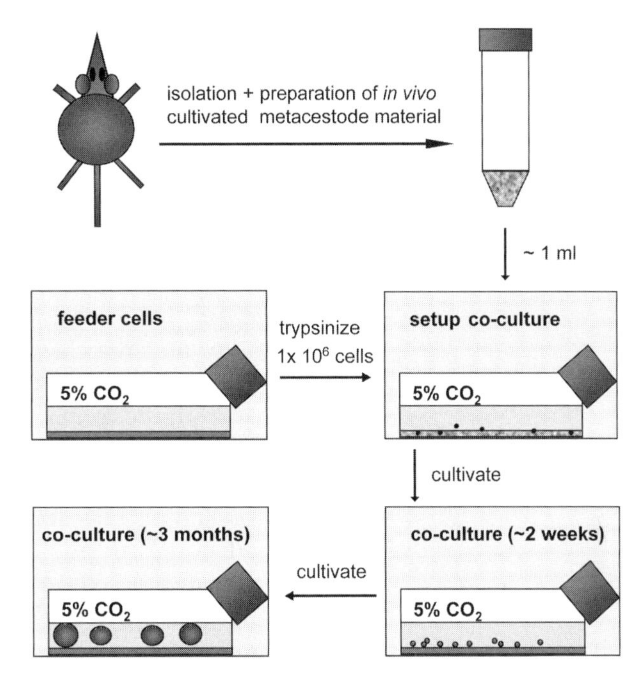

Fig. 17.1. Flowchart summarizing the setup and maintenance of *E. multilocularis* metacestode co-cultures with host feeder cells.

1. Parasite material isolated from a laboratory host (*see* **Section 3.1.4**) is first sieved through a plastic tea strainer to remove larger particles (a plastic tea strainer is used because it is meshed finer than metal tea strainer). Add parasite material to the mesh and pipette up and down several times. Rinse with 1x PBS but ensure that the total volume of parasite material plus PBS does not exceed 50 mL. Pour the small parasite particles mixed with PBS in a 50-mL tube and allow the *Echinococcus* material to pellet for ~20 min. Carefully remove supernatant with a pipette.

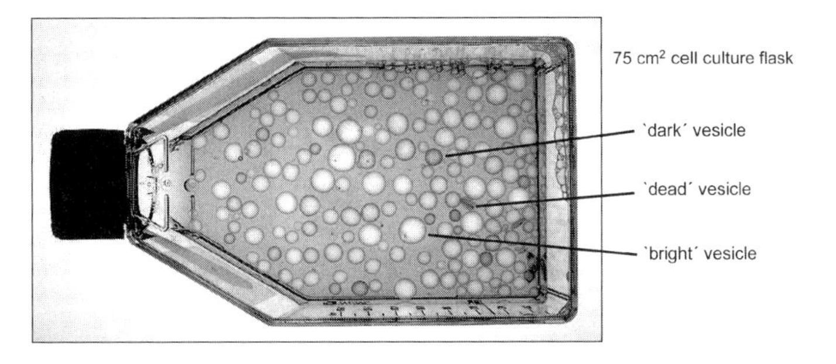

Fig. 17.2. *E. multilocularis* metacestode vesicles in 75 cm^2 cell culture flask after 8 weeks of co-culture with host feeder cells. Indicated are collapsed ("dead") vesicles that can still contain living cells but that are unsuitable for axenic cultivation. Likewise, vesicles with "dark" appearance are not optimal for axenic cultivation. Best results are obtained with "bright," floating vesicles.

2. Culture setup: To every 1 mL of the fine metacestode pellet add 50 mL DMEM (including FBS and antibiotics) and transfer to a 75 cm^2 cell culture flask, then add 1×10^6 freshly trypsinized Reuber RH^- cells per flask (for trypsin treatment of rat Reuber RH^- cells, *see* **Note 4**).
3. Medium change: Change the medium for the first time about 1 to 2 weeks after culture setup. Earlier change is necessary if the medium becomes acidified (indicator switch to orange, pH <6.5). Always wait until the latest possible time point for medium change (i.e., never change before the medium turns orange). For the medium exchange, remove the culture from the culture flask by decanting into a 50-mL tube and let stand for ~20 min. Metacestode material will precipitate. Discard supernatant and replace with fresh DMEM (including FBS and antibiotics) to 50 mL volume. Add new 1×10^6 freshly trypsinized RH^- cells and pour into the previously used cell culture flask (you can reuse the cell culture flasks several times). Repeat the process every week (remember to always add 1×10^6 freshly trypsinated RH^- cells!).
4. Splitting cultures: As soon as the metacestode vesicles in the flask have reached a volume of 5 mL, transfer half of it to a new culture flask and add 1×10^6 freshly trypsinized RH^- cells. Further splitting must be done at volumes of 10, 15, and 20 mL (*see* **Note 5**). Parasite growth is impaired by too many vesicles per flask, mostly because protoscolex production is induced (*see* **Note 6**).
5. Culture sieving: If very "clean" metacestode vesicles are required (e.g., for an axenic cultivation), the vesicles need to be sieved with a plastic tea strainer to remove small particles (i.e., small parasite vesicles and host tissue). The vesicles must be at least 5 weeks of age and should have a size of at least 3 to 4 mm. Pour the vesicles very carefully into a plastic tea strainer, from the strainer into a beaker, and then into a 50-mL tube add some DMEM (including FBS and antibiotics) to facilitate pouring. Bring the volume to 50 mL by adding DMEM (including FBS and antibiotics). Add 1×10^6 freshly trypsinized RH^- cells, then pour the vesicles into a fresh 75 cm^2 cell culture flask.

3.3. Axenic In Vitro Culture

Growth and differentiation of *E. multilocularis* metacestode vesicles under axenic conditions only occurs in the presence of host cell (hepatocyte)-conditioned medium *(6)*. Prior to starting an axenic culture, this medium has to be prepared in sufficient amounts. A flowchart for medium preparation and vesicle isolation is displayed in **Fig. 17.3**.

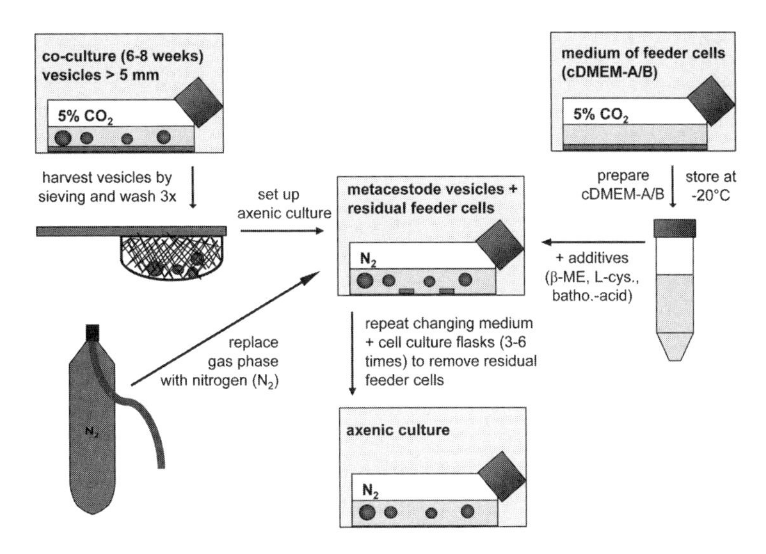

Fig. 17.3. Flowchart summarizing setup and maintenance of *E. multilocularis* metacestodes in axenic (host-cell free) culture.

3.3.1. Preparing Conditioned Medium for Axenic Cultivation

1. Depending on the experimental need, prepare either cDMEM-A or cDMEM-B medium.
   (a) cDMEM-A medium is used for fast metacestode vesicle growth. However, up to 30% of the vesicles will degenerate within 6 weeks of axenic cultivation. For supernatant preparation, seed 1×10^6 RH^- cells in a 175 cm^2 cell culture flask with 50 mL DMEM (including FBS, antibiotics) and incubate for 7 days at 37°C with 5% CO_2.
   (b) cDMEM-B medium is used for slower metacestode growth. However, more than 90% of the vesicles will survive the first 6 weeks of axenic cultivation. Seed 2×10^7 RH^- cells in a 175 cm^2 cell culture flask with 50 mL DMEM (including FBS, antibiotics) and incubate for 3 days at 37°C with 5% CO_2.
2. At the end of incubation, remove medium from hepatocytes, centrifuge for 10 min at $3000 \times g$, and pool the supernatant of several flasks in a new container. Sterilize medium by sterile filtration using an 0.2-μm filter (if the volume exceeds 50 mL, bottle top filters are comfortable), then aliquot (40 mL cDMEM-A/B per 50-mL tube) and store at −20°C (shelf life at least 3 months; freeze tubes in upright position!).
3. Keep medium frozen until use. Thaw conditioned medium 30 min before culture setup and add 1 μL/mL β-mercaptoethanol (stock solution 1:100, diluted in H_2O), 1 μL/mL bathocuproine disulfonic acid (stock solution 10 mM), 1 μL/mL L-cysteine (stock solution 100 mM), and antibiotics

(penicillin/streptomycin, stock 10 mg/mL each, use 10 μL/mL medium). cDMEM-A/B medium including reducing substances and antibiotics will subsequently be referred to as Ax-cDMEM-A/B.

3.3.2. Setting Up an Axenic In Vitro Culture

1. The choice of co-cultivated metacestode vesicles is very important. Use vesicles that have been cultivated at least for 6 to 8 weeks in co-culture (*see* **Section 3.2.5**) with a diameter of at least 5 mm.
2. Sieve the co-cultivated metacestode vesicles carefully into a plastic tea strainer. Pour them carefully into a 250-mL beaker (turn the tea strainer upside down, put it close to the inner wall of the beaker, and let the vesicles carefully slide down the beaker wall).
3. Pour 1x PBS over the back side of the strainer to wash off any remaining vesicles. Fill beaker to 200 mL with 1x PBS. Sway beaker softly and decant again into the tea strainer. Repeat this step 3 times to remove feeder cells.
4. After the last washing step, pour metacestode vesicles carefully into a new beaker (wash off any vesicles remaining in the sieve with several milliliters of Ax-cDMEM-A/B).
5. Place metacestode vesicles from the beaker into a 50-mL tube by decanting. Determine total volume and split the vesicles into several tubes if required. In the case of small vesicles (~5 mm in diameter), start with 10 mL of total parasite vesicle volume per 50 mL tube. In the case of medium-sized vesicles (5 to 7 mm), use 15 mL total vesicle volume per 50-mL tube, and in the case of larger vesicles (>7 mm), use 20 mL per 50-mL tube (the number of parasite cells per tube is critical, not the volume). Fill the tube(s) with Ax-cDMEM-A/B to 50 mL and pour carefully into 75 cm^2 cell culture flasks. Replace the air in the cell culture flasks with nitrogen (put a sterile 10-mL pipette in the flexible tube from the nitrogen bottle, open the valve slightly, and dip the tip of the pipette into the medium in the cell culture flask. Keep the pipette in the medium until the emerging bubbles fill the flask completely). Close the lid and incubate at 37°C.
6. Anaerobic growth of metacestode vesicles during axenic cultivation acidifies the medium very fast. A switch of the indicator to orange within 2 days is a good sign for vesicle proliferation. Under optimal conditions, medium must be changed every 2 days. Pour the vesicles from the cell culture flask into a 50-mL tube and drain the medium with a metal tube and medical gauze (*see* **Note 7**). Fill the tube with fresh Ax-cDMEM-A/B and pour into a clean 75 cm^2 flask. Fill in nitrogen (*see* **Section 3.3.2.3**) and incubate at 37°C for

2 more days. Fill approximately 20 mL DMEM (including FBS, antibiotics) into the previously used flask and incubate at 37°C (the culture is *axenic* if no adherent hepatocytes grow in the "old" bottle after a few days). Repeat medium change until the culture is axenic (about three to six changes). Cell culture flasks can be re-used once the culture is free of feeder cells (*see* **Note 8**). Particularly in axenic cultures, which require significant resources (material and time) for setup and maintenance, strict care has to be taken to avoid any contamination (*see* **Note 6**).

3.4. Studying the Influence of Defined Growth Factors on Echinococcus In Vitro Development

The influence of defined factors (e.g., growth factors or inhibitors) on growth, differentiation, or vesicle survival is difficult to assess in cell culture flasks. The following experimental setup allows easier measurement of vesicle volume and protoscolex development in axenic culture (**Fig. 17.4**). For this, axenically precultivated vesicles are transferred to either 15-mL tubes (with volume scale), which allows facilitated measurement of vesicle volume (e.g., in growth fact or studies) or to 24-well cell culture trays, which facilitates microscopic analysis (e.g., to measure the effects of inhibitors). The type of vesicles you use for these experiments is critical. Do not use vesicles with less than 5 mm. Reducing conditions have to be maintained throughout the cultivation procedures.

1. Fill 1 mL of hepatocyte-conditioned medium (Ax-cDMEM-A/B) into every 15-mL tube or every well of the cell culture tray. The metacestode vesicles will be individually picked and transferred to the tube (or well).
2. First, sieve the metacestode vesicles carefully over a plastic tea strainer and transfer them carefully into a beaker. Add 1x PBS to cover the vesicles. Damaged vesicles appear pink, yellow, or red. These are not suitable. Intact vesicles are translucent and should be chosen. Pour the vesicles from the beaker carefully into a Petri dish and pick individual vesicles carefully

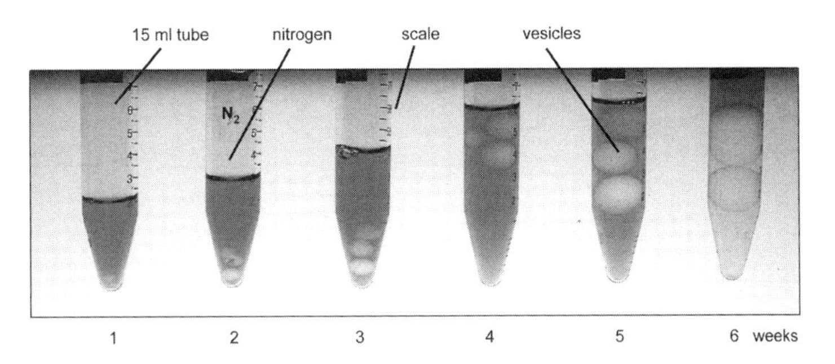

Fig. 17.4. *E. multilocularis* metacestode vesicles axenically cultivated under growth-promoting conditions (with conditioned hepatocyte medium) in 15-mL tubes for 6 weeks.

into the Falcon tube (or culture well) using a truncated (blue) 1000-μL automatic pipette tip (*see* **Note 9**).

3. Each tube with parasite needs a nitrogen gas-phase. Put a sterile 10-mL pipette into the flexible tube of the nitrogen gas canister, open valve slightly, and hold the tip of the pipette for about 3 s in a distance of ~1 cm above the fluid in the tube. Close tubes tightly and incubate at 37°C for 48 h. In the case of 24-well cell culture trays, a modified procedure has to be applied: place the trays into a Ziploc freezer plastic bag, fill the bag with nitrogen, close the Ziploc, and incubate at 37°C for 48 h.
4. Prepare media for positive control, negative control, and test tubes in sufficient amounts (1 mL per Falcon tube or well; the number of tubes/wells depends on the number of substances to be tested). As a positive control for vesicle growth, use Ax-cDMEM-A/B. As a negative control use DMEM (without FBS and additives but with antibiotics). Vesicles in this control medium should survive and remain intact for at least 7 days (survival is supported by hydatid fluid within the vesicles). If the vesicles die earlier, damage had been inflicted during the transfer process, and the experiment has to be started anew. For test media, use Ax-cDMEM-A/B and include the test substance (e.g., growth factor, hormone, or inhibitor) in relevant concentrations.
5. Check metacestode vesicles in control and test tubes microscopically for damage or contamination and use only tubes with intact, noncontaminated vesicles. Try to keep the groups comparable (same vesicle size distribution in the groups). After removing "old" medium from the vessel, fresh control or test medium is added. Do not change medium from more than 20 vessels simultaneously to avoid that vesicles are too long without covering. To remove the previous medium, put a (blue) 1000-μL pipette tip on a 10-mL plastic pipette and remove medium carefully by pipetting. Keep the tube tilted and remove medium only at the fluid surface. It is not crucial to remove the medium completely. Do not get close to the vesicles to avoid damage. Add 1 mL of the prepared medium (depending on test group) to every vessel, again changing the pipette for every different medium. Afterwards, add nitrogen (*see* **Section 3.3.2.3**). Close the tubes or the Ziploc freezer bag tightly and incubate again at 37°C for 48 h. Change medium every second day until the end of the experiment. Total vesicle volume and the production of protoscoleces or brood capsules should be measured throughout the experiment.

3.5. Isolation and In Vitro Cultivation of E. multilocularis Primary Cells

E. multilocularis metacestode primary cells that are not overgrown by host cells can only be established on the basis of axenically cultured vesicles (due to the absence of host cells). For long-term primary cell maintenance, it is important to apply reducing conditions as already outlined above for axenic cultivation. Under oxygen-rich culture conditions, primary cells will degenerate within 2 to 3 days of culture. Primary cell cultivation also requires hydatid fluid of axenically cultured vesicles as the growth medium (i.e., a medium in which *Echinococcus* cells reside *in vivo*). This medium has to be prepared in sufficient amounts before setup of primary cell cultures (**Fig. 17.5**).

3.5.1. Preparation of Hydatid Fluid as Growth Medium

Only hydatid fluid of fast-growing metacestode vesicles from axenic culture should be used. First, sieve the axenically cultivated vesicles from several flasks carefully over a plastic tea strainer and place the vesicles into a 250-mL beaker. Fill with 1x PBS to 200 mL, sway

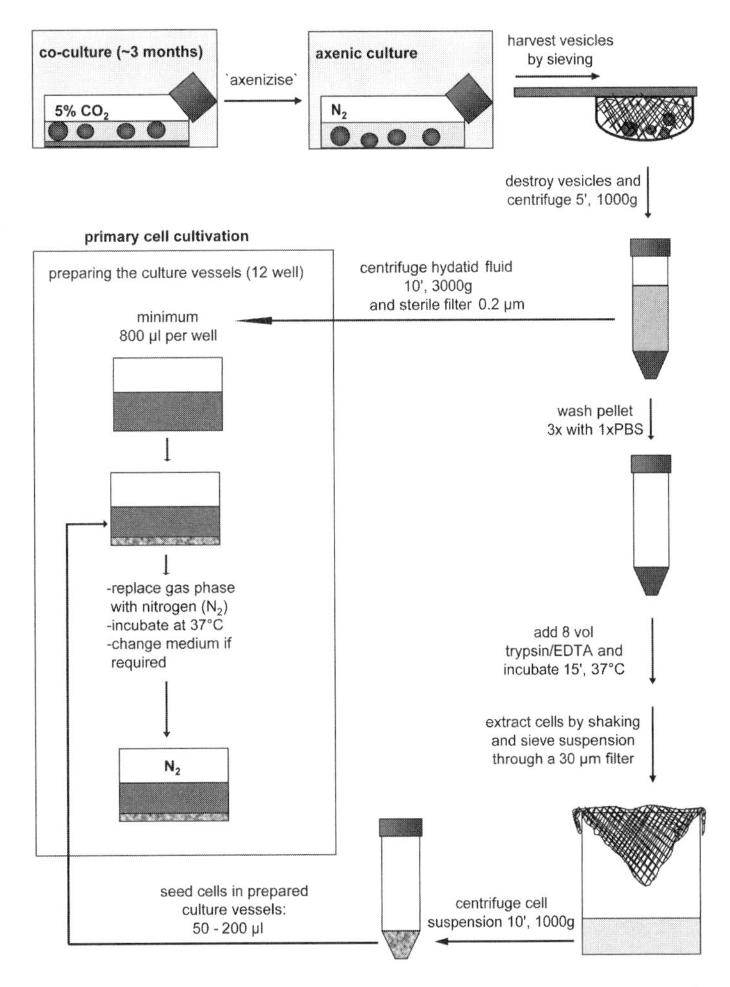

Fig. 17.5. Flowchart summarizing setup and maintenance of *E. multilocularis* primary cell culture.

softly, and sieve again with the strainer. Pour the washed vesicles (without 1x PBS) into a fresh beaker. Use a sterile 10-mL plastic pipette to transfer the vesicles into 50-mL tubes. Larger vesicles will be physically damaged through this treatment releasing hydatid fluid. Centrifuge Falcon tubes (containing hydatid fluid, damaged vesicles, and smaller, intact vesicles) for 5 min at 2000 × *g*, then pour supernatant (hydatid fluid) slowly into a fresh 50-mL tube. The supernatant will still contain smaller vesicles that are intact. These can be separated from the hydatid fluid by another filtration step using a tea strainer. The isolated hydatid fluid must then be sterile filtered (0.2-μm filter) and can be stored for several months at −80°C. Upon thawing, add penicillin/streptomycin stock solution (10 μL/mL) and use the hydatid fluid directly as growth medium (i.e., no further additives such as β-mercaptoethanol or L-cysteine have to be added).

3.5.2. Choosing Vesicles for Primary Cell Culture

The choice of suitable vesicles is a very critical issue in primary cell cultivation. Best results are obtained with "old" vesicles that have been kept in co-culture with hepatocytes for at least 3 months. These vesicles have to be in active growth phase (i.e., should increase in volume about 1.5-fold within 1 week and have a diameter of more than 5 mm) and should not contain brood capsules or protoscoleces.

3.5.3. Axenization of Vesicles

Suitable vesicles of co-cultures are manually picked using a truncated (blue) 1000-mL automatic pipette tip (*see* **Note 8**) and are subjected to axenization as described above (*see* **Section 3.3**). Only use vesicles for primary cell culture that are fully devoid of host cells as these would rapidly overgrow the parasite cells.

3.5.4. Primary Cell Isolation

Similar to the procedure described above for hydatid fluid preparation (*see* **Section 3.5.1**), use a 10-mL pipette to transfer the vesicles into 50-mL Falcon tubes. Larger vesicles will be physically damaged through this procedure, resulting in "vesicle ghosts" that contain parasite cells (of the germinal layer) attached to the outer laminated layer (smaller vesicles may sometimes require blue 1000-μL pipette tips for physical destruction). Centrifuge the tubes for 5 min at 2000 × *g*, then pour the supernatant slowly into a fresh 50-mL tube (this hydatid fluid, sterile filtered as mentioned above [*see* **Section 3.5.1**], can be used as growth medium for the first days of primary cell culture). Vesicle ghosts will remain at the bottom as a pellet.

3.5.5. Washing Vesicle Pellets

Fill tube containing the vesicle ghosts to 50 mL with 1x PBS and sway softly to loosen the pellet. Let stand for 5 min at room temperature, then centrifuge for 5 min at 1000 × *g*. Discard supernatant and repeat washing step twice.

3.5.6. Trypsin Digestion of Vesicle Ghosts

Add 8 volumes of prewarmed Trypsin/EDTA solution (37°C) to the pellet. Sway softly until the pellet is loosened and incubate for 15 min at 37°C. Repeat the swaying several times during the incubation, then shake the tube very softly, turning it slowly overhead and back for approximately 3 min. The first cells will detach from the laminated layer (turning the solution turbid). Centrifuge the tube for 2 min at 400 × *g* and decant the supernatant into a fresh 50-mL tube (primary cell preparation A). Fill the tube containing the remaining vesicles to 35 mL with 1x PBS and shake with medium intensity for 3 min. Repeat the centrifugation step and decant the supernatant (primary cell preparation B). Fill the tube once more to 35 mL with 1x PBS and shake vigorously until only transparent ghosts of the metacestode vesicles (mostly swimming on top) are visible (primary cell preparation C).

3.5.7. Cell Sieving

Sieve the collected supernatants (primary cell preparations A, B, and C) through a 30-μm sieve (polyester tissue conically attached in a beaker). Use 1x PBS to rinse as required. Distribute the collected fluid containing primary cells to 50-mL Falcon tubes and centrifuge at 1000 × *g* for 5 to 10 min. Decant the supernatant (which should be almost clear). Resuspend the primary cell pellets (very!) carefully with a 1000-μL pipette and pool them. The primary cells can now be seeded into 12-well cell culture trays that have been preloaded with 800 μL hydatid fluid (per well).

3.5.8. Seeding Primary Cells

Add 50 to 200 μL of the primary cell suspension in each well of the cell culture tray (fill empty wells with water or 1x PBS, otherwise the culture wells will dry very fast). Put culture tray in a Ziploc freezer plastic bag and fill with nitrogen. Close Ziploc bag and incubate at 37°C.

3.5.9. Growth and Medium Change

Echinococcus primary cells will not adhere to the culture vessel but remain in suspension (anchorage independent growth). Within 1 week after seeding, the cells should form aggregates, which is a good indicator that the cell extraction was successful. For further cultivation, add 0.5 mL of hydatid fluid to each well weekly (always refill the Ziploc bags with nitrogen!). After about 3 to 4 weeks, remove two thirds of the accumulated hydatid fluid and replace by fresh 0.5 mL hydatid fluid. During this procedure, try to remove as few primary cells as possible (if necessary, these can be recollected from the removed hydatid fluid through centrifugation as outlined above).

4. Notes

1. Different *E. multilocularis* isolates (from natural hosts or from humans) vary significantly in their capacity to produce protoscoleces. Usually, fresh isolates produce large amounts of protoscoleces during maintenance in laboratory mice. However, for as yet unknown reasons, these isolates tend to loose the capacity of protoscolex production after several years of passage in laboratory hosts. The problem can be circumvented by a method introduced by Kamiya and Sato *(12)* in which protoscoleces of such isolates are fed to immunosuppressed jirds where they grow out to egg-producing adult worms. Utilization of these eggs for novel infection of the jird as intermediate host yields isolates that have a restored capacity of protoscolex production *(12)*.

2. Apart from rat Reuber RH^- cells, the following cell lines can be used and give good results: Chinese hamster ovary (CHO) cells (CCL-61), HeLa human epidermoid carcinoma cells (CCL-2), human embryonic kidney (HEK293) cells (CRL-1573), and SH-SY5Y human neuroblastoma cells (CRL-2266) (*see also* Ref. *6*).

3. The capacity of *E. multilocularis* isolates to produce protoscoleces *in vitro* matches that of *in vivo* protoscolex production. In general, isolates should be used that still produce protoscoleces as non–protoscolex producers (such as isolate MP1) must be considered developmental mutants *(5)*.

4. Trypsinization of RH^- cells: Remove medium from confluently grown cell culture flasks and tilt for 5 to 10 min in order to accumulate the remaining medium. Draw off medium with a 10-mL pipette, add Trypsin/EDTA solution (3 mL for a 75 cm^2 flask, 4 mL for a 175 cm^2 flask), and incubate for 10 min at 37°C. Trypsinization is finished when cells can be detached by flicking on the bottle. Remove cells with a 10-mL pipette, resuspend carefully several times, transfer to a 50-mL tube, and add 1 vol DMEM (including FBS and antibiotics) to stop trypsinization. Cells may be stored prior to use for about 1 h at room temperature.

5. Proper splitting is critical to gain large amounts of parasite vesicles. Once the vesicles in a culture flask (75 cm^2) have reached a total volume of 5 mL, they should be distributed to two flasks (2.5 mL each) with feeder cells. Once the vesicles in each of these flasks have reached a volume of 10 mL, they should be distributed to two novel flasks (5 mL each). The next splitting steps are then done when the vesicles per

flask have reached 15 mL (2 × 7.5 mL) and 20 mL. From one initial co-culture flask of 1 mL meshed parasite material, this protocol gives 16 culture flasks of 25 mL parasite vesicles each, after about 10 weeks.

6. Apart from early protoscolex production, contamination of the culture with bacteria or fungi is the most prominent reason for nongrowing parasite vesicles. Because the parasite material originally derives from laboratory animals, contamination with a variety of organisms that do not usually occur in cell culture laboratories can occur. Cultures with nongrowing vesicles should therefore be carefully checked for a variety of different contaminations (e.g., mycoplasma or fungi). Because several antifungal substances are also active against *Echinococcus (13)*, it is not recommended to use these in addition to antibiotics for the prevention of contaminations. Particularly in axenic culture, contaminations with anaerobic organisms (e.g., *Corynebacterium* spp.) have to be considered.
7. Growing metacestode vesicles frequently do not sediment to the bottom of the tube but are floating. In this case, medium cannot be removed by decanting. The problem can be circumvented by using a lab-made sieving device. Cover one end of a short metal pipe (length ~5 cm, diameter ~4 cm, stainless steel) with medical gauze (fix with tape) and autoclave. For the removal of the medium, press the culture tube onto the gauze and pour the medium off. The vesicles will fall back into the tube or, in the case that some remain in the gauze, can be brought back using the rim of the tube.
8. After several weeks of vesicle growth in axenic culture, the absence (or presence) of host cells can also be measured by PCR-based techniques. For this, genomic DNA is isolated from vesicles using commercially available kits (e.g., DNeasy; Qiagen) and PCR reactions are performed for parasite-specific and host cell–specific genes. For parasite-specific PCRs, primers that hybridize to the spliced leader gene emsl are recommended. Host-cell control PCRs can be run with primers specific for β-tubulin encoding genes. Respective primer sequences are available *(14, 15)*.
9. Picking metacestode vesicles: cut off the pointed end of a 1000-μL (blue) automatic pipette tip using a red-hot scalpel (Bunsen burner). This step is important to obtain smooth edges on the pipette tip that will usually not damage metacestode vesicles during picking. Place the truncated tip onto a 1000-μL automatic pipette, choose vesicle, and apply vacuum to suck the vesicle into the tip. Transfer vesicle to new tube (or well) and release vacuum.

Acknowledgments

The authors would like to thank Dirk Radloff and Lucas Sesterhenn for excellent technical assistance as well as Dennis Tappe and Henning Schröder for many helpful suggestions. This work was supported by the Deutsche Forschungsgemeinschaft (SFB 479).

References

1. Brehm, K., Spiliotis, M., Zavala-Gongora, R., Konrad, C. and Frosch, M. (2006) The molecular mechanisms of larval cestode development: first steps into an unknown world. *Parasitol. Int.* **55**, S15–S21.
2. Siles-Lucas, M. and Hemphill, A. (2002) Cestode parasites: application of in vivo and in vitro models for studies on the host-parasite relationship. *Adv. Parasitol.* **51**, 133–230.
3. Hemphill, A. and Gottstein, B. (1996) Immunological and morphological studies on the proliferation of in vitro cultivated *Echinococcus multilocularis* metacestodes. *Parasitol. Res.* **81**, 605–614.
4. Jura, H., Bader, A., Hartmann, M., Maschek, H. and Frosch, M. (1996) Hepatic tissue culture model for study of host-parasite interactions in alveolar echinococcosis. *Infect. Immun.* **64**, 3484–3490.
5. Brehm, K., Wolf, M., Beland, H., Kroner, A. and Frosch, M. (2003) Analysis of differential gene expression in *Echinococcus multilocularis* larval stages by means of spliced leader differential display. *Int. J. Parasitol.* **33**, 1145–1159.
6. Spiliotis, M., Tappe, D., Sesterhenn, L. and Brehm, K. (2004) Long term in vitro cultivation of *Echinococcus multilocularis* metacestodes under axenic conditions. *Parasitol. Res.* **92**, 430–432.
7. Spiliotis, M., Tappe, D., Brückner, S., Mösch, H.U. and Brehm, K. (2005) Molecular cloning and characterization of Ras- and Raf-homologues from the fox-tapeworm *Echinococcus multilocularis*. *Mol. Biochem. Parasitol.* **139**, 225–237.
8. Zavala-Gongora, R., Kroner, A., Bernthaler, P., Knaus, P. and Brehm, K. (2006) A member of the transforming growth factor-b receptor family from *Echinococcus multilocularis* is activated by human bone morphogenetic protein 2. *Mol. Biochem. Parasitol.* **146**, 265–271.
9. Spiliotis, M., Konrad, C., Gelmedin, V., Tappe, D., Brückner, S., Mösch, H.U. and Brehm, K. (2006) Characterization of EmMPK1, an ERK-like MAP kinase from Echinococcus multilocularis which is activated in response to human epidermal growth factor. *Int. J. Parasitol.* **36**, 1097–1112.
10. Howell, M. J. and Matthaei, K. (1988) Points in question: in vitro culture of host or parasite cells? *Int. J. Parasitol.* **18**, 883–884.
11. Yamashita, K., Uchino, J., Sato, N., Furuya, K. And Namieno, T. (1997) Establishment of a primary culture of *Echinococcus multilocularis* germinal cells. *J. Gastroenterol.* **32**, 344–350.
12. Kamiya, M. and Sato, H. (1990) Complete life cycle of the canid tapeworm, *Echinococcus multilocularis*, in laboratory rodents. *FASEB J.* **4**, 3334–3339.
13. Reuter, S., Merkle, M., Brehm, K., Kern, P. and Manfras, B. (2003) Effect of amphotericin B on larval growth of *Echinococcus multilocularis*. *Antimicrob. Agents Chemother.* **47**, 620–625.
14. Brehm, K., Jensen, K. and Frosch, M. (2000) mRNA trans-splicing in the human parasitic cestode *Echinococcus multilocularis*. *J. Biol. Chem.* **275**, 38311–38318.
15. Konrad, C., Kroner, A., Spiliotis, M., Zavala-Gongora, R. and Brehm, K. (2003) Identification and molecular characterization of a gene encoding a member of the insulin receptor family in *Echinococcus multilocularis*. *Int. J. Parasitol.* **33**, 301–312.

Chapter 18

Culture of Exoerythrocytic Stages of the Malaria Parasites *Plasmodium falciparum* and *Plasmodium vivax*

Liwang Cui, Namtip Trongnipatt, Jetsumon Sattabongkot, and Rachanee Udomsangpetch

Abstract

The two most prevalent human malaria parasites, *Plasmodium falciparum* and *Plasmodium vivax*, cause the majority of malaria-related morbidity and mortality. Compared with our knowledge about the erythrocytic stages, we understand little about the liver exoerythrocytic (EE) stages of the human malaria parasites. Our recent development of a hepatocyte line from normal human liver tissue is crucial for successful culturing of the liver stages of both *P. falciparum* and *P. vivax*. This technical advancement should be an important tool for directly studying developmental biology of the EE stages of the human malaria parasites and developing drugs against parasite liver stages.

Key words: Malaria culture, erythrocytic stage, exoerythrocytic development, host cell invasion, drug resistance, reticulocyte, cord blood.

1. Introduction

Hepatocyte infection is the first natural step in the establishment of malaria infections. Once being injected from a female anopheline mosquito, *Plasmodium* sporozoites travel to the liver sinusoid. The sinusoidal cell layer is composed of endothelial cells interspersed with Kupffer cells, resident macrophages of the liver. Before invading a hepatocyte and developing into the next infective stages, a sporozoite traverses through a Kupffer cell and perhaps several other hepatocytes *(1, 2)*. This migration through host cells activates exocytosis of the sporozoite apical organelles, a prerequisite for the successful formation of parasitophorous vacuole *(3)*. Once inside the vacuole of an infected hepatocyte, the sporozoite transforms into a trophozoite, which undergoes schizogony to produce

Steffen Rupp, Kai Sohn (eds.), *Host-Pathogen Interactions*, DOI: 10.1007/978-1-59745-204-5_18,

numerous merozoites. For some malaria parasites such as *Plasmodium vivax* and *Plasmodium ovale*, the development of certain trophozoites is arrested at earlier stages to form dormant cells termed *hypnozoites*, which are responsible for relapses of the disease.

Our knowledge about the exoerythrocytic (EE) development of human malaria parasites is rather limited; most studies on EE stages have used animal malaria models. This is largely due to the absence of a convenient system to culture the EE stages *in vitro*. Complete *in vitro* EE development of human malaria parasites has been achieved using primary hepatocyte cultures *(4–6)*, but primary hepatocytes do not grow continuously in culture and need to be isolated from liver. The human hepatoma cell line HepG2-A16 has been used to culture the EE stages of several strains of *P. vivax (7, 8)*. However, *Plasmodium falciparum* could not achieve complete maturation in this cell line *(9)*. Recently, we established a hepatocyte line, HC-04, from normal human liver tissue that supports the complete EE cycle for both *P. falciparum* and *P. vivax (10)*. This cell line should be useful for studying EE development *in vitro*, determining the gene expression repertoires of two parasites, monitoring drug resistance, and elucidating the mechanism of hypnozoite development.

2. Materials

2.1. Cell Culture

1. HC-04 cell line (ATCC patent deposit no. PTA-3441; ATCC, Manassas, VA).
2. Culture medium (CM): mix equal volumes of Minimum Essential Medium (MEM) and F-12 Nutrient mix (Hams) (Invitrogen, Carlsbad, CA) supplemented with 10% fetal bovine serum (FBS; Invitrogen), 15 mM HEPES (Sigma, St. Louis, MO), 20 mM sodium bicarbonate ($NaHCO_3$), and 15 μM phenol red (Sigma) (*see* **Note 1**).
3. Penicillin (10,000 units/mL) and streptomycin (10,000 μg/mL) (Invitrogen) are added to CM at final concentrations of 100 units/mL and 100 μg/mL, respectively.
4. Phosphate-buffered saline (PBS): dissolve 8 g NaCl, 0.2 g KCl, 0.24 g KH_2PO_4, and 1.44 g Na_2HPO_4 in 800 mL of distilled water, adjust pH to 7.4 with HCl, add distilled water to 1 L, and sterilize by autoclave.
5. 0.25% Trypsin-EDTA solution (Invitrogen).
6. Trypan blue (Invitrogen) is made in PBS at 0.1% (w/v).
7. Culture flasks (25 cm^2) and 15-mL culture tubes (Corning, Acton, MA) and CO_2 incubator used at 37°C, with 5% CO_2 (Thermo Fisher Scientific, Waltham, MA).

8. Hemocytometer (Fisher, Pittsburgh, PA).
9. Dimethyl sulfoxide (DMSO) (Sigma).
10. Cryovial (1.8 mL) (Costar, Cambridge, MA).

2.2. Mosquito Feeding and Collection of Sporozoites

1. Mosquito feeding apparatus (**Fig. 18.1**): circulated water bath (Thermo Fisher Scientific), water-jacketed glass feeders, Baudruche membrane (Joseph Long, Belleville, NJ) or Parafilm, mosquito collection paper cups (John W. Hock, Gainesville, FL), and small elastic bands.
2. 15-mL plastic tubes.
3. Normal human AB serum is obtained from a blood bank.
4. Heparinized tubes.
5. Mosquito sterilization solutions: 70% ethanol; 400 units/mL penicillin, and 400 μg/mL streptomycin solution (Invitrogen); and 2.5 μg/mL amphotericin B (Sigma).
6. Dissection medium: serum-free CM with 400 units/mL penicillin and 400 μg/mL streptomycin.
7. 1.5-mL Eppendorf tubes, plastic tissue grinders (Daigger, Vernon Hills, IL).

2.3. Infection of Hepatocyte Line

1. Invasion medium is prepared from CM supplemented with 10% human serum (heat-inactivated at 56°C for 30 min) and 200 units/mL penicillin and 200 μg/mL streptomycin.
2. 96-well tissue culture plates (Corning).

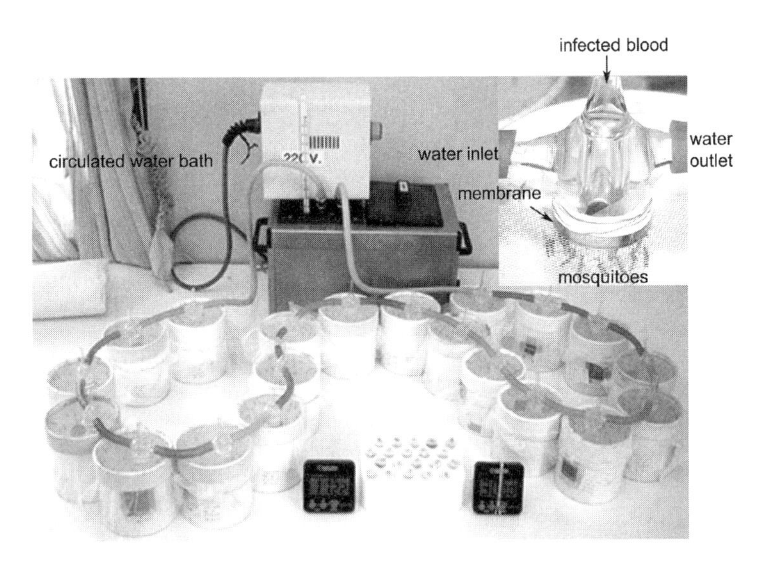

Fig. 18.1. Membrane feeders for mosquitoes. A series of water-jacketed membrane feeders are connected by a water tube to a circulated water bath. In each feeder, an aliquot of infected blood is fed to ~100 mosquitoes, which are kept in a mosquito container. Inset is an enlarged view of a membrane feeder.

2.4. Giemsa Staining

1. Shandon Cytospin Cytocentrifuge and Shandon TPX white filter cards (Thermo Fisher Scientific).
2. Methanol (Sigma), Giemsa Solution (Fisher), phosphate buffer for Giemsa stain (Fisher).
3. Microscope slides (Fisher), immersion oil type B (Carqille Laboratories, Cedar Grove, NJ).

2.5. Indirect Immunofluorescent Assay (IFA)

1. Acetone (Sigma).
2. Blocking buffer: 5% (w/v) skim milk prepared in PBS.
3. Primary monoclonal antibody: anti-circumsporozoite protein (CSP) of *P. vivax* variant VK210 and anti-CSP of *P. falciparum* (Kirkegaard and Perry Laboratories, Gaithersburg, MD) is used at a concentration of 1 μg/mL (*see* **Note 2**). Other specific antibodies such as anti–*P. falciparum* heat shock protein 70 (Hsp70) can be obtained from the malaria reagent depository MR4.
4. Secondary antibodies: FITC-conjugated anti-mouse (Dako Cytomation, Carpinteria, CA) or anti-human IgG (Sigma).
5. Evan's Blue solution (Sigma) for counter staining slides: 0.01% (w/v) solution is prepared in blocking buffer.
6. 4′-6-Diamidino-2-phenylindole (DAPI) (Sigma) for nuclei staining: 2.5 μg/mL solution prepared in blocking buffer.
7. Mounting medium: Prolong antifade kit (Invitrogen).
8. Microscope cover slips (22 × 40 × 0.15 mm) (Fisher).

2.6. Preparation of Reticulocyte-Enriched Blood for Erythrocytic Invasion by P. vivax

1. Complete culture medium (CCM): RPMI-1640 supplemented with L-Glu (Invitrogen) and carbohydrate (Invitrogen), 370 μM hypoxanthine (Sigma), 25 mM HEPES, 10 μg/mL gentamicin (Invitrogen), and 0.225% $NaHCO_3$ (w/v) (Sigma).
2. CF-11 columns are prepared by pouring the CF-11 cellulose powder (Whatman, Middlesex, UK) into 10- to 20-mL syringe tubes. The columns are packed by tapping the side of the syringe tubes, wrapped with aluminum foil, and autoclaved in the upright position. Before use, it needs to be equilibrated with PBS.
3. Plasmodipur filter (Euro-Diagnostica BV, Arnhem, The Netherlands) can be used as an alternative for CF-11 columns.

3. Methods

3.1. Maintenance of HC-04 Cell Line

1. The hepatocyte cell line HC-04, originally established from healthy liver biopsy, is maintained continuously in 25 cm^2 culture flasks in CM supplemented with penicillin and

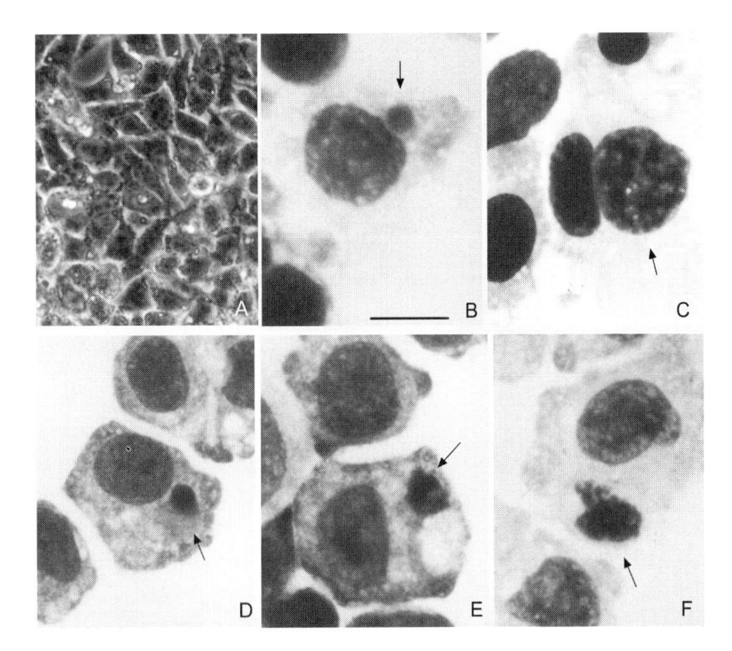

Fig. 18.2. (**A**) HC-04 cells and infection by (**B, C**) *P. falciparum* sporozoites and (**D–F**) *P. vivax* sporozoites. (**A**) A phase contrast view of the HC-04 cells showing morphology of liver parenchymal cells. (**B–F**) Giemsa staining of infected HC-04 cells. Note the flattened cell morphology as the result of the cytospin procedure. (**B**) Four-day old *P. falciparum* trophozoite. (**C**) Fourteen-day-old *P. falciparum* schizont. (**D, E**) Four-day old *P. vivax* trophozoites. (**F**) Fourteen-day-old *P. vivax* schizont. Scale bar = 10 μM. Arrows indicate the parasites. (From Sattabongkot, J., Yimamnuaychoke, N., Leelaudomlipi, S., Rasameesoraj, M., Jenwithisuk, R., Coleman, R. E., Udomsangpetch, R., Cui, L., and Brewer, T. G. 2006. Establishment of a human hepatocyte line that supports in vitro development of the exo-erythrocytic stages of the malaria parasites *Plasmodium falciparum* and *P. vivax*. *Am. J. Trop. Med. Hyg.* **74**, 708–715. Reproduced and modified with permission from The American Society of Tropical Medicine and Hygiene.)

streptomycin. All culture procedures are done under sterile conditions in a laminar flow hood. Daily observation of cell growth is done using an inverted compound microscope (**Fig. 18.2A**).

2. Cells are passaged when approaching 80% to 90% confluence. First, the cell monolayer is washed with PBS 3 times and PBS is aspirated by using a serologic pipette. To loosen cell attachment, 0.25% trypsin-EDTA solution is added and incubated at 37°C for 6 min. The trypsinized cells are collected in 15-mL culture tubes and centrifuged at 70 × *g* for 5 min. The cell pellet is resuspended in CM, and the cells are seeded into a culture flask at a density of 20,000 cells/cm^2. Medium is changed every 48 h (*see* **Note 3**).
3. Cell numbers are quantified by using a hemocytometer under a compound microscope at 100× magnification. Briefly, 1 volume of resuspended cells in CM is mixed with 1 volume of 0.1% trypan blue in PBS and loaded onto a hemocytometer.

The cells are counted in four of the nine large squares to obtain the average number of cells in one large square. The total number is multiplied by 2 (dilution factor) and 10,000 to get the number of cells/mL. Cells stained blue are dead.

4. For storage of the hepatocyte line in liquid nitrogen, cells are trypsinized, washed, and counted. Then, 3×10^6 to 5×10^6 cells are resuspended in 1 mL of 10% DMSO in CM and transferred to a sterile 1.8-mL cryovial. The cells are immediately frozen at −80°C for 4 to 24 h and then moved to liquid nitrogen.
5. To start a culture from the frozen cells, the cryovial is removed from liquid nitrogen tank, uncapped under sterile conditions to release air, and recapped. The cells are completely thawed in a 37°C water bath. The cryovial is rinsed briefly with 70% ethanol and wiped dry with sterile paper towel in the hood. Under the hood, the cell suspension is transferred into a 50-mL centrifuge tube. CM is added slowly to a final volume of 10 mL and mixed thoroughly by pipetting up and down. Cell number and viability is counted using a hemocytometer. Cells are transferred to a 25 cm^2 culture flask at the density of 2×10^4 cells/cm^2. The medium in a 25 cm^2 culture flask is ~5 mL, which is changed every 48 h.

3.2. Mosquito Infection and Collection of Plasmodium Sporozoites

1. *Anopheles dirus* has been maintained in colony at the Armed Forces Research Institute of Medical Sciences (AFRIMS) in Bangkok, Thailand.
2. Infections by *P. falciparum* or *P. vivax* are normally diagnosed by examining Giemsa-stained thick and thin blood smears using a compound microscope under oil immersion (magnifications 1000×). For confirmed cases, 5 mL of blood is withdrawn by using sterile techniques into a heparinized tube. Blood is centrifuged at 1500 × *g* for 5 to 10 min, and the plasma is removed. The pellet is washed with 10 mL PBS and reconstituted with naïve type AB serum to the original volume for mosquito feeding. Reconstitution of the blood is necessary because patient serum usually has transmission-blocking activities *(11)*.
3. The mosquito feeding apparatus consists of a series of water-jacketed glass membrane feeders connected by plastic tubes to a circulating water bath (**Fig. 18.1**). Each feeder has 0.3 to 0.5 mL of infected blood to feed ~100 5- to 7-day-old female mosquitoes for 30 min. Unfed mosquitoes are removed from the container, and engorged mosquitoes are kept in an insectary at 24°C for sporozoite collection. On days 7 to 10, up to 10 mosquitoes are dissected to view the presence of oocysts on the midgut to confirm infections.

4. Sporozoite dissection is performed on a clean bench under a dissection scope. Infected mosquitoes 15 days after feeding are chilled on ice for 15 min and their legs are removed with a pair of forceps. They are sterilized by sequential, brief immersion in three solutions: 70% ethanol, penicillin/streptomycin, and amphotericin B. The cleaned mosquitoes are wiped on a piece of sterile gauze and transferred to a few drops of dissection medium on a sterile slide. Mosquitoes are dissected by separating the head from the body using a pairs of fine forceps to view the transparent salivary glands, which are collected into a microcentrifuge tube.
5. The collected salivary glands are centrifuged in a microcentrifuge for 10 min at 10,000 × *g* and 4°C and washed twice with the dissection medium.
6. The salivary glands are ground briefly in 50 μL invasion medium, and the number of sporozoites is estimated by using a hemocytometer (see **Section 3.1**).

3.3. Infection of HC-04 Cells

1. The HC-04 experimental cells are cultured in 96-well culture plates. Two days before infection, 50,000 HC-04 cells in ~100 μL of CM are seeded in each well.
2. Approximately 2×10^4 sporozoites in 25 μL of invasion medium are added into each well and incubated in a 5% CO_2 incubator at 37°C for 4 h. The medium is removed using a pipette, and 100 μL of fresh invasion medium is added. The sporozoite-infected HC-04 cells are incubated at 37°C under 5% CO_2, and the medium is changed daily.

3.4. Detection and Quantitation of EE Stages by Giemsa Staining

1. The HC-04 cells are collected by washing with PBS and trypsin digestion (30 μL/well) (see **Section 3.1**) and resuspended in 200 μL of invasion medium.
2. HC-04 cell monolayers are prepared on microscope slides by centrifuging the cells in a Shandon Cytospin Cytocentrifuge at 800 rpm for 2 min (*see* **Note 4**).
3. The Giemsa HC-04 slides are fixed with methanol briefly and stained with Giemsa diluted with 6 volumes of phosphate buffer for 20 min. The stained slides are rinsed with fresh water for a few seconds, dried at room temperature, and visualized under a compound microscope at 1000× magnification (**Fig. 18.2B–F**).

3.5. Detection and Quantitation of EE Stages by IFA

1. Cytospin is performed as in **Section 3.4**. The HC-04 slides are fixed with cold acetone for 10 min and kept at −70°C until use.
2. The HC-04 slide is thawed and dried in a desiccator at room temperature. A few drops of blocking buffer are added to

cover the surface area with cells. The slide is incubated at 37°C for 30 min in a humidified chamber. The blocking buffer is removed, and primary monoclonal antibody for CSP diluted in blocking buffer 1 μg/mL is added to cover the sample surface. After incubation at 37°C for 1 h, the slide is gently washed in cold PBS for 5 min (*see* **Note 5**).

3. Secondary antibodies diluted 40× in blocking buffer containing 0.01% Evan's Blue solution and 2.5 μg/mL of DAPI are added onto the slide and incubated at 37°C for 1 h. Finally, the slide is washed gently with cold PBS for 5 min and mounted using Prolong Antifade kit and a coverslip. The slide is viewed under an epifluorescent microscope using a FITC filter at 1000× magnification (**Fig. 18.3**) (*see* **Note 6**).

3.6. Preparation of Red Blood Cells for Erythrocytic Invasion

1. The umbilical cord blood from normal full-term delivery is used as a reticulocyte-enriched blood source. After centrifugation at 500 × *g* for 10 min, the plasma is removed, and the cells are resuspended in two original volumes of PBS if a CF-11 column is used for the removal of white blood cells (WBCs). Alternatively, the cells are resuspended in PBS to 20 mL for the removal of WBCs using Plasmodipur columns.
2. To remove WBCs from the cord blood, a CF-11 column is pre-equilibrated with PBS. The blood cell suspension is added to the column, and the flow-through is collected into a new 50-mL tube. Alternatively, red blood cells (RBCs) are centrifuged at 250 × *g* for 5 min and washed once with RPMI-1640 medium (*see* **Note 7**).

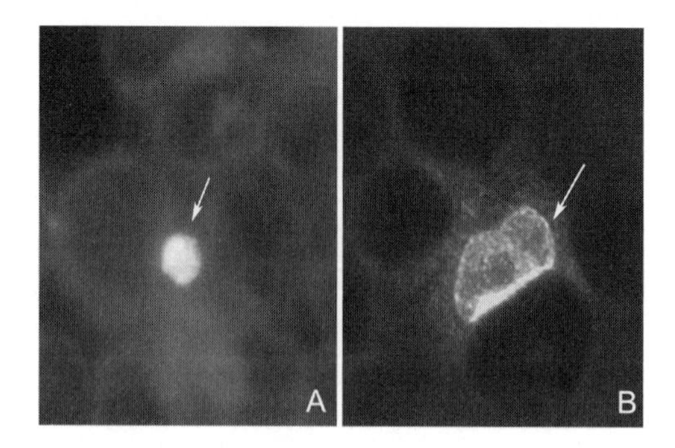

Fig. 18.3. IFA staining of (**A**) *P. falciparum* trophozoite and (**B**) *P. vivax* trophozoite in infected HC-04 cells. Primary antibodies are (**A**) anti-PfHsp70 and (**B**) anti-PvCSP VK210. The slides are counterstained with Evan's Blue, which gives the pictures a reddish background. (From Sattabongkot, J., Yimamnuaychoke, N., Leelaudomlipi, S., Rasameesoraj, M., Jenwithisuk, R., Coleman, R. E., Udomsangpetch, R., Cui, L., and Brewer, T. G. 2006. Establishment of a human hepatocyte line that supports in vitro development of the exo-erythrocytic stages of the malaria parasites *Plasmodium falciparum* and *P. vivax*. *Am. J. Trop. Med. Hyg.* **74**, 708–715. Reproduced and modified with permission from The American Society of Tropical Medicine and Hygiene.)

3.7. Infection of RBCs by Merozoites Developed from Infected Hepatocytes

1. To determine whether *P. vivax* EE stages developed in the *in vitro* culture can mature and invade reticulocytes, reticulocyte-enriched cord blood is added at 2% to 5% hematocrit in CCM supplemented with 10% to 25% human AB serum. This co-incubation is normally done 6 to 20 days after hepatocyte infection. For *P. falciparum* infection, normal RBCs are used.
2. At 24 to 30 h after the co-incubation, RBCs are collected and used to prepare both thick and thin blood films. The films are stained with Giemsa for the observation of asexual blood stages. Alternatively, parasites can also be visualized by the aforementioned IFA procedure using specific antibodies (*see* **Note 8**).

4. Notes

1. We normally prepare MEM and F-12 Nutrient mix (Hams) separately, and they can be stored at 4°C for up to a month. CM normally is used within a week. During storage, the pH of the medium will increase, and phenol red is added as a pH indicator. It will change from orange to pink color when pH arises.
2. The PvCSP VK210 is the predominant variant in Thailand. For areas with prevalent VK247 variant, the primary antibody should also include this variant for IFA detection.
3. The HC-04 cells resembles liver parenchymal cells in morphology. Under optimal culture conditions, it has a doubling time of ~24 h. The cells may lose some characteristics of hepatocytes during continuous culture, which might be important for parasite invasion. We have monitored the synthesis and secretion of hepatocyte proteins such as albumin, α-fetoprotein, and transferrin in different passages. We simply collect the spent culture medium and perform Western blots using antibodies against these proteins (Sigma). Detailed reagents and methods have been described *(10)*.
4. We normally make two cell spots on a microscope slide. For each well of the 96-well culture, we make 2, 4, and 6 spots for days 0 to 1, 2 to 4, and >5-day culture to obtain good monolayers. Note that the cell morphology changes dramatically after the cytospin procedure, which allows better visualization of the parasite trophozoites or schizonts in infected cells.
5. For anti-CSP antibodies, fluorescence is observed during early transformation from sporozoites to trophozoites, but the fluorescence diminishes later. At later stages, other antibodies should be used. For *P. falciparum*, we used anti-Hsp70 and a hyperimmune serum from a *P. falciparum* patient that reacts strongly with the EE stages.

6. For *P. falciparum*, green fluorescent protein (GFP)-labeled transgenic parasites can be used to allow better parasite visualization and quantification. GFP can be directed by constitutive promoters such as those for HSP86 and elongation factor-1α. In recognition of the potential of this system for selecting drugs targeting EE stages, we are currently working on better quantification methods using GFP-labeled parasites and quantitative polymerase chain reaction.
7. Cord blood prepared in this way typically has 3% to 8% reticulocytes. They are stored at 4°C and used within 4 days. Under these storage conditions, a ~2.2% decrease in reticulocyte numbers per day is observed. More detailed methods for culturing *P. vivax* blood stages using cord blood have been described *(12)*. For *P. falciparum* infection of erythrocytes, normal blood from a blood bank is routinely used.
8. We have observed continuous parasite maturation up to 28 days after sporozoite infections, suggesting that the parasite development in this *in vitro* system is asynchronous and not optimal. This is especially true for *P. vivax*, as we have observed small (~5 μm) parasites even after 20 days. These parasites are reminiscent of hypnozoites, which suggests that this *in vitro* system could be used to study the mechanisms of hypnozoite development in *P. vivax*.

Acknowledgments

We want to thank Nongnuch Yimamnuaychoke, Ratawan Ubalee, and Rachaneeporn Jenwithisuk for technical support and acknowledge grant support from the U.S. Army and the Fogarty International Center (1 D43 TW000657-A1).

References

1. Pradel, G., and Frevert, U. (2001) Malaria sporozoites actively enter and pass through rat Kupffer cells prior to hepatocyte invasion. *Hepatology* **33**, 1154–1165.
2. Mota, M. M., Pradel, G., Vanderberg, J. P., Hafalla, J. C., Frevert, U., Nussenzweig, R. S., Nussenzweig, V., and Rodriguez, A. (2001) Migration of *Plasmodium* sporozoites through cells before infection. *Science* **291**, 141–144.
3. Mota, M. M., Hafalla, J. C., and Rodriguez, A. (2002) Migration through host cells activates *Plasmodium* sporozoites for infection. *Nature Med.* **8**, 1318–1322.
4. Mazier, D., Beaudoin, R. L., Mellouk, S., Druilhe, P., Texier, B., Trosper, J., Miltgen, F., Landau, I., Paul, C., Brandicourt, O., et al. (1985) Complete development of hepatic stages of *Plasmodium falciparum* in vitro. *Science* **227**, 440–442.
5. Mazier, D., Collins, W. E., Mellouk, S., Procell, P. M., Berbiguier, N., Campbell, G. H., Miltgen, F., Bertolotti, R., Langlois, P., and Gentilini, M. (1987) *Plasmodium ovale*: in vitro development of hepatic stages. *Exp. Parasitol.* **64**, 393–400.
6. Mazier, D., Landau, I., Druilhe, P., Miltgen, F., Guguen-Guillouzo, C., Baccam, D.,

Baxter, J., Chigot, J. P., and Gentilini, M. (1984) Cultivation of the liver forms of *Plasmodium vivax* in human hepatocytes. *Nature* **307**, 367–369.

7. Hollingdale, M. R., Collins, W. E., and Campbell, C. C. (1986) In vitro culture of exoerythrocytic parasites of the North Korean strain of *Plasmodium vivax* in hepatoma cells. *Am. J. Trop. Med. Hyg.* **35**, 275–276.
8. Hollingdale, M. R., Collins, W. E., Campbell, C. C., and Schwartz, A. L. (1985) In vitro culture of two populations (dividing and nondividing) of exoerythrocytic parasites of *Plasmodium vivax*. *Am. J. Trop. Med. Hyg.* **34**, 216–222.
9. Hollingdale, M. R., Nardin, E. H., Tharavanij, S., Schwartz, A. L., and Nussenzweig, R. S. (1984) Inhibition of entry of *Plasmodium falciparum* and *P. vivax* sporozoites into cultured cells; an in vitro assay of protective antibodies. *J. Immunol.* **132**, 909–913.
10. Sattabongkot, J., Yimamnuaychoke, N., Leelaudomlipi, S., Rasameesoraj, M., Jenwithisuk, R., Coleman, R. E., Udomsangpetch, R., Cui, L., and Brewer, T. G. (2006) Establishment of a human hepatocyte line that supports in vitro development of the exo-erythrocytic stages of the malaria parasites *Plasmodium falciparum* and *P. vivax*. *Am. J. Trop. Med. Hyg.* **74**, 708–715.
11. Sattabongkot, J., Maneechai, N., Phunkitchar, V., Eikarat, N., Khuntirat, B., Sirichaisinthop, J., Burge, R., and Coleman, R. E. (2003) Comparison of artificial membrane feeding with direct skin feeding to estimate the infectiousness of *Plasmodium vivax* gametocyte carriers to mosquitoes. *Am. J. Trop. Med. Hyg.* **69**, 529–535.
12. Udomsangpetch, R., Somsri, S., Panichakul, T., Chotivanich, K., Sirichaisinthop, J., Yang, Z., Cui, L., and Sattabongkot, J. (2007) Short-term in vitro culture of field isolates of *Plasmodium vivax* using umbilical cord blood. *Parasitol. Int.* **56**, 65–69.

Chapter 19

Modulation of Caspase Activation by *Toxoplasma gondii*

Diana Hippe, Andrea Gais, Uwe Gross, and Carsten G. K. Lüder

Abstract

Apoptosis plays crucial roles for the outcome of infection with various infectious agents. The host's apoptotic program may be modulated after infection in order to combat the pathogen or to restrict the immune response. In addition, distinct microorganisms alter the apoptotic program of the host in order to meet the requirements for their further distribution. The activation of caspases (i.e., cysteine proteases with specificity for aspartic acid residues) preludes the disassembly of the cell in response to apoptosis-inducing stimuli. This depends on the proteolytic cleavage of inactive proforms into catalytically active subunits. Analyses of the proteolysis and the enzymatic activity of caspases therefore represent valuable tools to study apoptotic programs during infection. The apicomplexan parasite *Toxoplasma gondii* interferes with the caspase cascade of its host cell in order to facilitate intracellular survival. The modulation of caspase activation by *T. gondii* is determined by SDS-PAGE and immunoblotting with caspase-specific antibodies. Furthermore, the impact of the parasite on caspase activity is fluorimetrically determined by measuring the cleavage of caspase-specific substrate analogues.

Key words: Apoptosis, caspase activation, substrate cleavage, Western blot, SDS-PAGE, fluorimetry, *Toxoplasma gondii*

1. Introduction

Within the past decade, apoptosis, a form of programmed cell death (PCD), has been recognized as a crucial determinant of the host-pathogen interaction *(1, 2)*. This particularly relates to the indispensable roles of apoptosis in shaping the host's immune response during infection *(3)* and as an important effector mechanism of both the innate and adaptive immunity *(4)*. Activation of a class of cysteine proteases (i.e., caspases) preludes the final commitment to death after cells have encountered intrinsic or extrinsic apoptotic stimuli *(5)*. Caspases are synthesized as inactive proforms (i.e., zymogens). Proapoptotic stimuli result in the cleavage and

Steffen Rupp, Kai Sohn (eds.), *Host-Pathogen Interactions*, DOI: 10.1007/978-1-59745-204-5_19,

concomitant activation of caspases in a cascade-like manner, which ultimately triggers the disintegration of the cell *(6)*.

Extrinsic apoptotic stimuli like ligation of Fas/CD95 lead to the formation of the death-inducing signaling complex (DISC) and the consecutive activation of the initiator caspase 8 *(7)*. Caspase 8 in turn directly activates caspases 3, 6, and 7. These so-called effector caspases form the "point of no return" within the apoptotic pathway, which is beyond regulatory events *(5, 8)*. Intrinsic apoptotic stimuli such as DNA damage, growth factor deprivation, toxins, and, importantly also, intracellular infection activate the mitochondrial pathway, in which different signaling cascades (e.g., Akt/PKB-pathway or JNK) lead to the release of cytochrome c from the mitochondria *(9, 10)*. Cytosolic cytochrome c forms a complex with Apaf-1 and the initiator caspase 9, which acquires autocatalytic function in the presence of ATP *(11)*. Active caspase 9 activates the effector caspases 3, 6, and 7, which are responsible for the cleavage of death substrates and the subsequent fragmentation of DNA.

The modulation of apoptotic pathways of the host cell is a widespread characteristic of intracellular pathogens *(12)*. *Toxoplasma gondii* is an obligatory intracellular parasite that inhibits apoptosis of its host cell in order to ensure its intracellular survival *(13)*. We have shown that *T. gondii* prevents activation of caspases 9 and 3 after induction of apoptosis with different proapoptotic stimuli including the kinase inhibitor staurosporine *(14)* (Hippe and Lüder, unpublished data). Furthermore, the parasite is able to diminish activation of caspase 8 after ligation of Fas/CD95 *(15)*. Proform and cleavage products of the different caspases were detected by immunostaining. Furthermore, the activity of caspases was fluorimetrically determined by measuring the cleavage of AMC-conjugated peptide substrates.

2. Materials

2.1. Cell Culture and Lysis

1. Cell culture medium for SKW6.4 cells and Jurkat E6.1 cells: Roswell Park Memorial Institute (RPMI) 1640 supplemented with 10% heat-inactivated fetal calf serum (FCS), 100 U/mL penicillin, and 100 μg/mL streptomycin (all reagents from Biochrom, Berlin, Germany).
2. Cell culture medium for L929 cells: Dulbecco's Modified Eagle's Medium (DMEM) supplemented with 1% FCS, 1 mM sodium pyruvate, nonessential amino acids, 100 U/mL penicillin, and 100 μg/mL streptomycin.
3. Cell culture medium for propagation of *T. gondii*: RPMI 1640 supplemented with 1% heat-inactivated FCS, 100 U/mL penicillin, and 100 μg/mL streptomycin.

4. NP40 lysis buffer: 1% Nonidet P40, 150 mM NaCl, 50 mM Tris-HCl, pH 8.0 supplemented with 25 μL complete protease inhibitor cocktail per mL lysis buffer (Roche, Mannheim, Germany) (*see* **Note 1**). The protease inhibitors are purchased as EDTA-free tablets (cat. no. 1836170) and are dissolved in 400 μL H_2O/tablet. Store at −20°C and avoid repeated thawing and freezing.

2.2. Induction of Apoptosis

1. Staurosporine (from *Streptomyces* sp., Sigma-Aldrich, Taufkirchen, Germany) is dissolved at 1 mM in dimethyl sulfoxide (DMSO) and stored at −20°C.
2. Agonistic mouse anti-Fas/CD95 (human) IgM antibody (clone CH11) is purchased from Upstate Biotechnology, New York, NY) and stored at −20°C.

2.3. SDS-Polyacrylamide Gel Electrophoresis (SDS-PAGE)

1. Separating buffer stock solution: 2 M Tris-HCl, pH 8.8.
2. Stacking buffer stock solution: 0.5 M Tris-HCl, pH 6.8.
3. 10% SDS stock solution.
4. 30% acrylamide/bisacrylamide solution with an acrylamide/bisacrylamide ratio of 37.5:1 (equivalent to C = 2.6%) (Roth, Karlsruhe, Germany). Unpolymerized acrylamide is neurotoxic and should be handled with caution.
5. A 10% ammonium persulfate (APS) stock solution should be stored in aliquots at −20°C. After thawing, it can be stored for 1 week at 4°C. *N,N,N,N'*-tetramethyl-ethylenediamine (TEMED) is stored at 4°C. It is stable for extended periods (months to years), however, in order to avoid contamination of the stock solution, aliquots should be used.
6. SDS-PAGE running buffer: 25 mM Tris, 192 mM glycine, 0.1% SDS, pH 8.5 (*see* **Note 2**).
7. SDS-sample buffer (5x) 312.5 mM Tris-HCl, pH 6.8, 10% SDS, 50% glycerol; 162.5 mM dithiothreitol (DTT), 0.05% bromphenol blue. Store in aliquots at −20°C and avoid repeated thawing and freezing.
8. Prestained molecular weight markers: Mr range 175 to 6.5 kDa (New England Biolabs, Frankfurt/Main, Germany).

2.4. Western Blot and Immunostaining of Caspases 8, 9, and 3

1. Hybond ECL nitrocellulose membrane (NC; Amersham Biosciences, Freiburg, Germany).
2. Fivefold concentrated stock solutions for semidry Western blotting transfer: Anode 1: 1.5 M Tris-HCl, pH 10.4; Anode 2: 125 mM Tris-HCl, pH 10.4; Cathode: 200 mM 6-aminocaproic acid, pH 7.6; methanol.
3. Ponceau S solution: 0.1% Ponceau S in 5% acetic acid.

4. Blocking solution: 5% (w/v) dried skimmed milk, 0.2% (v/v) Tween-20, 0.02% NaN_3 in Ca^{2+} and Mg^{2+}-free Dulbecco's phosphate-buffered saline (PBS) (prepared from Instamed PBS powder, Biochrom, Berlin, Germany). Adjust the pH to 7.4 (*see* **Note 3**). Because of the preservative NaN_3, the solution is stable for several weeks at 4°C.
5. Incubation buffer: 5% (w/v) dried skimmed milk, 0.05% (v/v) Tween-20 in PBS, pH 7.4 (*see* **Notes 3** and **4**).
6. Washing buffer: 0.05% Tween-20 in PBS, pH 7.4.
7. Primary antibodies: rabbit IgG anti–caspase 3 (Becton Dickinson/Pharmingen, cat. no. 552785); rabbit anti–caspase 9 (Becton Dickinson/Pharmingen, cat. no. 68086E); mouse IgG1 anti–caspase 8 (clone 1C12; Cell Signaling Technology, Beverly, MA), mouse IgG1 anti-actin (clone C4; kindly provided by J. L. Lessard, Cincinnati, OH).
8. Secondary antibodies: horseradish peroxidase (HRPO)-conjugated goat anti-mouse IgG or donkey anti-rabbit IgG (Dianova, Hamburg, Germany).
9. Enhanced chemiluminescence (ECL) detection: ECL detection reagents and Hyperfilm ECL (Amersham Biosciences, Freiburg, Germany).

2.5. Stripping of the Nitrocellulose Membrane

1. Stripping buffer: 0.2 M NaOH.

2.6. Fluorimetric Measurement of Caspase Activity

1. 96-well microtiter plates (Greiner, Frickenhausen, Germany).
2. MDB buffer: 10 mM HEPES, pH 7.0, 40 mM β-glycerophosphate, 50 mM NaCl, 2 mM $MgCl_2$, and 5 mM ethylene glycol-bis(2-aminoethyl-ether)-*N,N,N′,N′*-tetraacetic acid (EGTA). After sterile filtration (0.2 μm), the buffer can be stored at room temperature.
3. 10% 3-8(cholamidopropyl)dimethylammonio]-1-propanesulfonate (CHAPS) in MDB buffer. Store at 4°C.
4. Bovine serum albumin (BSA): 10 mg/mL in MDB buffer. Store in aliquots at −20°C.
5. Caspase cleavage buffer: 0.1% CHAPS, 0.1 mg/mL BSA in MDB buffer supplemented with one of the following caspase substrates: 10 μM Ac-DEVD-7-amino-4-methylcoumarin (AMC) (caspase 3/7 substrate; Bachem, Weil am Rhein, Germany), 50 μM Ac-IETD-AMC (caspase 8 substrate; Alexis, Grünberg, Germany), or 50 μM Ac-LEHD-AMC (caspase 9 substrate; Bachem, Weil am Rhein, Germany) (*see* **Note 5**).

3. Methods

Jurkat and SKW6.4 represent well characterized and widely used cell types for the analysis of signaling pathways during apoptosis and its modulation by pathogens *(16, 17)*. Apoptosis can be easily induced in these cells by different proapoptotic triggers, and this considerably facilitates the discovery of an antiapoptotic effect of a certain pathogen. They can also be used as models for different apoptotic pathways. In SKW6.4 cells, activation of Fas/CD95 leads to apoptosis via activation of the extrinsic pathway without the need of a mitochondrial amplification loop. In contrast, triggering of Fas/CD95 or treatment with staurosporine predominately signals via the intrinsic, that is, mitochondrial apoptotic pathway in Jurkat cells. It has to be stressed, however, that intracellular pathogens may not easily or even not at all infect Jurkat and SKW6.4 cells. In such cases, appropriate host cells for the certain microorganism have to be chosen, and the efficacies of several proapoptotic stimuli have to be evaluated. The methods described herein for the analyses of caspases are nevertheless also adaptable to other pathogen-host cell systems. Because microorganisms can both induce and block apoptosis, the impact of infection (or a pathogen-derived molecule) should be tested on spontaneous caspase activation (i.e., in the absence of a proapoptotic stimulus) as well as on cells that have been treated with a proapoptotic stimulus (e.g., anti-Fas/CD95 or staurosporine).

All the antibodies listed above for the analyses of caspases recognize the denatured proteins and are thus reliable for use after separation of cellular lysates by SDS-PAGE and Western blot transfer. Importantly, they detect both the inactive precursor caspases as well as catalytic subunits. The caspase 8–specific antibody detects the proform, the intermediate cleavage products, and the 18-kDa active subunit. Anti–caspase 9 recognizes both the proform and the 35-kDa cleavage product, but not the 12- and 10-kDa subunits. Likewise, anti–caspase 3 binds the proform and the cleavage products at 20 and 17 kDa, but not the 12-kDa small subunit. SDS-PAGE and immunoblotting using these antibodies are therefore well suited to analyze the level of caspase cleavage (i.e., its activation). The ECL detection method allows consecutive immunostainings for different caspases. Furthermore, it is recommended to probe the membrane also with an antibody recognizing actin to confirm equal protein loading onto each lane.

Another key feature for the impact of caspases on the host-pathogen interplay is their enzymatic activity. This is particularly important because caspases that have been activated via proteolytic cleavage are not necessarily enzymatically active due to various natural or synthetic inhibitors *(18)*. The fluorimetric measurement of caspase activities is based on the cleavage of distinct fluorochrome-conjugated peptides. Upon cleavage of such caspase-specific substrate analogues, free fluorochromes emit light after being excited at a specific wavelength. This method is thus applicable for the quantification of caspase activity. In order to allow the comparison of enzymatic activities, equal amount of host cells from different samples rather than equal protein contents have to be used as microbial infection will increase the protein concentrations and would considerably distort the results.

3.1. Cell Culture Procedures, Infection Assays, and Induction of Apoptosis

1. Human B lymphoblastoid SKW6.4 cells (kindly provided by P. H. Krammer, Heidelberg, Germany) and human-derived Jurkat T leukemia cells (ECACC, Salisbury, UK) are subcultured in fresh medium 2 to 3 times per week (*see* **Note 6**). All cell cultures were kept at 37°C and 5% CO_2 in saturated humidity.
2. For experiments, SKW 6.4 or Jurkat cell suspensions are centrifuged at 400 × *g* for 5 min, washed once in culture medium, and then counted in a Neubauer hemocytometer grid using 0.1% trypan blue to distinguish viable and dead cells (*see* **Note 7**). For each experimental sample, 1×10^6 viable cells per well are then seeded in 12-well tissue culture plates in a total volume of 1 mL.
3. Tachyzoites of the *T. gondii* strain NTE *(19)* are propagated in murine L929 fibroblasts as host cells. Parasite–host cell co-cultures are set up twice a week in 12-well tissue culture plates at different parasite to host cell ratios. For this, confluent L929 cells are detached from the tissue culture dishes using a cell scraper, are thoroughly resuspended, and are then mixed with parasite suspensions isolated from co-cultures after initiation of cell lysis (*see* **Note 8**). After addition of fresh medium, co-cultures are kept at 37°C, 5% CO_2 in saturated humidity until host cell lysis occurred (regularly within 6 to 10 days).
4. For infection of Jurkat and SKW6.4 cells, tachyzoites are isolated from L929 co-cultures after the initiation of host cell lysis by differential centrifugation. For this, parasite-L929 co-culture suspensions are centrifuged at 35 × *g* for 5 min to pellet contaminating host cells (*see* **Note 9**). The supernatant is then centrifuged at 1350 × *g* for 10 min. The parasites are washed twice in Jurkat/SKW6.4 culture medium, counted in a Neubauer hemocytometer grid, and adjusted to 1×10^8 parasites/mL Jurkat/SKW6.4 culture medium. Parasites are then added to the host cells at different parasite to host cell ratios (regularly 5:1 to 50:1; *see* **Note 10**).

5. In order to induce apoptosis, Jurkat and SKW6.4 cells are treated for 3 h with 1 μM staurosporine or 250 ng/mL agonistic anti-Fas/CD95 antibody, respectively. Control samples in the absence of proapoptotic stimuli are cultured in parallel (*see* **Note 11**). The time point of treatment with proapoptotic stimuli depends on the question under investigation and may vary from 30 min to 45 h postinfection (p.i.) (*see* **Note 12**).

3.2. Preparation of NP40 Cell Lysates

1. After centrifugation of the cell suspensions at 3500 × *g* for 3 min, cells are washed once in PBS, pH 7.4.
2. After centrifugation (as above), cells are resuspended in 50 μL NP40 lysis buffer per sample (*see* **Note 13**) and incubated for 15 min on ice. They are then vigorously vortexed (at least 15 s at maximum speed).
3. The samples are then centrifuged for 5 min at 21,000 × *g* and 4°C and the supernatant stored at −80°C until further use.

3.3. SDS-PAGE

1. These instructions assume the use of a BioRad Protean 3 Mini gel system (8 cm × 7.5 cm vertical slab gels, 0.75 mm thick).
2. Prepare a 12% separating gel by mixing 0.94 mL separating buffer stock solution, 50 μL SDS stock solution, 1.98 mL H_2O, 2 mL acrylamide/bisacrylamide, 20 μL APS, and 10 μL TEMED. Pour a gel with a height of 5.5 cm into the gel casting chamber.
3. Carefully overlay the separating gel with water-saturated butanol and let the gel polymerize (approximately 1 h). Remove the butanol and rinse gel surface extensively with H_2O. Either directly pour the stacking gel or store the gel overnight at 4°C overlaid with separating buffer/SDS (final concentrations as in the gel).
4. Prepare a 4.5% stacking gel by mixing 625 μL stacking buffer stock solution, 25 μL SDS stock solution, 1.47 mL H_2O, 360 μL acrylamide/bisacrylamide, 5 μL saturated bromphenol blue solution, 10 μL APS, and 10 μL TEMED. Insert the comb and let the stacking gel polymerize for at least 1 h.
5. Assemble the gel unit and add running buffer to the cathode and anode chambers. Carefully remove the comb and wash the sample pockets with running buffer.
6. Mix 20 μL cell extract from each sample with 5 μL 5x SDS sample buffer. Boil samples and molecular weight marker protein mixture for 5 min at 99°C (*see* **Note 14**).

7. Load 20 μL per sample into the pockets of the stacking gel using a Hamilton microliter syringe. Wash the syringe with running buffer from the anode chamber between applying the different samples.
8. Run the gel at 25 mA for approximately 1 h until the dye front has reached the end of the gel.

3.4. Western Blotting and Immunostaining

1. Proteins are transferred to nitrocellulose membranes by semidry blotting using a discontinuous buffer system. For this, 20 mL of each of the transfer stock solutions are mixed with 60 mL H_2O and 20 mL methanol. For the transfer unit, six qualitative Whatman filter papers (no. 3; cut into pieces of 8 × 5.5 cm) are soaked in anode transfer buffer 1, three filter papers are soaked in anode transfer buffer 2, and nine filter papers are soaked in cathode transfer buffer. Also soak a nitrocellulose membrane of the same size in anode transfer buffer 2.
2. Disassemble the electrophoresis unit. Carefully remove the stacking gel as it would firmly stick to the nitrocellulose membrane.
3. Assemble a sandwich transfer unit in the following order: cathode – filters soaked in cathode buffer – polyacrylamide gel – nitrocellulose membrane – filters soaked in anode 2 buffer – filters soaked in anode 1 buffer – anode. During assembly of the transfer unit, avoid air bubbles entrapped between the different layers, and drain excessive buffer solution.
4. Transfer proteins at 0.8 mA/cm^2 gel size (i.e., 35 mA) for 90 min.
5. In order to visualize efficient transfer of the proteins, the NC membrane is incubated for 2 min in ponceau S and washed twice in H_20 for 2 min each.
6. After removal of the lane with the marker proteins (these would fade during the staining procedure), the NC membrane is incubated in blocking solution for 2 h at room temperature.
7. After being washed once for 5 min in washing solution, the membrane is incubated overnight at 4°C (*see* **Note 15**) with anti–caspase 9 (1:1000), anti–caspase 8 (1:1000), or anti–caspase 3 (1:2000) diluted in incubation buffer (*see* **Note 16**).
8. Wash the membrane 3 times for 5 min each in washing solution.
9. The NC membrane is then incubated for 90 min at room temperature with appropriate HRPO-conjugated secondary antibodies diluted at 1:10,000 in incubation buffer.

10. The membrane is washed 3 times in washing solution for 5 min each and twice for 15 min each (*see* **Note 17**).
11. Immunocomplexes are visualized by ECL detection. For this, the ECL detection reagent is prepared as recommended by the manufacturer and the NC membrane overlaid on a glass plate with the solution for 1 min. Thereafter, excessive detection reagent is drained off and the membrane covered with transparency foil. A light-sensitive ECL Hyperfilm is then exposed to the membrane (**Fig. 19.1**; *see* **Note 18**).
12. If desired, signals can be densitometrically quantified using a BioDoc II imaging system (Biometra, Göttingen, Germany).

3.5. Stripping and Reprobing Blots for Actin as Loading Control

1. After the first immunostaining, the membrane is stripped of the signal and is then reprobed using an antibody that either recognizes another caspase (or another protein depending on the experiment) or actin as control protein to ensure equal loading of each lane (*see* **Note 19**).
2. After being rinsed in H_2O, the membrane is incubated for 5 min in stripping solution.

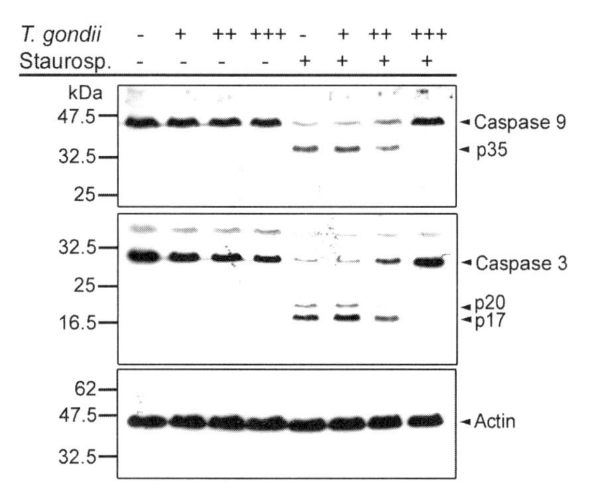

Fig. 19.1. Proteolytic cleavage of caspases 9 and 3 in *T. gondii*–infected and noninfected Jurkat cells treated or not with proapoptotic staurosporine. Jurkat cells were infected with *T. gondii* at parasite to host cell ratios of 5:1 (+), 15:1 (++), or 30:1 (+++) or were left noninfected. After 21 h, cells were treated for 3 h with 1 μM staurosporine as indicated. Cell lysates were then separated by SDS-PAGE, transferred to a nitrocellulose membrane, and analyzed for the presence of proforms and cleavage products of caspase 9 and caspase 3 by immunostaining and enhanced chemiluminescence detection. After the membrane was stripped of the signals, it was reprobed with anti-actin and an appropriate secondary antibody in order to visualize an equal loading of each lane.

3. The membrane is then washed twice in H_2O for 5 min each.
4. After stripping, unspecific binding sites of the NC membrane are again blocked and the NC membrane subsequently immunostained as described above. Actin is visualized by incubation with anti-actin diluted 1:10,000 in incubation buffer and HRPO-conjugated anti-mouse IgG secondary antibody.

3.6. Fluorimetric Measurement of Caspase Activation

1. Ten microliters of each sample are distributed in triplicate in a 96-well plate.
2. Add 90 μL of freshly prepared caspase cleavage buffer to each well and mix by gentle rotation (*see* **Note 20**).
3. Immediately after addition of the substrate, the fluorescence intensity is measured at excitation and emission wavelengths of 380 and 460 nm, respectively, and 37°C using a Victor V multilabel counter (Perkin Elmer, Boston, MA) or an equivalent reader. The fluorescence is recorded every 5 min over a period of 60 min.
4. After calculating the mean fluorescence intensity of each triplicate, the kinetics of the fluorescence intensities in each sample are graphically visualized (**Fig. 19.2A**). The substrate cleavage (i.e., the caspase activity) is then determined by calculating the increase in the fluorescence intensity from 5 to 60 min (**Fig. 19.2B**; *see* **Note 21**).

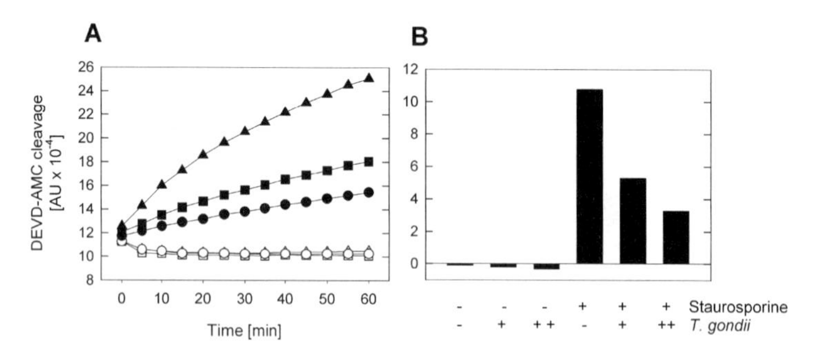

Fig. 19.2. Activity of caspase 3/7 in noninfected and *T. gondii*–infected Jurkat cells treated or not with staurosporine. At 12 h postinfection, Jurkat cells infected with *T. gondii* at parasite to host cell ratios of 15:1 (**A**, ■; **B**, +) or 30:1 (**A**, ●; **B**, ++) and noninfected controls (**A**, ▲; **B**, −) were induced to undergo apoptosis by treatment with 1 μM staurosporine for 3 h (closed symbols) or were left untreated (open symbols). Cell lysates were fluorimetrically analyzed for the cleavage of the caspase 3/7-specific substrate analogue DEVD-AMC over a period of 60 min (**A**). The absolute increase of DEVD-AMC cleavage from 5 to 60 min was then calculated as a determinant of the caspase activity (**B**).

4. Notes

1. Because of the loss of inhibitor activity during prolonged storage, the protease inhibitors have to be added freshly not longer than 30 min before usage.
2. Do not adjust the pH of the running buffer with HCl because this will considerably decrease the resolution of protein separation. Normally, 192 mM glycine should suffice to adjust the pH to approximately 8.5. In the case of pH >8.6, decrease it by the addition of glycine.
3. Adjust pH with HCl after solubilization of skimmed milk because this will considerably alter the pH.
4. Because NaN_3 substantially inhibits the enzymatic activity of horseradish peroxidase (HRPO), the antibody incubation solution (at least for HRPO-conjugated secondary antibodies) is prepared without sodium azide. Stocks of the incubation solution can nevertheless be prepared and stored as aliquots at −20°C. After thawing, it can be stored at 4°C for not longer than 1 week.
5. AMC-conjugated caspase substrates are prepared as 10 mM stock solutions in H_2O and stored in aliquots at −20°C. They are light sensitive and rather unstable at higher temperatures. Therefore, removal of the solutions from the deep-freezer should be minimized.
6. Jurkat and particularly SKW6.4 cells require rather high cell densities for optimal growth. Therefore, cultures have to be checked regularly and should only be diluted when the culture medium already becomes yellowish. Jurkat and SKW6.4 suspensions can then be subcultured at dilutions of 1:6 or 1:3, respectively.
7. Viable cells exclude trypan blue, whereas dead cells are stained blue.
8. Because *T. gondii* tachyzoites loose their infectivity during prolonged extracellular culture, they should be used shortly after the initiation of host cell lysis and when potential new host cells of the parasite are still present.
9. In order to increase the parasite yield, pelleted host cells can be again resuspended in culture medium and lysed by forcing them through a 26-gauge syringe needle. The resulting parasite suspension is again centrifuged at $35 \times g$ for 5 min and the supernatant pooled with that from the first centrifugation step.
10. Jurkat and SKW6.4 cells are suspension cells and infection efficiency with *T. gondii* is rather low. Therefore, high parasite

to host ratios are required to guarantee sufficiently high infection rates. For example, addition of *T. gondii* to SKW6.4 cells at parasite to host ratios of 10:1 or 50:1 results in mean infection rates of 59% and 97% *(15)*.

11. Because *T. gondii* has also been shown to promote apoptosis in leukocytes from parasite-infected mice *(13)*, determining the effect of parasitic infection on spontaneous apoptosis of leukocytes *in vitro* may also be of major interest. Therefore, noninfected and parasite-infected control cultures in the absence of staurosporine or anti-Fas/CD95 are run in parallel.
12. During extended periods of parasite infection, lysis of host cells may considerably interfere with the detection of caspase activation and activity. Such cell lysis can be controlled by comparing the host cell counts of infected and noninfected cell cultures. Routinely, we add proapoptotic stimuli at 21 hours p.i. (i.e., infect host cells for a total period of 24 h). The duration of infection should not be extended beyond 48 h due to considerable parasite-induced lysis of host cells.
13. For subsequent biochemical analyses, it is critical that equal amounts of host cells are extracted from each sample. This is particularly important because comparison and subsequent adjustment of the protein content of the different NP40 cell lysates is inappropriate due to the parasite proteins present in samples from *T. gondii*–infected host cells. Determination of host cell counts from different samples indicated that infection with *T. gondii* at parasite to host cell ratios up to 50:1 for 24 h does not significantly increase or decrease the number of Jurkat or SKW6.4 cells compared with noninfected controls. Under those experimental conditions described herein, we therefore extract each cell sample in the same amount of lysis buffer. However, when host cells are infected for extended periods or when using other host cell types, it may be necessary to determine the cell number for each sample and to adjust the amount of lysis buffer accordingly (2×10^7 host cells/mL NP40 lysis buffer).
14. Centrifuge the samples after boiling to avoid residual liquid in the lids.
15. Alternatively, the primary antibody can be applied for 2 h at room temperature. However, the sensitivity and specificity of antibody binding is increased after incubation overnight.
16. After addition of 0.02 % NaN_3 as a preservative and storage at 4°C, the primary antibody solution can be repeatedly used.
17. In the case of high background staining, the washing procedure can be extended for several hours.

18. Appropriate signals should be visible within 5 to 10 min. However, the exact duration of exposure has to be determined for each experiment and may vary between 30 s and 1 h.
19. NC membranes can be reprobed 4 to 5 times. However, because very strong signals may not be completely eliminated by the stripping procedure, stainings that give only weak signals should precede those that will result in strong signals.
20. Because of the light sensitivity and instability, the AMC-conjugated caspase substrate is added to the cleavage buffer after the samples have already been distributed into the microtiter plate and just before adding the cleavage buffer to the samples.
21. Because the fluorescence intensity in samples from non-apoptotic cells initially often slightly decreases, the fluorescence intensity at 0 min is not included into the calculation. Furthermore, if the fluorescence intensity reaches a plateau, only those values from the linear increase should be included in the subsequent calculation.

Acknowledgments

The authors would like to thank Peter H. Krammer, Heidelberg, Germany, and James L. Lessard, Cincinati, Ohio, for kindly providing SKW6.4 cells and anti-actin antibody, respectively. Diana Hippe is recipient of a Ph.D. scholarship from the Karl-Enigk-Stiftung, Hannover, Germany.

References

1. Liles, W. C. (1997) Apoptosis – role in infection and inflammation. *Curr. Opin. Infect. Dis.* **10**, 165–170.
2. Lüder C. G. K., Gross, U. and Lopes, M. F. (2001) Intracellular protozoan parasites and apoptosis: diverse strategies to modulate parasite-host interactions. *Trends Parasitol.* **17**, 480–486.
3. Opferman, J. T. and Korsmeyer, S. J. (2003) Apoptosis in the development and maintenance of the immune system. *Nature Immunol.* **4**, 410–415.
4. Williams, G. T. (1994) Programmed cell death: a fundamental protective response to pathogens. *Trends Microbiol.* **2**, 463–464.
5. Hengartner, M. O. (2000) The biochemistry of apoptosis. *Nature* **407**, 770–776.
6. Earnshaw, W. C., Martins, L. M. and Kaufmann, S. H. (1999) Mammalian caspases: structure, activation, substrates, and functions during apoptosis. *Ann. Rev. Biochem.* **68**, 383–424.
7. Krammer, P. H. (2000) CD95's deadly mission in the immune system. *Nature* **407**, 789–795.
8. Danial, N. N. and Korsmeyer, S. J. (2004) Cell death: critical control points. *Cell* **116**, 205–219.
9. Green, D. R. and Reed, J. C. (1998) Mitochondria and apoptosis. *Science* **281**, 1309–1312.
10. Anderson, P. (1997) Kinase cascades regulating entry into apoptosis. *Microbiol. Mol. Biol. Rev.* **61**, 33–46.
11. Li, P., Nijhawan, D., Budihardjo, I., Srinivasula, S. M., Ahmad, M., Alnemri, E. S. and Wang, X. (1997) Cytochrome c and dATP-dependent formation of

Apaf-1/caspase-9 complex initiates an apoptotic protease cascade. *Cell* **91**, 479–489.

12. Schaumburg, F., Hippe, D., Vutova, P. and Lüder, C. G. K. (2006) Pro- and antiapoptotic activities of protozoan parasites. *Parasitology* **132**, S69–S85.

13. Lüder, C. G. K. and Gross, U. (2005) Apoptosis and its modulation during infection with *Toxoplasma gondii*: Molecular mechanisms and role in pathogenesis. In: Griffin, D. E, ed. Role of apoptosis in infection. *Curr. Topics Microbiol. Immunol.* **289**, 219–238.

14. Goebel, S., Gross, U. and Lüder, C. G. K. (2001) Inhibition of host cell apoptosis by *Toxoplasma gondii* is accompanied by reduced activation of the caspase cascade and alterations of poly(ADP-ribose) polymerase expression. *J. Cell Sci.* **114**, 3495–3505.

15. Vutova, P., Wirth, M., Hippe, D., Gross, U., Schulze-Osthoff, K., Schmitz, I. and Lüder, C. G. K. (2007) *Toxoplasma gondii* inhibits Fas/CD95-triggered cell death by inducing aberrant processing and degradation of caspase 8. *Cell. Microbiol.* **9**, 1556– 1570.

16. Scaffidi, C., Fulda, S., Srinivasan, A., Friesen, C., Li, F., Tomaselli, K. J., Debatin, K.-M., Krammer, P. H. and Peter, M. E. (1998) Two CD95 (APO-1/Fas) signaling pathways. *EMBO J.* **17**, 1675–1687.

17. Ferrari, D., Stepczynska, A., Los, M., Wesselborg, S. and Schulze-Osthoff, K. (1998) Differential regulation and ATP requirement for caspase-8 and caspase-3 activation during CD95- and anticancer drug-induced apoptosis. *J. Exp. Med.* **188**, 979–984.

18. Riedl, S. J. and Shi, Y. (2004) Molecular mechanisms of caspase regulation during apoptosis. *Nature Rev. Mol. Cell Biol.* **5**, 897–907.

19. Gross, U., Müller, W. A., Knapp, S. and Heesemann, J. (1991) Identification of a virulence-associated antigen of *Toxoplasma gondii* by use of a mouse monoclonal antibody. *Infect. Immun.* **59**, 4511–4516.

Part IV
Host Responses

Chapter 20

Introduction: Host Responses

Martin Schaller, Günther Weindl, and Bernhard Hube

The primary objective for pathogens is the ability to overcome host defense mechanisms and establish an infection in the host. During this complex process, it is critically important for a pathogen to recognize and respond to its host environment and adapt to changing microenvironments in the host. However, in addition to the identification and characterization of virulence factors and survival systems expressed by different pathogens, investigation of host response mechanisms and host pathways that are subverted by these microorganisms is required for a profound understanding of infectious diseases. Furthermore, the interaction between symbiotic, commensal, and pathogenic microorganisms and host defense mechanisms plays a central role in determining whether colonization remains harmless or leads to localized and systemic infections.

In vertebrates, a successful host response to infection is mediated by an immunologic defense consisting of innate and adaptive (or acquired) immunity. The innate immune system acts as a first line of defense, serving as a physical and chemical barrier and responding to a wide spectrum of pathogens by using a limited repertoire of germ-line–encoded proteins. Yet, it cannot control invading pathogens, which have developed evasion strategies against this defense system. The initiation of humoral and cell-mediated adaptive responses enables specific recognition and elimination of pathogenic organisms. The adaptive immune system generates a random and wide variety of antigen receptors by somatic gene rearrangement, followed by selection and expansion of cell clones expressing receptors with relevant specificities, as well as establishment of immunologic memory. A well-coordinated interaction of innate and adaptive immune responses provides a highly effective regulatory network to combat infectious diseases.

Steffen Rupp, Kai Sohn (eds.), *Host-Pathogen Interactions*, DOI: 10.1007/978-1-59745-204-5_20,

However, the redundancy and functional plasticity of the immune system complicates the experimental study of its components. The availability of various animal and human *in vitro* models greatly accelerates our ability to understand the host immune response, but the transfer between species and/or models involves assumptions that require thorough evaluation. Recent breakthroughs in immunology have revealed numerous new targets and mechanisms and have revolutionized the concepts of innate and adaptive immunity. Together with the continuous development of novel technologies and research tools, exciting insights into host response have been gained. Through the use of new technologies, it is becoming much easier to accumulate and acquire this information, but data are often more difficult to interpret.

Although many questions still remain unanswered, studying host responses will provide information that facilitates our understanding of molecular mechanisms of pathogenesis and host-pathogen interaction and will be helpful to develop molecular tools for the diagnosis of infection susceptibilities in patients. Increased knowledge of pathogen escape strategies and the cooperation of innate and adaptive immune system will undoubtedly offer novel therapeutic strategies for immunomodulation and pharmacologic targeting, which might be a new and promising way to prevent or treat infections.

Chapter 21

Fungal and Bacterial Killing by Neutrophils

David Ermert, Arturo Zychlinsky, and Constantin Urban

Abstract

Neutrophils are professional phagocytes of the innate immune system that are essential to control bacterial and fungal infections. These cells engulf and kill invading microbes. Additionally, activated neutrophils are able to release neutrophil extracellular traps (NETs). These fibers consist of chromatin decorated with antimicrobial proteins to trap and kill microbes. Appropriate quantitative methods are required to understand the nature of interactions of neutrophils with pathogens. Here we present assays to measure killing mediated by phagocytosis, by NETs, by a combination of both, and by granular extract. As examples, we use *Candida albicans* for fungal and *Shigella flexneri* for bacterial pathogens.

Key words: Neutrophils, phagocytosis, neutrophil extracellular traps, killing, survival, *Candida albicans, Shigella flexneri.*

1. Introduction

1.1. Neutrophils

Polymorphonuclear neutrophils (neutrophils or PMNs) are essential cells of the innate immune system that form a first line of defense against pathogenic microbes such as bacteria and fungi *(1, 2)*. The important role of neutrophils is underscored by the fact that individuals with decreased quantity or quality of neutrophils suffer from recurrent, life-threatening infections *(3)*.

Neutrophils differentiate in the bone marrow from pluripotent hematopoietic progenitor cells into mature neutrophils. They are the most abundant innate immune cells in the blood (65% to 75% of all white blood cells). Mature neutrophils are terminally differentiated cells with a life span of only a few hours in circulation *(4)*.

The main characteristics of neutrophils are the trilobulated nucleus and the numerous granules that can be found in the

Steffen Rupp, Kai Sohn (eds.), *Host-Pathogen Interactions*, DOI: 10.1007/978-1-59745-204-5_21,

cytoplasm. The granule contents are synthesized mainly during maturation in the bone marrow. There are three different granule types, which are called azurophilic, specific, and gelatinase granules. Azurophilic (primary) granules contain acidic hydrolases, such as neutrophil elastase, oxidases (e.g., myeloperoxidase), and antimicrobial peptides, such as defensins. Specific (secondary) and gelatinase (tertiary) granules contain an overlapping set of antimicrobials, such as lysozyme and matrix degrading enzymes (e.g., gelatinase), which are essential for migration of neutrophils through tissue *(5)*.

The recruitment of neutrophils from the bloodstream to the site of infection is initiated by chemokines and cytokines. For instance, infected epithelia secrete interleukin 8 (IL-8), which attracts neutrophils from the bloodstream to the site of infection *(6)*. This process is called extravasation and is regulated by selectins, integrins, and other cell adhesion molecules. Once the neutrophil reaches the infection site, microbes and microbial compounds such as lipopolysaccharide (LPS) activate the neutrophils via transmembrane receptors. For laboratory purposes, phorbol esters, such as phorbol myristate acetate (PMA), are widely used to activate neutrophils directly through intracellular pathways *(7)*. Recruited and activated neutrophils in turn release effector proteins (e.g., TNF-α or IL-8), which attract additional neutrophils and other immune cells.

At the site of infection, neutrophils engulf microbes. The vesicles containing the pathogens, called phagosomes, fuse with primary and secondary neutrophil granules, and the antimicrobial contents are discharged into the lumen of the phagosome, which is then called the phagolysosome *(8)*. During phagocytosis, neutrophils consume large amounts of oxygen. This non-mitochondrial respiration is termed *respiratory burst*. A multicomponent leukocyte oxidase mediates this burst transferring electrons to molecular oxygen *(9)*. The resulting reactive oxygen intermediates (ROI) dismutate first to hydrogen peroxide and then to hypochlorous acid. The latter is formed by the action of myeloperoxidase. The combination of antimicrobials and ROI kill phagocytosed microorganisms *(10)*.

Recently a second antimicrobial mechanism of neutrophils was described. Activated neutrophils eliminate microbes in the extracellular space by releasing chromatin decorated with granular antimicrobials *(11)*. The neutrophil extracellular traps (NETs) capture and kill microbes such as *Shigella flexneri, Staphylococcus aureus*, and *Candida albicans*. Pathogens themselves can induce NET formation in neutrophils *(12)*. During that process, the chromatin decondenses, the nuclear membrane dissolves, and nuclear material mixes with granular components. At last, the plasma membrane ruptures, and the granule decorated chromatin spills out *(13)*.

In this chapter, we introduce four different assays to measure the antimicrobial activity of neutrophils: (i) intracellular, (ii) extracellular, (iii) both combined, and (iv) granular extract. For this purpose, we use two pathogens that are investigated in our laboratory: *Candida albicans* serves as an example for a eukaryotic pathogen, and *Shigella flexneri* represents a prokaryotic pathogen.

1.2. Candida

The genus *Candida* belongs to the order Saccharomycetales (phylum Ascomycota, class Endomycetes). The yeast *C. albicans* is fastidious diploid, and proliferates asexually *(14)*. *C. albicans* can grow in different morphologic forms: as yeast-form cells, as a filamentous form called hyphae, or as a mixture of both called pseudohyphae. In its yeast form, the diameter of the cells is 3 to 6 μm, and daughter cells bud from a mother cell. Hyphae emerge through continuous apical growth of single cells. True hyphae show no septation between mononuclear cells, which can then branch *(15)*. Pseudohyphae are clusters of elongated cells with clear septation. Environmental conditions regulate the morphology of *C. albicans.* Temperatures of more than 30°C, low nitrogen, serum or pH of 7 induce hyphal growth *(16)*. Usually, *C. albicans* is a commensal colonizing the skin and the oral, gastrointestinal, and genital tracts of humans. However, as a facultative pathogen, *C. albicans* is the causative agent for more than 50% of all candidiasis. These infections range from mild superficial infections of mucosa to severe life-threatening systemic infections. Particularly in immunocompromised persons, for example after long-term antibiotic treatments, in AIDS or transplantation patients, candidal infections can disseminate and become systemic. *C. albicans* has recently emerged as the fourth most frequent cause of nosocomial bloodstream infections *(17)*.

1.3. Shigella

Shigella are Gram-negative, rod-shaped, nonmotile bacteria discovered by Kiyoshi Shiga in 1898. *Shigella* cause bacillary dysentery, a severe form of diarrhea. Each year, 160 million infections occur worldwide causing 1 million deaths *(18)*. Especially children and older people die by *Shigella* infection. In developing countries, *S. flexneri* is the most prevalent strain. Beside *S. flexneri*, there are three more strains, *S. boydii, S. sonnei*, and *S. dysentery*, which have different geographic distribution. *Shigella* is transmitted via the fecal-oral route. Only 10 to 100 bacteria are sufficient to cause dysentery. The bacteria are taken up by M-cells in the Peyer's patches to cross the intestinal barrier. Once in the lamia propria of the colon, they invade epithelial cells via the basolateral side *(19)*. Macrophages as well as neutrophils engulf *S. flexneri* at the site of infection. In macrophages, *S. flexneri* escapes from the phagosome and induces macrophage apoptosis. In contrast, *S. flexneri* is not able to escape from neutrophilic phagosomes as neutrophil elastase cleaves *S. flexneri* virulence factors essential for phagosomal escape. Eventually *S. flexneri* is killed in the phagolysosome of neutrophils *(20)*.

2. Materials

2.1. Culture of Microorganisms

2.1.1. Candida albicans

1. YPD: 1% yeast extract, 2% bacto peptone (both Becton Dickinson, Heidelberg, Germany), and 2% glucose (Sigma-Aldrich, Steinheim, Germany).
2. RPMI medium (Invitrogen, Karlsruhe, Germany), substituted 2% glucose (Sigma-Aldrich).

2.1.2. Shigella flexneri

1. TSB: Bacto Tryptic Soy-broth (Becton Dickinson).

2.2. Ficoll Purification of Human Neutrophils

1. Vacutainer K2E 18 mg (Becton Dickinson) color code: purple.
2. RPMI-HEPES: RPMI (Invitrogen) without phenol red, substituted with 10 mM HEPES (Invitrogen).
3. 20% human serum albumin (HSA; Grifols, Langen, Germany).
4. Dextran (MP Biomedicals Inc, Eschwege, Germany).
5. Hypaque Ficoll (GE Healthcare, Uppsala, Sweden).
6. Plastic Pasteur pipette.
7. 15-mL plastic centrifuge tubes.
8. Cold double-distilled H_2O (ddH_2O), pyrogen free.

2.3. Preparation of Human Neutrophil Granular Extract from Buffy Coats

1. Human buffy coats (available at blood banks).
2. Hanks Balanced Salt Solution without Calcium and Magnesium ($HBSS^-$; Invitrogen).
3. 15-mL plastic centrifuge tubes.
4. Sonicator Bandelin Sonopuls HD2070.
5. Sulfuric acid.
6. Double-distilled H_2O (ddH_2O), pyrogen free.
7. Slide-A-Lyzer Dialysis Cassette, 3.5-kDa molecular weight cutoff, 0.5 to 3 mL capacity (Pierce, Rockford, IL).
8. Sodium acetate (Sigma-Aldrich), for preparation of 20 mM solution at pH 4.0 (*see* **Section 4.2**).
9. 2 mL siliconized reaction tubes.

2.4. Killing Assays with Neutrophils

2.4.1. Intracellular Killing (Phagocytosis)

1. *C. albicans* or *S. flexneri* culture.
2. RPMI-HEPES: RPMI (Invitrogen) without phenol red, substituted with 10 mM HEPES (Invitrogen).
3. 24-well tissue culture test plates (TPP; Trasadingen, Switzerland).
4. Heat-inactivated pooled human serum (for preparation *see* **Section 4.3**).
5. 20% human serum albumin (HSA; Grifols).
6. Phosphate-buffered saline (PBS: 140 mM NaCl pH 7.0, 2.7 mM KCl, 9 mM Na_2HPO_4, 1.5 mM KH_2PO_4).
7. Double-distilled H_2O (ddH_2O), pyrogen free.
8. 16-cm cell scraper (Sarstedt, Newton, NC).
9. 1.5-mL reaction tubes.
10. YPD-Agar plates (*C. albicans*) or TSB-Agar plates (*S. flexneri*).

2.4.2. Extracellular Killing (NETs)

Use materials listed in **Section 2.4.1** and add:
1. Phorbol 12-myristate 13-acetate (PMA; Sigma-Aldrich).
2. RNase-free and protease-free deoxyribonuclease 1 (DNase-1) (Worthington, Lakewood, NJ).

2.4.3. Combined Killing (Phagocytosis and NETs)

Use materials listed in **Section 2.4.1** and add:
1. Phorbol 12-myristate 13-acetate (PMA; Sigma-Aldrich).
2. RNase-free and protease-free deoxyribonuclease 1(DNase-1) (Worthington, Lakewood, NJ).
3. Cytochalasin D (CytD; Sigma-Aldrich).

2.5. Killing Assays with Human Neutrophil Granular Extract

1. *C. albicans* or *S. flexneri* culture.
2. Human neutrophil granular extract (hNGE) (for preparation of hNGE, *see* **Section 3.3**).
3. Hanks Balanced Salt Solution without Calcium and Magnesium ($HBSS^-$; Invitrogen) containing 3% casamino acids (Roth, Karlsruhe, Germany).
4. YPD-Agar plates (*C. albicans*) or TSB-Agar plates (*S. flexneri*).

3. Methods

3.1. Culture of Microorganisms

3.1.1. Candida albicans

1. Inoculate 10 mL of YPD with *Candida albicans* from cryo stocks and shake culture overnight at 30°C.
2. For experiments, dilute culture to an optical density at wavelength 600 nm (OD_{600}) of 0.1 and shake for 4 h. Incubate

in YPD at 30°C to induce yeast-form cells and in RPMI with 2% glucose at 37°C to induce hyphae.

3.1.2. Shigella flexneri

1. Inoculate 5 mL of TSB with *Shigella flexneri* and shake culture overnight at 37°C.
2. For experiments, dilute 50 μL of the overnight culture in 5 mL fresh TSB and shake culture for 2 h at 37°C.

3.2. Ficoll Purification of Human Neutrophils

1. Dissolve 3% dextran in RPMI-HEPES.
2. Take a 10 mL whole blood sample with a K2E Vacutainer. Ten milliliters of whole blood contain approximately 1×10^7 to 2×10^7 PMNs. This is sufficient for a killing assay with 5 to 10 conditions in triplicate.
3. Add 2 volumes of collected blood with 1 volume 3% dextran and invert twice. Change lid and incubate at room temperature for approximately 30 min until red blood cells separates from serum containing other blood cells like neutrophils and lymphocytes.
4. After separation, transfer 1 volume of supernatant with a plastic Pasteur pipette on top of 1 volume Hypaque Ficoll in a 15-mL centrifugation tube.
5. Centrifuge tubes 30 min at $300 \times g$ at room temperature.
6. Remove supernatant and resuspend pellet in 1 mL RPMI-HEPES. Pool all cells from the same donor in a 15-mL centrifugation tube and fill up to 15 mL with RPMI-HEPES.
7. Centrifuge tubes 10 min at $300 \times g$ at room temperature.
8. Remove supernatant and lyse red blood cells by adding 1 mL ddH_20. After 30 to 60 s, add 10 mL RPMI-HEPES to the tube.
9. Centrifuge tubes 10 min at $300 \times g$ at room temperature.
10. Remove supernatant and resuspend pellet in 1 mL RPMI-HEPES and count cells.
11. Dilute cells to 2×10^6 cells/mL in RPMI-HEPES.

For an alternative method of neutrophil purification, *see* **Section 4.1**.

3.3. Preparation of Human Neutrophil Granular Extract from Buffy Coats

1. For the separation of neutrophils from buffy coats use the Ficoll purification protocol (**Section 3.2, steps 1** to **8**). After the red blood cell lysis step (**step 8**), resuspend the cells in 5 mL HBSS$^-$, and count and transfer 5×10^8 cells to 15-mL plastic centrifuge tubes. Centrifuge to pellet cells at $250 \times g$ for 5 min at 4°C. Discard the supernatant, quick freeze in liquid nitrogen, and store pellet at −80°C until use. A total of 5×10^8 neutrophils equals a volume of 0.5 mL of packed cells. Expect approximately 5×10^8 neutrophils per unit of

buffy coat. To proceed, thaw cells on ice and add ice-cold water to adjust to a cell concentration of 4.5×10^8cells/mL.

2. Sonicate cells in the centrifugation tubes in a beaker with ice 2 times for approximately 30 s until the granules are released. This occurs when the samples look milky white and are homogenous.
3. Prepare 200 mM sulfuric acid with pyrogen-free ddH_2O and add to the homogenized cells to give a final concentration of 80 mM sulfuric acid. Incubate on ice for 30 min, mixing every 5 min.
4. Transfer the solution to siliconized 2 mL reaction tubes and spin at 4°C at $16{,}000 \times g$ for 20 min.
5. Dialyze in Spectraphor 3.5-kDa membrane against 2 l of 20 mM sodium acetate buffer pH 4 for 2 days at 4°C. Change buffer once every 24 h.
6. Remove the solution from the dialysis bags and transfer to 2 mL reaction tubes. Spin at $16{,}000 \times g$ for 5 min. Transfer supernatant to a fresh tube and discard the pellet. Store granular extract (hNGE) at 4°C. Do not freeze.

3.4. Killing Assays with Neutrophils

3.4.1. *C. albicans*

3.4.1.1. Intracellular Killing (Phagocytosis)

1. Resuspend human neutrophils at 2×10^6/mL and allow them to adhere to 24-well tissue culture test plates in 500 μL RPMI-HEPES per well at 37°C with 5% CO_2. For measuring killing of opsonized *C. albicans*, use RPMI-HEPES containing 1% heat inactivated pooled human serum. For measuring killing of nonopsonized *C. albicans*, use RPMI-HEPES containing 1% HSA. Each sample should be tested in triplicates.
2. Wash a sufficient amount of *C. albicans* cells with PBS and adjust the number of *C. albicans* cells for infection of neutrophils. Determine the *C. albicans* concentration, centrifuge at $2000 \times g$ for 5 min at 4°C and resuspend the *C. albicans* in PBS. Using yeast-form *C. albicans* measure the optical density at 600 nm wavelength (OD_{600}) and correlate 3×10^7 *C. albicans*/mL to 1 OD_{600}. Using hyphae count in a Neubauer chamber to determine the hyphal concentration.
3. Dilute the *C. albicans* suspensions accordingly to the desired multiplicity of infection (MOI, *C. albicans* cells per neutrophil) in PBS. MOIs between 1 and 0.01 are usual. To calculate the correct dilutions for different MOIs see **Table 21.1**.
4. Add 50 μL of the adjusted *C. albicans* suspension to each well containing neutrophils, centrifuge plates with $700 \times g$ for 10 min at 37°C, and incubate the plates for the desired time point at 37°C with 5% CO_2. For example: Dilute *C. albicans* for an MOI of 0.1 and 0.01, add the *C. albicans* suspension, centrifuge and incubate for 30 min, 60 min, and 120 min.

Table 21.1
Correlation of MOI with concentration of microbes to use in killing assays.

Total assay volume is 500 μL per sample in 24-well tissue culture test plate

Desired MOI	Neutrophils in 500 μL medium	Microbes in 50 μL
1	2×10^6/mL	2×10^7/mL
0.1	2×10^6/mL	2×10^6/mL
0.01	2×10^6/mL	2×10^5/mL

5. Prepare identical samples without adding neutrophils. These samples serve as controls.
6. After incubation carefully remove the supernatants to 1.5-mL reaction tubes and store on ice. Lyse neutrophils with 500 μL ice-cold distilled water per well. Incubate for 5 min on ice and scrape the bottom of each well thoroughly with a mini cell scraper to remove all remaining *C. albicans* and neutrophils. Pipette vigorously up and down and add the suspension to the corresponding supernatants collected in the 1.5-mL reaction tubes.
7. Dilute each sample 10^{-1}, 10^{-2}, 10^{-3}, and plate 50 μL of each dilution on YPD-Agar plates and incubate the plates for 24 h at 30°C.
8. Count the colony forming units (CFU) on the plates. Plates with CFU counts between 30 and 600 can be evaluated.
9. Calculate the percent survival of each sample using the formula:

$$\%_{\text{survival}} = \frac{\text{df} \times \text{CFU}_{\text{with neutrophils}}}{\text{df} \times \text{CFU}_{\text{without neutrophils}}} \times 100$$

To calculate percent phagolysosomal killing use:

$$\%_{\text{phagolysosomal killing}} = \frac{(\text{df} \times \text{CFU}_{\text{without neutrophils}} - \text{df} \times \text{CFU}_{\text{with neutrophils}})}{\text{df} \times \text{CFU}_{\text{without neutrophils}}} \times 100$$

Calculate mean values from triplicates and determine the mean deviation (df = dilution factor).

3.4.1.2. Extracellular Killing (NETs)

1. Resuspend human neutrophils at 2×10^6/mL and allow them to adhere to 24 well tissue culture test plates in 500 μL RPMI-HEPES with 1% HSA per well at 37°C with 5% CO_2. To induce NET formation, add 500 μL RPMI medium with 1% HSA containing 20 nM PMA to each sample and incubate for 4 h at 37°C with 5% CO_2.
2. Centrifuge the plates at $300 \times g$ for 5 min at room temperature and aspirate the medium carefully in order to not remove or destroy the fragile NETs. Immediately add fresh medium. For measuring killing of opsonized *C. albicans* use 450 μL RPMI-HEPES containing 1% heat inactivated pooled human serum. For measuring killing of nonopsonized *C. albicans*, use RPMI-HEPES containing 1% HSA.
3. To adjust the number of *C. albicans* cells for infection of neutrophils determine the concentration of the *C. albicans* culture and dilute as described in **Section 3.4.1.1** (**steps 2** and **3**) and in **Table 21.1**.
4. Dilute your *C. albicans* suspensions to an MOI of 0.01 in PBS.
5. Add 50 μL of the adjusted *C. albicans* suspension to each well containing the neutrophils, centrifuge plates with $700 \times g$ for 10 min at 37°C, and incubate the plates for the desired time point at 37°C with 5% CO_2. As an example: Use an MOI of 0.01 and incubate for 30 min, 60 min, and 120 min.
6. Prepare identical samples and add DNase-1 at a final concentration of 100 U/mL. These samples serve as 100% controls as DNase-1 efficiently degrades the NETs and prevents NET-mediated killing in these samples.
7. After incubation, neutrophil lysis is not necessary as NET formation causes cell death. Directly scrape the bottom of each well thoroughly with a mini cell scraper to remove all remaining *C. albicans* and NETs. Pipette vigorously up and down and transfer the suspension to a 1.5-mL reaction tube.
8. Dilute each sample 10^{-1}, 10^{-2}, 10^{-3}, and plate 50 μL of each dilution on YPD-Agar plates and incubate the plates for 24 h at 30°C.
9. Count the colony forming units (CFU) on the plates. Plates with CFU counts between 30 and 600 can be evaluated.
10. Calculate the percent survival of each sample using the formula:

$$\%_{\text{survival}} = \frac{\text{df} \times \text{CFU}_{\text{without DNase}}}{\text{df} \times \text{CFU}_{\text{with DNase}}} \times 100$$

To calculate percent NET killing use:

$$\%_{\text{NET killing}} = \frac{(\text{df} \times \text{CFU}_{\text{with DNase}} - \text{df} \times \text{CFU}_{\text{without DNase}})}{\text{df} \times \text{CFU}_{\text{with DNase}}} \times 100$$

Calculate mean values from triplicates and determine the mean deviation (df = dilution factor).

3.4.1.3. Combined Killing (Phagocytosis and NETs)

1. Resuspend human neutrophils at 2×10^6/mL, and allow them to adhere to 24-well tissue culture test plates in 500 μL RPMI-HEPES with 1% HSA per well at 37°C with 5% CO_2 with addition of 20 nM PMA to induce NET formation and incubate for 20 min.
2. Centrifuge the plates at 300 × *g* for 5 min at room temperature and carefully replace the medium with RPMI-HEPES containing 1% heat-inactivated pooled human serum with or without Cytochalasin D (Cyt D, 10 μg/mL) and incubate further for 15 min before infection with *C. albicans*. Cyt D inhibits phagocytosis and does not affect the NETs.
3. To adjust the number of *C. albicans* cells for infection of neutrophils, determine the concentration of the *C. albicans* culture and dilute as described in **Section 3.4.1.1** (**steps 2** and **3**) and in **Table 21.1**.
4. Dilute your *C. albicans* suspensions to an MOI of 0.01 in PBS.
5. Preincubated control samples with 100 U/mL of DNase-1 to degrade NETs and prevents NET-mediated killing before the addition of *C. albicans*.
6. Add 50 μL of the adjusted *C. albicans* suspension to each well containing the neutrophils, centrifuge plates with 700 × *g* for 10 min at 37°C, and incubate the plates for the desired time point at 37°C with 5% CO_2. As an example: Dilute *C. albicans* for an MOI of 0.01, add the *C. albicans* suspension, centrifuge and incubate for 30 min, 60 min, and 120 min.
7. Prepare identical samples (i) without adding neutrophils and (ii) adding 100 U/mL DNase-1. These samples serve as controls.
8. After incubation, carefully remove the supernatant to a 1.5-mL reaction tube and store on ice. Lyse neutrophils with 500 μL ice-cold distilled water per well. Incubate for 5 min on ice and scrape the bottom of each well thoroughly with a mini cell scraper to remove all remaining *C. albicans* and neutrophils. Pipette vigorously up and down and add the suspension to the according supernatants collected in the 1.5-mL reaction tubes.

9. Dilute each sample 10^{-1}, 10^{-2}, 10^{-3}, and plate 50 μL of each dilution on YPD-Agar plates and incubate the plates for 24 h at 30°C.
10. Count the colony forming units (CFU) on the plates. Plates with CFU counts between 30 and 600 can be evaluated.
11. The percent killing can be determined separately for (a) total killing, (b) NET-mediated killing, and (c) intracellular killing. First calculate the survival as follows:

(a) total: $$\%_{\text{survival (total)}} = \frac{\text{df} \times \text{CFU}_{\text{with neutrophils}}}{\text{df} \times \text{CFU}_{\text{without neutrophils}}} \times 100;$$

(b) NET mediated:

$$\%_{\text{survival (net)}} = \frac{\text{df} \times \text{CFU}_{\text{with neutrophils and Cyt D}}}{\text{df} \times \text{CFU}_{\text{without neutrophils and with Cyt D}}} \times 100;$$

(c) intracellular:

$$\%_{\text{survival (intracellular)}} = \frac{\text{df} \times \text{CFU}_{\text{with neutrophils and DNase}}}{\text{df} \times \text{CFU}_{\text{without neutrophils and with DNase}}} \times 100.$$

Calculate individual percent killing: 100% – percent survival (df = dilution factor). Calculate mean values from triplicates and determine the mean deviation.

3.4.2. *S. flexneri*

3.4.2.1. Intracellular Killing (Phagocytosis)

1. Resuspend human neutrophils at 2×10^6/mL, and allow them to adhere to 24-well tissue culture test plates in 500 μL RPMI-HEPES per well at 37°C with 5% CO_2. For measuring killing of opsonized *S. flexneri*, use RPMI-HEPES containing 1% heat-inactivated pooled human serum. For measuring killing of nonopsonized *S. flexneri*, use RPMI-HEPES containing 1% HSA. Each sample should be tested in triplicate.
2. To adjust the number of *S. flexneri* cells for infection of neutrophils, determine the *S. flexneri* concentration. Centrifuge at 2000 × *g* for 5 min at 4°C and resuspend the *S. flexneri* cell pellet in PBS. Measure the optical density at 600-nm wavelength (OD_{600}) and correlate 4×10^8 *S. flexneri*/mL to 1 OD_{600}.
3. Dilute the *S. flexneri* suspensions accordingly to the desired multiplicity of infection (MOI; *S. flexneri* cells per neutrophil) in PBS. MOIs between 1 and 0.01 are usual. To calculate the correct dilutions for different MOIs, see **Table 21.1**.
4. Add 50 μL of the adjusted *S. flexneri* suspension to each well containing the neutrophils, centrifuge plates with 700 × *g* for 10 min at 37°C, and incubate the plates for the desired time

point at 37°C with 5% CO_2. For example: Dilute *S. flexneri* for an MOI of 1 and 0.1, add the *S. flexneri* suspension, centrifuge and incubate for 30 min, 60 min, and 120 min.

5. Prepare identical samples without adding neutrophils. These samples serve as controls.
6. After incubation, carefully remove the supernatant to a 1.5-mL reaction tube and store on ice. Lyse neutrophils with 500 μL ice-cold distilled water per well. Incubate for 5 min on ice and scrape the bottom of each well thoroughly with a mini cell scraper to remove all remaining *S. flexneri* and neutrophils. Pipette vigorously up and down and add the suspension to the corresponding supernatants collected in the 1.5-mL reaction tubes.
7. Dilute each sample 10^{-1}, 10^{-2}, 10^{-3}, and plate 50 μL of each dilution on TSB-Agar plates and incubate the plates for 24 h at 37°C.
8. Count the colony forming units (CFU) on the plates. Plates with CFU counts between 30 and 600 can be evaluated.
9. Calculate the percent of survival each sample using the formula:

$$\%_{\text{survival}} = \frac{\text{df} \times \text{CFU}_{\text{with neutrophils}}}{\text{df} \times \text{CFU}_{\text{without neutrophils}}} \times 100$$

To calculate percent phagolysosomal killing use:

$$\%_{\text{phagolysosomal killing}} = \frac{(\text{df} \times \text{CFU}_{\text{without neutrophils}} - \text{df} \times \text{CFU}_{\text{with neutrophils}})}{\text{df} \times \text{CFU}_{\text{without neutrophils}}} \times 100$$

Calculate mean values from triplicates and determine the mean deviation (df = dilution factor).

3.4.2.2. Extracellular Killing (NETs)

1. Resuspend human neutrophils at 2 × 10^6/mL, and allow them to adhere to 24-well tissue culture test plates in 500 μL RPMI-HEPES with 1% HSA per well at 37°C with 5% CO_2. To induce NET formation, add 500 μL RPMI-HEPES with 1% HSA containing 40 nM PMA to each sample and incubate for 4 h at 37°C with 5% CO_2.
2. Centrifuge the plates at 300 × *g* for 5 min at room temperature and aspirate the medium carefully in order to not remove or destroy the fragile NETs. Immediately add fresh medium. For measuring killing of opsonized *S. flexneri*, use 450 μL RPMI-HEPES containing 1% heat-inactivated pooled human serum. For measuring killing of nonopsonized *S. flexneri*, use RPMI-HEPES containing 1% HSA.

3. To adjust the number of *S. flexneri* cells for infection of neutrophils, determine the concentration of the *S. flexneri* culture and dilute as described in **Section 3.4.2.1** (**steps 2** and **3**) and in **Table 21.1**.
4. Dilute your *S. flexneri* suspensions to an MOI of 0.01 in PBS.
5. Add 50 μL of the adjusted *S. flexneri* suspension to each well containing the neutrophils, centrifuge plates with 700 × *g* for 10 min at 37°C, and incubate the plates for the desired time point at 37°C with 5% CO_2. As an example: Use an MOI of 0.01 and incubate for 30 min, 60 min, and 120 min.
6. Prepare identical samples and add DNase-1 at a final concentration of 100 U/mL. These samples serve as 100% controls as DNase-1 efficiently degrades the NETs and prevents NET-mediated killing in these samples.
7. After incubation, neutrophil lysis is not necessary as NET formation causes cell death. Directly scrape the bottom of each well thoroughly with a mini cell scraper to remove all remaining *S. flexneri* and NETs. Pipette vigorously up and down and transfer the suspension to a 1.5-mL reaction tube.
8. Dilute each sample 10^{-1}, 10^{-2}, 10^{-3}, and plate 50 μL of each dilution on TSB-Agar plates and incubate the plates for 24 h at 37°C.
9. Count the colony forming units (CFU) on the plates. Plates with CFU counts between 30 and 600 can be evaluated.
10. Calculate the percent of survival each sample using the formula:

$$\%_{\text{survival}} = \frac{\text{df} \times \text{CFU}_{\text{without DNase}}}{\text{df} \times \text{CFU}_{\text{with DNase}}} \times 100$$

To calculate percent NET killing use:

$$\%_{\text{NET killing}} = \frac{(\text{df} \times \text{CFU}_{\text{with DNase}} - \text{df} \times \text{CFU}_{\text{without DNase}})}{\text{df} \times \text{CFU}_{\text{with DNase}}} \times 100$$

Calculate mean values from triplicates and determine the mean deviation (df = dilution factor).

3.4.2.3. Combined Killing (Phagocytosis and NETs)

1. Resuspend human neutrophils at 2 × 10^6/mL, and allow them to adhere to 24-well tissue culture test plates in 500 μL RPMI-HEPES with 1% HSA per well at 37°C with 5% CO_2 with addition of 20 nM PMA to induce NET formation and incubate for 20 min.
2. Centrifuge the plates at 300 × *g* for 5 min at room temperature and carefully replace the medium with RPMI-HEPES

containing 1% heat-inactivated pooled human serum with or without Cytochalasin D (Cyt D, 10 μg/mL) and incubate further for 15 min before infection with *S. flexneri*. Cyt D treatment completely inhibits phagocytosis and does not affect the NETs.

3. To adjust the number of *S. flexneri* cells for infection of neutrophils, determine the concentration of the *S. flexneri* culture and dilute as described in **Section 3.4.2.1** (**steps 2** and **3**) as well as in **Table 21.1**.
4. Dilute your *S. flexneri* suspensions to an MOI of 0.01 in PBS.
5. Preincubate control samples with 100 U/mL of DNase-1 to degrade NETs and prevent NET-mediated killing before the addition of *S. flexneri*.
6. Add 50 μL of the adjusted *S. flexneri* suspension to each well containing the neutrophils, centrifuge plates at 700 × *g* for 10 min at 37°C, and incubate the plates for the desired time point at 37°C with 5% CO_2. As an example: Dilute *S. flexneri* for an MOI of 0.01, add the *S. flexneri* suspension, centrifuge and incubate for 30 min, 60 min, and 120 min.
7. Prepare identical samples (i) without adding neutrophils and (ii) adding 100 U/mL DNase-1. These samples serve as controls.
8. After incubation, carefully remove the supernatant to a 1.5-mL reaction tube and store on ice. Lyse neutrophils with 500 μL ice-cold distilled water per well. Incubate for 5 min on ice and scrape the bottom of each well thoroughly with a mini cell scraper to remove all remaining *S. flexneri* and neutrophils. Pipette vigorously up and down and add the suspension to the according supernatants collected in the 1.5-mL reaction tubes.
9. Dilute each sample 10^{-1}, 10^{-2}, 10^{-3}, and plate 50 μL of each dilution on TSB-Agar plates and incubate the plates for 24 h at 30°C.
10. Count the colony forming units (CFU) on the plates. Plates with CFU counts between 30 and 600 can be evaluated. The percent killing can be determined separately for (a) total killing, (b) NET-mediated killing, and (c) intracellular killing. First calculate the survival as follows:

(a) total: $$\%_{\text{survival (total)}} = \frac{\text{df} \times \text{CFU}_{\text{with neutrophils}}}{\text{df} \times \text{CFU}_{\text{without neutrophils}}} \times 100;$$

(b) NET mediated:

$$\%_{\text{survival (net)}} = \frac{\text{df} \times \text{CFU}_{\text{with neutrophils and Cyt D}}}{\text{df} \times \text{CFU}_{\text{without neutrophils and with Cyt D}}} \times 100;$$

(c) intracellular:

$$\%_{\text{survival (intracellular)}} = \frac{\text{df} \times \text{CFU}_{\text{with neutrophils and DNase}}}{\text{df} \times \text{CFU}_{\text{without neutrophils and with DNase}}} \times 100.$$

Calculate individual percent killing: 100%-percent survival (df = dilution factor). Calculate mean values from triplicates and determine the mean deviation.

3.5. Killing Assays with Human Neutrophil Granular Extract

3.5.1. C. albicans

1. Incubate *Candida albicans* (4×10^5/mL) with human neutrophil granule extract (hNGE) in a total volume of 50 μL for 30 to 120 min at 37°C with shaking in $HBSS^-$ with 3% casamino acids. In order to kill *C. albicans*, dilute the hNGE to 2 to 6 mg/mL total protein. Include one control without hNGE in the same buffer.
2. After incubation spot 5-μL aliquots of serial dilutions (1:3) onto YPD-Agar plates and incubate for 24 h at 30°C (**Fig. 21.1**).
3. Examine the plates for growth to define the minimal incubation or minimal concentration required for killing of *C. albicans*.

3.5.2. S. flexneri

1. Incubate *S. flexneri* (1×10^6/mL) with hNGE in a total volume of 100 μL for 30 to 120 min at 37°C with shaking in $HBSS^-$ with 3% casamino acids. In order to kill *S. flexneri*, dilute the hNGE to 0.1 to 2 mg/mL total protein. Include one control without hNGE in the same buffer.
2. After incubation, dilute each sample 10^{-1}, 10^{-2}, 10^{-3}, and plate 50 μL of each dilution on TSB-Agar plates and incubate the plates for 24 h at 37°C.
3. Count the colony forming units (CFU) on the plates.

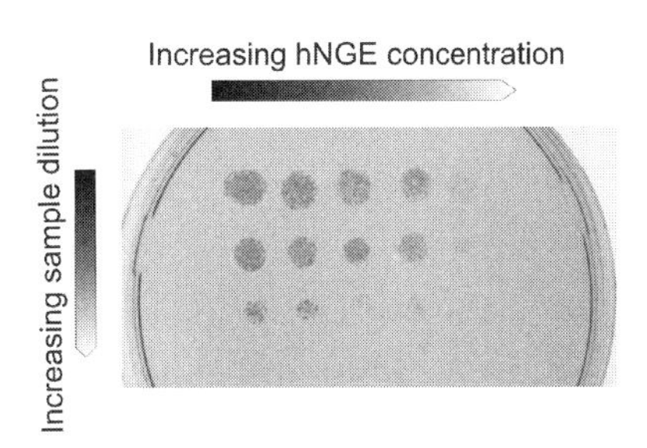

Fig. 21.1. YPD-Agar plated with spots of 5-μL aliquots of *C. albicans* suspensions in different dilutions after incubation in different concentrations of human neutrophil granular extract (hNGE). The plate was incubated for 24 h at 30°C.

4. Calculate the percent survival of each sample using the formula:

$$\%_{\text{survival}} = \frac{\text{df} \times \text{CFU}_{\text{with hNGE}}}{\text{df} \times \text{CFU}_{\text{without hNGE}}} \times 100$$

To calculate percent hNGE mediated killing use:

$$\%_{\text{hNGE killing}} = \frac{(\text{df} \times \text{CFU}_{\text{without hNGE}} - \text{df} \times \text{CFU}_{\text{with hNGE}})}{\text{df} \times \text{CFU}_{\text{without hNGE}}} \times 100$$

Calculate mean values from triplicates and determine the mean deviation (df = dilution factor).

4. Notes

4.1. Notes to Section 3.2

Ficoll purification is based on Ref. *21*. Do not extend **step 8** of the red blood cell lysis over 60 s. This step will also harm and, at least to a certain degree, activate the neutrophils. Do not repeat this step even in case you have little remains of red blood cells. Those will not interfere with your assay.

Here we present an alternative method to purify neutrophils *(22)*. The advantage of this method is that **step 8**, the red blood cell (RBC) lysis, is not required. RBC lysis stresses the neutrophils and can preactivate them. Preactivated neutrophils might change in function and less efficiently kill microbes compared with naive cells. However, the Ficoll preparation is suitable for the experiments described here.

Materials for Histopaque/Percoll Purification

1. Vacutainer K2E 18 mg (Becton Dickinson) color code: purple.
2. Percoll (GE Healthcare).
3. RPMI-HEPES: RPMI (Invitrogen) without phenol red, substituted with 10 mM HEPES (Invitrogen).
4. Histopaque 119 (Sigma-Aldrich).
5. 10x phosphate-buffered saline (PBS): 1.4 M NaCl pH 7.0, 27 mM KCl, 90 mM Na_2HPO_4, 15 mM KH_2PO_4.
6. 1x PBS containing 0.5% human serum albumin (HSA): 140 mM NaCl pH 7.0, 2.7 mM KCl, 9 mM Na_2HPO_4, 1.5 mM KH_2PO_4 supplemented with 0.5% HSA (Grifols).

Method of Histopaque/Percoll Purification

1. Prepare Percoll solutions for discontinuous Percoll gradient. (Solutions can be stored at 4°C overnight).

Mix 13.5 mL Percoll with 1.5 mL 10x PBS (100% isotonic Percoll solution).

Mix 375 μL RPMI with 2125 μl 100% Percoll (85% solution).

Mix 500 μL RPMI with 2000 μL 100% Percoll (80% solution).

Mix 625 μL RPMI with 1875 μL 100% Percoll (75% solution).

Mix 750 μL RPMI with 1750 μL 100% Percoll (70% solution).

Mix 875 μL RPMI with 1625 μL 100% Percoll (65% solution). Store it at room temperature untill use in step 9.

2. Carefully layer 2 mL of each solution with a plastic Pasteur pipette in a 15-mL centrifugation tube starting with 85% solution. Store it at room temperature until use in step 9.
3. Take a 10 mL whole blood sample with a K2E Vacutainer. Ten milliliters whole blood corresponds with 1×10^7 to 2×10^7 PMNs, which is sufficient for a killing assay with 5 to 10 conditions in triplicate.
4. Layer 5 mL of collected blood on 5 mL Histopaque 1119 with a plastic Pasteur pipette in a 15-mL centrifugation tube.
5. Centrifuge tubes 20 min at $800 \times g$ and room temperature.
6. Carefully discard the interphase and collect the diffuse red phase of Histopaque 1119 above the red blood cell pellet.
7. Wash 5 mL of collected cells in 10 mL PBS with 0.5% HSA in a new centrifugation tube.
8. Spin down the cells for 10 min at $300 \times g$ and room temperature.
9. Remove supernatant and resuspend cell pellet in 2 mL PBS with 0.5% HSA and load cells on top of Percoll gradient.
10. Centrifuge gradient 20 min at $800 \times g$ and room temperature.
11. Collect distinct white layer between clear 70% and 75% Percoll layers in a 15-mL centrifugation tube (**Fig. 21.2**).
12. Fill up to 15 mL PBS with 0.5% HSA and spin down cells for 10 min at $300 \times g$ at room temperature.
13. Resuspend cell pellet in 1 mL PBS with 0.5% HSA and count cells.
14. Dilute cells to 2×10^6 cells / mL in RPMI-HEPES.

Frequently at **step 11**, additional neutrophil bands appear between the 75% and 80% as well as between the 80% and 85% Percoll layers. The bands at higher densities contain neutrophils that are already approaching an apoptosis program. Particularly, neutrophils that accumulate between the layers of 80% and 85% are close to apoptosis (**Fig. 21.2**).

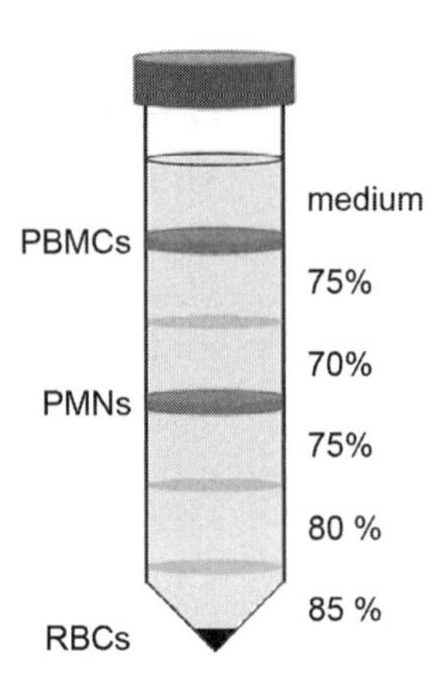

Fig. 21.2. Percoll gradient for neutrophil purification. Concentration of Percoll is indicated in percent. PBMCs, peripheral blood mononuclear cells; PMNs, polymorphonuclear neutrophils; RBCs, red blood cells.

4.2. Notes to Section 3.3

Preparation of human neutrophil granular extract is based on Ref. *23*.

1. Usually 1 unit of buffy coat is about 50 to 80 mL of blood enriched in leukocytes from 500 mL peripheral blood.
2. Use siliconized tubes and tips to minimize loss of cationic granular proteins due to adhesion to plastic surfaces.
3. Sonication of neutrophils (**step 2**) to release intact granules is a crucial step to purify granule proteins. This step must be optimized for your own purposes. Use different pulse duration and intensity for a maximal protein yield.
4. If the cells settle upon standing again, then you need to pulse again. Be careful to not overheat the cells.
5. Preparation of 20 mM sodium acetate at pH 4.0: Prepare 0.2 M sodium acetate and 0.2 M acetic acid. Mix 180 mL of 0.2 M sodium acetate with 820 mL 0.2 M acetic acid. The resulting solution is 10-fold concentrated. Dilute this solution appropriately before use.

4.3. Notes to Sections 3.4.1 and 3.4.2

1. To obtain pooled human serum, collect 5 mL peripheral blood from three different volunteers in tubes without anticoagulants, such as EDTA. Let the blood coagulate (in about 30 to 60 min). To accelerate coagulation, use centrifuge tubes containing a clot activator (serum tube PP, nerbe plus; Winsen/Luhe, Germany). After coagulation, centrifuge tubes at $3000 \times g$ for 30 min and collect supernatant (serum). Pool the sera from different donors and heat inactivate at 56°C in a water bath for 30 min to destroy complement components. The pooled human serum can be stored in aliquots at −20°C.

2. An essential step in **Sections 3.4.1** and **3.4.2** is to centrifuge the plates after infection of the neutrophils with *C. albicans* or *S. flexneri* respectively (**step 4** in **Sections 3.4.1.1** and **3.4.2.1; step 5** in **Sections 3.4.1.2** and **3.4.2.2; step 6** in **Sections 3.4.1.3** and **3.4.2.3**). Particularly, the bacteria do not settle so fast that all added microbes will be in close vicinity to the neutrophils that adhered to the bottom of the wells.
3. *C. albicans* and to a minor extent *S. flexneri* adhere to the surfaces of the tissue culture plates used in the killing assays. When removing the samples, many fungal or bacterial cells might be lost by this means. This in turn falsifies your results. To remove *C. albicans* properly, we add 20 μg/mL proteinase K for 10 min before scraping (**step 6** in **Section 3.4.1.1, step 7** in **Section 3.4.1.2**, and **step 8** in **Section 3.4.1.3**). The addition of protease increases the yield of *C. albicans* especially from control plates without neutrophils.
4. To overcome the problem of hyphae clumping, measure six values for each condition. To alleviate the distortion caused by extreme values, calculate the trimmed mean omitting the highest and lowest values. Additionally, dilute samples such that 200 to 500 CFU result per plate.
5. Slight sonication after removing the microbes from the wells might help to resuspend clumped fungal or bacterial cells for better results. At low energy levels, the procedure does not harm *C. albicans* or *S. flexneri*. Nevertheless, the adequate sonication dose should be determined experimentally for your own purposes and performed with all samples equally.
6. For higher dilutions of your samples (10^{-2} and 10^{-3}), use only one half of an agar plate per dilution to safe material. This is possible as you expect less CFU.

4.4. Notes to Section 3.5

1. Always add hNGE as the last component to each of your samples and use siliconized tubes to minimize loss on plastic surfaces prior to incubation.
2. We plate *C. albicans* in serial dilutions of 5-μL aliquots to determine minimal lethal concentrations of hNGE because this approach provides more reproducible results (**Fig. 21.1**) than does plating, as described for *S. flexneri*.

References

1. Borregaard, N. & Cowland, J.B. (1997) Granules of the human neutrophilic polymorphonuclear leukocyte. *Blood* **89,** 3503–3521.
2. Nathan, C. (2006) Neutrophils and immunity: challenges and opportunities. *Nat Rev Immunol* **6,** 173–182.
3. Levy, O. (2000) Antimicrobial proteins and peptides of blood: templates for novel antimicrobial agents. *Blood* **96,** 2664–2672.
4. Pericle, F., Liu, J.H., Diaz, J.I., Blanchard, D.K., Wei, S., Forni, G. & Djeu, J.Y. (1994) Interleukin-2 prevention of

apoptosis in human neutrophils. *Eur J Immunol* **24,** 440–444.
5. Faurschou, M. & Borregaard, N. (2003) Neutrophil granules and secretory vesicles in inflammation. *Microbes Infect* **5,** 1317–1327.
6. Weber, C. (2003) Novel mechanistic concepts for the control of leukocyte transmigration: specialization of integrins, chemokines, and junctional molecules. *J Mol Med* **81,** 4–19.
7. Cox, J.A., Jeng, A.Y., Blumberg, P.M. & Tauber, A.I. (1987) Comparison of subcellular activation of the human neutrophil NADPH-oxidase by arachidonic acid, sodium dodecyl sulfate (SDS), and phorbol myristate acetate (PMA). *J Immunol* **138,** 1884–1888.
8. Cohn, Z.A. & Hirsch, J.G. (1960) The influence of phagocytosis on the intracellular distribution of granule-associated components of polymorphonuclear leucocytes. *J Exp Med* **112,** 1015–1022.
9. Clark, R.A., Leidal, K.G., Pearson, D.W. & Nauseef, W.M. (1987) NADPH oxidase of human neutrophils. Subcellular localization and characterization of an arachidonate-activatable superoxide-generating system. *J Biol Chem* **262,** 4065–4074.
10. Hampton, M.B., Kettle, A.J. & Winterbourn, C.C. (1998) Inside the neutrophil phagosome: oxidants, myeloperoxidase, and bacterial killing. *Blood* **92,** 3007–3017.
11. Brinkmann, V., Reichard, U., Goosmann, C., Fauler, B., Uhlemann, Y., Weiss, D.S., Weinrauch, Y. & Zychlinsky, A. (2004) Neutrophil extracellular traps kill bacteria. *Science* **303,** 1532–1535.
12. Urban, C.F., Reichard, U., Brinkmann, V. & Zychlinsky, A. (2006) Neutrophil extracellular traps capture and kill Candida albicans yeast and hyphal forms. *Cell Microbiol* **8,** 668–676.
13. Fuchs, T.A., Abed, U., Goosmann, C., Hurwitz, R., Schulze, I., Wahn, V., Weinrauch, Y., Brinkmann, V. & Zychlinsky, A. (2007) Novel cell death program leads to neutrophil extracellular traps. *J Cell Biol* **176,** 231–241.
14. Magee, B.B. & Magee, P.T. (2000) Induction of mating in Candida albicans by construction of MTLa and MTLalpha strains. *Science* **289,** 310–313.
15. Odds, F.C. (1994) Pathogenesis of Candida infections. *J Am Acad Dermatol* **31,** S2–5.
16. Ernst, J.F. (2000) Transcription factors in Candida albicans – environmental control of morphogenesis. *Microbiology* **146** (Pt 8), 1763–1774.
17. Pfaller, M.A., Jones, R.N., Messer, S.A., Edmond, M.B. & Wenzel, R.P. (1998) National surveillance of nosocomial blood stream infection due to Candida albicans: frequency of occurrence and antifungal susceptibility in the SCOPE Program. *Diagn Microbiol Infect Dis* **31,** 327–332.
18. Kotloff, K.L., Winickoff, J.P., Ivanoff, B., Clemens, J.D., Swerdlow, D.L., Sansonetti, P.J., Adak, G.K. & Levine, M.M. (1999) Global burden of Shigella infections: implications for vaccine development and implementation of control strategies. *Bull World Health Organ* 77, 651–666.
19. Ogawa, M. & Sasakawa, C. (2006) Intracellular survival of Shigella. *Cell Microbiol* **8**, 177–184.
20. Weinrauch, Y., Drujan, D., Shapiro, S.D., Weiss, J. & Zychlinsky, A. (2002) Neutrophil elastase targets virulence factors of enterobacteria. *Nature* **417,** 91–94.
21. Boyum, A. (1968) Isolation of mononuclear cells and granulocytes from human blood. Isolation of monuclear cells by one centrifugation, and of granulocytes by combining centrifugation and sedimentation at 1 g. *Scand J Clin Lab Invest* **97,** 77–89.
22. Aga, E., Katschinski, D.M., van Zandbergen, G., Laufs, H., Hansen, B., Muller, K., Solbach, W. & Laskay, T. (2002) Inhibition of the spontaneous apoptosis of neutrophil granulocytes by the intracellular parasite Leishmania major. *J Immunol* **169,** 898–905.
23. Weiss, J., Elsbach, P., Olsson, I. & Odeberg, H. (1978) Purification and characterization of a potent bactericidal and membrane active protein from the granules of human polymorphonuclear leukocytes. *J Biol Chem* **253,** 2664–2672.

Chapter 22

Endothelial Cell Stimulation by *Candida albicans*

Quynh T. Phan and Scott G. Filler

Abstract

The opportunistic fungal pathogen *Candida albicans* enters the bloodstream and causes hematogenously disseminated infection in hospitalized patients. During the initiation of a hematogenously disseminated infection, endothelial cells are one of the first host cells to come in contact with *C. albicans*. Endothelial cells can significantly influence the local host response to *C. albicans* by expressing leukocyte adhesion molecules and proinflammatory cytokines. Thus, it is of interest to investigate the response of endothelial cells to *C. albicans in vitro*. We describe the use of real-time PCR and enzyme immunoassays to measure the effects of *C. albicans* on the endothelial cell production of E-selectin and tumor necrosis factor α *in vitro*.

Key words: endothelial cell, *Candida albicans*, E-selectin, IL-8, EIA, real-time PCR.

1. Introduction

Endothelial cells form the lining of the blood vessels. In addition to regulating vascular permeability and vascular tone, endothelial cells are important for initiating and regulating the local host response to infection. Endothelial cells are capable of synthesizing a variety of factors that influence the activity of neutrophils and other leukocytes. These factors include proinflammatory cytokines such as tumor necrosis factor α (TNF-α), interleukin 1 (IL-1), IL-6, IL-8, and monocyte chemotactic protein 1 (MCP-1). Endothelial cells also express leukocyte adhesion molecules such as P-selectin, E-selectin, intercellular adhesion molecule 1 (ICAM-1), and vascular cell adhesion molecule 1 (VCAM-1). In general, the expression and secretion of these immunomodulators by endothelial cells help recruit leukocytes to foci of infection. The proinflammatory cytokines also activate the microbicidal activities of these recruited leukocytes.

Steffen Rupp, Kai Sohn (eds.), *Host-Pathogen Interactions*, DOI: 10.1007/978-1-59745-204-5_22,

During the initiation of hematogenously disseminated candidiasis, *Candida albicans* cells in the bloodstream must traverse the endothelial cell lining of the vasculature to invade the deep tissues. Thus, endothelial cells are one of the first host cells that interact with this organism. Infection with *C. albicans in vitro* stimulates endothelial cells to synthesize and release a variety of leukocyte adhesion molecules and proinflammatory cytokines *(1, 2)*. For example, endothelial cells express E-selectin on their surface in response to *C. albicans* infection *(1, 2)*. This leukocyte adhesion molecule mediates rolling adherence of neutrophils and monocytes to endothelial cells, and its presence likely contributes to the recruitment of these leukocytes at foci of *C. albicans* infection. Endothelial cells also secrete large amounts of IL-8 in response to *C. albicans in vitro (1, 2)*. This cytokine is a potent neutrophil chemoattractant. Also, neutrophils that have been incubated in IL-8 have enhanced ability to kill *C. albicans (3)*.

The effects of *C. albicans* on endothelial cell expression of E-selectin and IL-8 can be measured at the level of mRNA accumulation using real-time reverse transcriptase PCR. The surface expression of E-selectin and secretion of IL-8 into the medium can also be measured by enzyme immunoassays (EIAs).

2. Materials

2.1. *C. albicans* Culture

1. Yeast extract peptone dextrose (YPD) agar: dissolve 5 g of yeast extract (Becton Dickinson, Sparks, MD), 10 g peptone (Becton Dickinson), 10 g glucose (Fisher Scientific, Pittsburgh, PA), and 10 g agar (Fisher Scientific) in 500 mL of water. Autoclave at 121°C for 15 min. Avoid overheating and remove medium from autoclave as soon as the sterilization cycle is completed to minimize caramelization.
2. Yeast nitrogen base (YNB) liquid medium: dissolve 6.7 g of YNB (Becton Dickinson) and 5 g of glucose (Fisher Scientific) in 100 mL sterile irrigation water and filter sterilize. This medium can be stored at 4°C for up to 1 month. For use, dilute 1:10 with endotoxin-free water (*see* **Note 1**).

2.2. Cell Culture

1. M199 tissue culture medium: M199 powder (Gibco/BRL, Bethesda, MD) is dissolved in 1 L endotoxin-free water. Next, add 100 mL of fetal bovine serum (FBS), 100 mL bovine calf serum (BCS) (both from Gemini Bio-products Woodland, CA; use serum lots with low endotoxin concentration), and 10 mL of a solution containing 1% penicillin, 1% streptomycin, and 2 mM L-glutamine (Gemini Bio-products).

The FBS and BCS can be stored in 50-mL aliquots at −20°C. Store the penicillin–streptomycin–L-glutamine solution in 5-mL aliquots at −20°C. Filter sterilize the medium and store at 4°C. Warm medium up to 37°C in a water bath before adding it to the endothelial cells.

2. 96-well and 24-well flat-bottom tissue culture plates, and Primaria 25 cm^2 tissue culture flasks are from Becton Dickinson (Franklin Lakes, NJ).
3. Phosphate-buffered saline without Ca^{++} or Mg^{++} (PBS) and Hanks balanced salt solution (HBSS) are from Mediatech, Inc (Herndon, VA) (*see* **Note 2**).
4. Solution of 10x trypsin (5 g/L) and ethylenediamine tetraacetic acid (EDTA) 2 g/L (Gemini Bioproducts) is stored in 4-mL aliquots at −20°C. Dilute 1:10 in PBS and warm to room temperature before use. The 1x solution can be stored at 4°C for 7 days.
5. Dilute 2% tissue culture quality gelatin stock solution (Sigma-Aldrich, St. Louis, MO) 1:10 in endotoxin-free water. Filter sterilize and store at 4°C for up to 1 month.
6. RPMI 1640 medium is from Irvine Scientific (Santa Ana, CA).

2.3. RNA Extraction and Real-time PCR

2.3.1. RNA Extraction (All Reagents Are Molecular Biology Grade)

1. TRI reagent from Ambion (Austin, TX).
2. Chloroform (Sigma-Aldrich).
3. Glycogen (Sigma-Aldrich).
4. RNase-free water (Ambion).
5. Isopropanol (Sigma-Aldrich).
6. 75% ethanol (Sigma-Aldrich).

2.3.2. DNase Treatment and RNA Cleanup

1. TURBO DNase I and 10X TURBO DNase buffer (Ambion).
2. RNA clean up kit from Zymo Research (Orange, CA).

2.3.3. Real-time RT-PCR

1. Retroscript (Ambion):
   (a) 10 units/μL RNase inhibitor
   (b) 10X reverse transcriptase buffer
   (c) 25 mM dNTP mixture
   (d) 100 units/μL M-MLV reverse transcriptase
   (e) 50 μM oligo(dT)
2. Taqman universal PCR master mix (Applied Biosystems, Foster City, CA).
3. IL-8 primers (Applied Biosystems).
4. E-selectin primer (Applied Biosystems).
5. GAPDH primer (Applied Biosystems).

6. 96-well microtiter reaction plate (Applied Biosystems).
7. Adhesive cover (Applied Biosystems).
8. Compressor pad (Applied Biosystems). The pad can be reused.

2.4. E-selectin EIA

1. 10X TBS-Glycine: Add 6.61 g Tris-HCl, 0.97 g Tris Base, 8.77 g NaCl, and 10.5 g glycine to ~99 mL water, adjust pH to 7.5 and volume to 100 mL, and filter sterilize. This solution can be stored at 4°C for up to 6 months. It should be diluted to 1X just before use.
2. The PBS used for rinsing fixed endothelial cells does not need to be endotoxin-free (*see* **Note 2**).
3. Blocking buffer: Add 1 mL of sheep serum (Sigma-Aldrich) and 3 g of bovine serum albumin (BSA; Sigma-Aldrich) to 99 mL of PBS, and filter sterilize. The solution can be stored at 4°C for up to 3 months.
4. 0.1% BSA in PBS: Add 1 g of BSA to 1 L of PBS, and filter sterilize. The solution can be stored at 4°C for up to 3 months.
5. 4% paraformaldehyde in PBS: Warm up 10 mL of PBS to about 90°C, and then add 0.4 g of paraformaldehyde. Vortex to mix. Paraformaldehyde is virtually impossible to dissolve in cold PBS. Do not inhale the vapors of this solution because they are toxic. Make this solution up the day of the experiment and always use when fresh.
6. M199 + 5% FBS should be made up right before.
7. Primary antibody: anti–E-selectin antibody (Biodesign, Saco, MN). Antibody can be aliquoted in 100 μL volumes and stored at −80°C. Thawed aliquots can be stored at 4°C for up to 2 weeks.
8. Secondary antibody: sheep anti-mouse IgG conjugated with horse radish peroxidase is purchased from Sigma-Aldrich.
9. *O*-phenylenediamine dihydrochloride (OPD) tablets are from Sigma-Aldrich.

3. Methods

The endothelial cell response to *C. albicans* can be analyzed using real-time PCR to measure the accumulation of mRNA for E-selectin and IL-8. The advantages of this technique are that it is sensitive and requires a relatively small amount of endothelial cells. Also, primers for virtually all human leukocyte adhesion molecules and cytokines are commercially available, and the mRNA expression of multiple immunomodulators can be determined using cDNA prepared from a single sample.

It is frequently useful to verify that increased mRNA expression is associated with increased protein expression. To determine this association, changes in the expression of leukocyte adhesion molecules on the endothelial cell surface can be quantified by EIA. Also, the conditioned medium above the endothelial cells can be collected and analyzed for cytokine content using either commercially available EIAs or flow cytometric bead arrays.

3.1. Growth and Preparation of *C. albicans*

1. *C. albicans* cells are maintained on YPD agar plates at room temperature. Approximately 16 h before the experiment, a single colony is used to inoculate 10 mL of endotoxin-free YNB broth containing 0.5% glucose, which is incubated in rotary shaker at ~150 rpm at 30°C (*see* **Note 3**).
2. After incubation, vortex the culture tube vigorously to break up the clumps. Next, harvest the organisms by centrifuging them at 1000 × *g* for 5 min at room temperature.
3. Decant supernatant. Vortex pellet vigorously and then resuspend in 10 mL sterile, endotoxin-free PBS. Centrifuge at 1000 × *g* for 5 min at room temperature. Repeat this step a total of 2 times. After the last centrifugation, resuspend the pellet in 10 mL HBSS. Vortex vigorously again.
4. Dilute stock suspensions of *C. albicans* 1:100 in PBS and count the number of cells with a hemacytometer (yield should be about 10^8 organisms per mL).
5. Make up a suspension of organisms in prewarmed M-199 + 20% serum (*see* **Notes 4** and **5**).
6. Warm all solutions to 37°C immediately before adding them to the endothelial cells.

3.2. Endothelial Cell RNA Extraction

1. Human umbilical vein endothelial cells are maintained in 25 cm^2 tissue culture flasks coated with 0.2 % gelatin. When the cells reach confluency, they are detached from the bottom of the flask using trypsin-EDTA (*see* **Note 6**). When cells are passaged for maintenance culture, they are split at a 1:3 ratio. For use in the real-time PCR experiments, the wells of a 24-well tissue culture plate are coated with 0.5 mL of 10 μg/mL fibronectin in PBS for at least 30 min. Next, aspirate the fibronectin solution and add endothelial cells to the wells. One flask of confluent endothelial cells can be split into all wells of the 24-well tissue culture plate (*see* **Notes 7** and **8**). To obtain sufficient RNA for the experiments, each condition should be tested in duplicate.
2. The night before the experiment, place some M-199 + 20% serum in the 5% CO_2 incubator for use in the experiment. Keep the cap on the container loose so that the tissue culture medium can equilibrate with the CO_2. The various stimuli

(e.g., *C. albicans* and TNF-α) should be prepared in this medium on the day of the experiment.

3. On the day of the experiment, aspirate the medium from two wells at a time and add the desired stimulus in a total volume of 500 μL (*see* **Note 9**). All experiments should include a set of wells that are incubated with medium alone. These wells are essential for using the $\Delta\Delta C_T$ method to calculate the extent of endothelial cell stimulation. It is also useful to include wells incubated with TNF-α (final concentration 400 pg/mL) as a positive control.
4. Incubate plate for 8 h at 37°C in 5% CO_2. At the end of the incubation period, examine wells with the inverted microscope to look for loss of endothelial cells and determine the morphology of *C. albicans*. Wild-type strains should form extensive hyphae.
5. Gently aspirate medium from each well and wash each well once with 500 μL of cold HBSS per well (*see* **Notes 10** and **11**).
6. Aspirate HBSS and add 250 μL of Tri Reagent to each well. Homogenize the cells by repeatedly passing them up and down with a P1000 Pipetman. Change pipet tip between samples.
7. Incubate plate at room temperature for 5 min.
8. Transfer the lysates to 1.5-mL Microfuge tubes, combining the lysates from each pair of wells that were exposed the same condition.
9. Samples can be either stored at −80°C or processed immediately.
10. Thaw lysates and incubate them at room temperature for 5 min to dissociate the nuclear proteins from the nucleic acids.
11. Add 100 μL of chloroform and vortex for 20 s.
12. Let sit at room temperature for 10 min.
13. Centrifuge samples in a Microfuge at 14,000 rpm for 15 min at 4°C.
14. Carefully remove tubes from centrifuge and aspirate the upper aqueous layer to about 3 mm above the interface. Transfer this upper phase to a fresh tube.
15. To each tube add 1 μL of glycogen (10 μg/μL in RNase-free water), 250 μL of RNase-free water, and 500 μL of isopropanol. Mix well using the vortex.
16. Incubate for 10 min at room temperature.
17. Centrifuge samples in a Microfuge at 14,000 rpm for 15 min at 4°C.

18. Remove supernatant and wash pellet by adding 500 μL of 75% ethanol and then vortexing.
19. Centrifuge samples in a Microfuge at 14,000 rpm for 10 min at 4°C.
20. Aspirate supernatant and dry pellet by inverting the Microfuge tubes on a paper towel for about 10 min. Do not let RNA dry out completely.
21. Resuspend pellets in 10 μL of RNase-free water.
22. Determine RNA concentration of the samples using a spectrophotometer.

3.3. Preparing cDNA

1. Remove DNA contamination by treating RNA samples with DNase I (2 units/μL) (*see* **Note 10**). Make up DNase I master mix by adding DNase I to DNase reaction buffer. Add 2 units of DNase I to 5 μg of each RNA sample and incubate at 37°C for 30 min.
2. Clean up the RNA samples with RNA clean-up kit (such as kit R1015, from Zymo Research, Inc., Orange, CA). The kit is designed to treat up to 5 μg RNA with a recovery efficiency of 80% to 95%. The purified RNA can be used immediately or stored at −70°C for future use. Keep RNA samples on ice if they are not going to be stored at −70°C.
3. For reverse transcription of the RNA, label two, 200-μL PCR tubes for each sample. One tube is for the reverse transcriptase reaction, and the other is for the no reverse transcriptase (no-RT) control.
   (a) Remove the secondary structure of the RNA and anneal the oligo(dT) primer to the RNA by adding oligo(dT) (final concentration of 5 μM) to 2 μg of RNA in RNase-free water. Each sample should be added to two separate tubes, and the final volume should be 12 μL per tube. Incubate the tubes at 70°C for 3 min and then rapidly cool on ice.
   (b) The reverse transcription reaction is performed in a final volume of 20 μL per sample, which will contain 8 μL of the reverse transcriptase master mix and 12 μL of RNA + oligo(dT). Make up the reverse transcriptase master mix so that it contains 12.5 μM oligo(dT), 12.5 mM dNTP mixture, 1.25 units/μL of RNase inhibitor, and 12.5 units/μL of reverse transcriptase in reverse transcriptase buffer. You will need 8 μL of master mix for each sample, so plan on making slightly more (about 9 μL per sample).
   (c) Add 8 μL of the reverse transcriptase master mix to one tube from each sample [which contains the RNA and

oligo(dT)] for a total volume of 20 μL. The other tube for each sample is for the no-RT control. Add 8 μL of RNase-free water to this tube; do not add the reverse transcriptase master mix to this tube.

(d) Incubate all tubes at 42°C for 1 h for reverse transcription to take place. Next, incubate the tubes at 92°C for 10 min to inactivate the reverse transcriptase.

(e) Place tubes on ice for immediate use or store at −20°C for later use.

3.4. Real-time PCR

1. These instructions assume the use of a real-time PCR thermocycler that can accept a 96-well microtiter plate. Also, the primers for E-selectin, IL-8, and GAPDH are proprietary Taqman primers obtained from Applied Biosystems. Proprietary primers for virtually all human and mouse genes are available from this and other companies. Alternatively, one can design one's own primers and monitor amplification with SYBR Green (Molecular Probes, Eugene, OR). GAPDH, which has fairly consistent expression, is used as the reference gene to control for differences in cDNA content among the various samples. Other reference genes, such as ACT1 can also be used (*see* **Note 12**).
2. Dilute all samples of cDNA, including the no-RT control, 1:30 using RNase-free water. Prepare a master mix of each primer set (E-selectin, IL-8, and GAPDH) in three clean microcentrifuge tubes. Add 10 μL of 2X Taqman PCR master mix, 1 μL of 20X primer mixture, and 9 μL of diluted cDNA to the wells of the microtiter plate. Each sample should be tested for each primer set in duplicate. Add only the GAPDH master mix to the no-RT controls as only a single set of primers is necessary for testing these controls.
3. Store any leftover diluted cDNA at −20°C.
4. Seal the plate with the adhesive pad, using the applicator to firmly seal the wells.
5. Place the sealed plate into the real-time thermocycler and determine the C_T for each well. The C_T of the no-RT control wells should be either undetectable or more than 30 cycles.
6. Use the C_T results from the duplicate wells to calculate the average the C_T for each condition.
7. To calculate the relative change in gene expression, use the $\Delta\Delta C_T$ method: For each condition, first calculate the ΔC_T

$$\Delta C_T = C_T(\text{target}) - C_T(\text{normalizer})$$

In this experiment, the target is either IL-8 or E-selectin, and the normalizer is GAPDH.

Next, calculate the $\Delta\Delta C_T$

$$\Delta\Delta C_T = \Delta C_T(\text{experimental}) - \Delta C_T(\text{control})$$

The experimental condition is endothelial cells exposed to either *C. albicans* or TNF-α, and the control condition is endothelial cells exposed to medium alone.

Finally, calculate the comparative expression level, which is $2^{-\Delta\Delta CT}$. Representative results of the real-time PCR experiment are shown in **Fig. 22.1**.

3.5. Whole Cell EIA to Detect E-selectin on the Endothelial Cell Surface

1. This protocol is designed to determine relative levels of E-selectin expressed on the surface of endothelial cells exposed to various stimuli. As an alternative to this protocol, one can purchase an EIA from companies such as R&D Systems for measuring the amount of soluble E-selectin that is shed from the endothelial cells into the medium.
2. This assay is performed using endothelial cell grown on the bottom of the wells of a 96-well tissue culture plate. Coat the wells by adding 50 μL of fibronectin (10 μg/mL in HBSS) for at least 30 min. Next aspirate the fibronectin and add endothelial cells. One 25 cm^2 flask of confluent endothelial cells should be sufficient to seed 18 wells of the 96-well plate (*see* **Notes 1** and **5** to **8**). Endothelial cells grow very slowly when in 96-well plates. Therefore, a relatively large number of cells should be added to each well so that the cells are at least 95% confluent within 3 to 4 h after they have been added. As in the real-time PCR experiment, the endothelial cells should be exposed to medium alone (negative control), *C. albicans* (experimental condition), and TNF-α (400 pg/mL; positive control). Six wells should be used for each condition; 3 wells will be incubated with

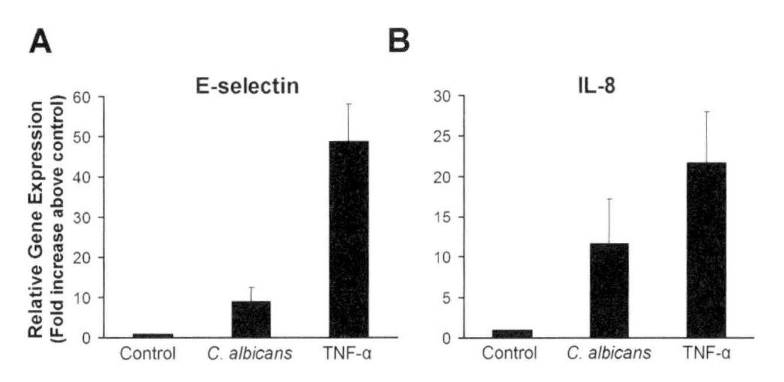

Fig. 22.1. Real-time PCR measurement of endothelial cell stimulation by *C. albicans*. Endothelial cells were incubated with tissue culture medium (control), *C. albicans*, or TNF-α for 8 h, after which the relative abundance of mRNA for (**A**) E-selectin and (**B**) IL-8 was measured using real-time PCR. Results are the mean ± SD of three independent experiments.

both the primary and secondary antibodies, and 3 wells will be incubated with the secondary antibody alone.

3. The day before the experiment, change medium above the endothelial cells, adding 200 μL of medium per well. Place sufficient M-199 + 20% serum for the experiment in the 5% CO_2 incubator to the night before the experiment. Keep the cap on the container loose so that the tissue culture medium can equilibrate with the CO_2. Set up the liquid culture of *C. albicans* as described in **Section 3.1**.
4. The next morning, aspirate the medium from the wells and add the various stimuli in a total volume of 85 μL per well (*see* **Note 9**). Process 6 wells at a time to avoid letting the endothelial cells dry up.
5. Incubate the 96-well plate in 5% CO_2 at 37°C for 8 h.
6. At the end of the incubation period, aspirate the medium from each well (*see* **Note 11**).
7. Gently add 200 μL of M199 + 5% FBS to each well using a multichannel pipettor at the slowest speed setting.
8. Remove the medium by gentle aspiration, and carefully add 200 μL of 4% paraformaldehyde to each well. Incubate plate at room temperature for 20 min to fix the cells.
9. Aspirate paraformaldehyde and add 200 μL of 1X TBS-glycine. Incubate for 5 min at room temperature.
10. Aspirate the TBS-glycine. From this point onward, aspiration can be done by inverting the plate and shake out the medium into a large beaker. Wash each well 3 times with 200 μL of PBS. Prepare anti–E-selectin primary antibody by diluting the E-selectin monoclonal antibody 1:100 in PBS + 0.1% BSA. Add 80 μL of the diluted antibody to 3 of the 6 wells exposed to each condition.
11. Add 80 μL of PBS + 0.1% BSA to the 3 wells of each condition that did not receive the anti–E-selectin primary antibody.
12. Place cover on plate and incubate at 4°C overnight (*see* **Note 13**).
13. Rinse each well 3 times with 200 μL PBS + 0.1% BSA and then 2 times with 200 μL of ddH_2O.
14. Block each well by adding 50 μL of PBS containing 3% BSA and 1% sheep serum. Incubate for 20 min at room temperature.
15. During this time, prepare the horse radish peroxidase labeled sheep anti-mouse secondary antibody by diluting it 1:2500 in PBS containing 3% BSA and 1% sheep serum.
16. Remove the blocking buffer from the well and add 50 μL of the secondary antibody to all wells.

17. Incubate plate for 1 h at room temperature.
18. Rinse each well 3 times with 200 μL PBS + 0.1% BSA and the 3 times with 200 μL ddH_2O.
19. Prepare OPD peroxidase substrate solution by adding one OPD tablet and one urea hydrogen peroxide phosphate-citrate buffer tablet into 20 mL water (*see* **Note 14**).
20. Add 150 μL of OPD solution to each well, and allow the color to develop at room temperature. Keep plate in the dark and check it every few minutes. The negative control well should turn faintly yellow, and positive control wells should be significantly more yellow. The usual incubation time is approximately 20 min.
21. Stop the reaction by adding 50 μL of 3 N HCl.
22. Transfer 150 μL from each well to the corresponding well of an empty 96-well flat bottom microtiter plate. This plated does not need to be sterile.
23. Determine the optical density (OD) of each with 490 nm using a spectrophotometer designed for 96-well plates.
24. Calculate the corrected OD for each condition by subtracting the OD of each no primary antibody well from the OD of the corresponding well that received primary antibody. Because no standard curve is included in this experiment, the amount of E-selectin expression induced by the various is usually just expressed as the corrected optical density. Alternatively, it can be expressed as the fold-stimulation relative to the endothelial cells exposed to medium alone. Representative results are shown in **Fig. 22.2**.

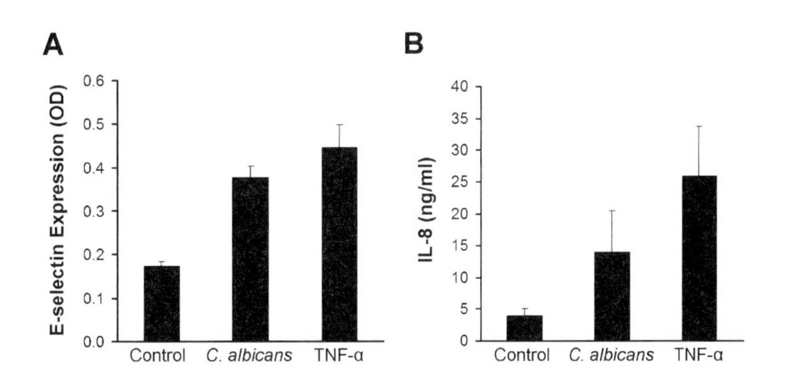

Fig. 22.2. E-selectin expression and IL-8 secretion by endothelial cells infected with *C. albicans*. Endothelial cells were incubated with tissue culture medium (control), *C. albicans*, or TNF-α for 8 h. (**A**) Surface expression of E-selectin, as measured by whole cell EIA. Results are the mean ± SD of a single experiment performed in triplicate. (**B**) Concentration of IL-8 in the medium above endothelial cells grown in a 24-well tissue culture plate as measured using a commercial EIA. Results are the mean ± SD of three experiments.

4. Notes

1. All containers, pipettes, reagents, and media that will be in contact with the endothelial cells should have as little endotoxin as possible. Most sterile, disposable plastic items are free of endotoxin. Glass and metal items can be made endotoxin-free by baking them at 180°C for 4 h. Reconstitute dry media using endotoxin-free water, such as pyrogen-free, sterile irrigation water.
2. PBS solution purchased commercially is endotoxin-free. Alternatively, PBS can be made from endotoxin-free ingredients as follows. Add 100 mg KCl, 100 mg KH_2PO_4, 4. g NaCl, and 1.08 g $Na_2HPO_4 \cdot 7H_2O$, to 450 mL water, and adjust the pH to 7.4 and the volume to 500 mL. Filter sterilize.
3. The organisms can be grown in a disposable sterile 50-mL polypropylene centrifuge tube with the cap screwed on loosely. To avoid having the caps fall off in the rotary shaker, apply a piece of tape running across the top of the tube and down one side.
4. Add the organisms to the M-199 just before use. If the organisms are allowed to sit in this medium at 37°C for very long, they will germinate and form aggregates of hyphae, which will result in variable stimulation of the endothelial cells.
5. When human umbilical vein endothelial cells are confluent, there are approximately 5×10^4 cells per cm^2. For the real-time PCR experiments using 24-well tissue culture plates, a multiplicity of infection (MOI) of 1 works well. For the whole cell EIAs performed using endothelial cells grown in 96-well plates, use an MOI of 2.
6. Trypsinizing and splitting endothelial cells: aspirate media from the C^{25} flask, and rinse flask with 1 mL of trypsin. Aspirate and add another 2 mL of trypsin. Wait for 10 to 15 s, then check the flask under the inverted microscope for the cell rounding. Once all endothelial cells are round up, slightly tap to lift cells off from the bottom of the flask. Neutralize the trypsin by adding 4 mL of M199 + 20% serum, and then transfer the cells to a 50-mL centrifuge tube. Centrifuge at $500 \times g$ for 5 min without braking. Gently aspirate supernatant without disturbing the pellet. Suspend pellet with M199 + 20% serum and resuspend the cells by gently aspirating and then expelling the mixture with a 10 mL pipette.
7. Most trypsin contains endotoxin. Therefore, to minimize stimulation caused by exposure to the trypsin, the endothelial cells should be split into the 24-well plate at least 48 h before use in the experiment.

8. The magnitude of endothelial cell stimulation by *C. albicans* is usually the greatest when endothelial cells are used at the first or second passage. Later passage cells usually have higher basal expression of leukocyte adhesion molecules and cytokines and lower induced expressions of these proinflammatory molecules.
9. To aspirate media without removing or damaging the endothelial cells, use a sterile Pasteur pipette and connected with rubber tubing to a vacuum flask. Use as low vacuum as possible to aspirate the medium slowly. Avoid touching the bottom of the wells with the pipette tip. Also, process only two wells at a time so that the endothelial cells will not dry out.
10. Use RNase-free pipette tips and centrifuge tubes, wear gloves, and keep samples on ice, except when specifically indicated.
11. Sometimes it is desirable to determine the cytokine concentration in the conditioned medium above the endothelial cells. In this case, collect the conditioned medium from each well and place it in a prelabeled, sterile Microfuge tube on ice. Centrifuge the Microfuge tube at 500 × *g* for 5 min at 4°C. Collect the supernatants and store them in 50- to 100-μL aliquots at −80°C. The cytokine concentration in the supernatants can be determined by commercially available EIA kits, such as from R&D Systems (Minneapolis, MN). Alternatively, the concentrations of multiple cytokines can be determined simultaneously the flow cytometric Cytokine Bead Array from R&D Systems.
12. When using the $\Delta\Delta C_T$ method of determining relative gene expression, it is important that the amplification efficiency of the primers for the reference gene is similar to that of the primers for the experimental genes. In our experience, some primers for ribosomal RNA have different efficiency compared with primers for nonribosomal RNA genes, and we therefore avoid using ribosomal RNA as the reference.
13. The wells can be incubated with the primary antibody at 4°C for up to 72 h.
14. OPD solution should be made within 5 min of use. The solution is carcinogenic; be careful when handling. The solution will eventually turn color if not used within 20 min. This solution should be disposed as a biohazard substance.

Acknowledgments

We thank Hyunsook Park for contributing data and expert advice. We also thank Norma Solis and the perinatal nurses at the Harbor-UCLA Medical Center Pediatric Clinical Research

Center for collecting and processing the umbilical cords. The work was supported in part by grants R01AI054928 and MO1RR00425 from the National Institutes of Health.

References

1. Filler SG, Pfunder AS, Spellberg BJ, Spellberg JP, Edwards JE, Jr. (1996) *Candida albicans* stimulates cytokine production and leukocyte adhesion molecule expression by endothelial cells. Infect Immun 64:2609–2617.
2. Orozco AS, Zhou X, Filler SG. (2000) Mechanisms of the pro-inflammatory response of endothelial cells to *candida albicans* infection. Infect Immun 68: 1134–1141.
3. Djeu JY, Matsushima K, Oppenheim JJ, Shiotsuki K, Blanchard DK. (1990) Functional activation of human neutrophils by recombinant monocyte-derived neutrophil chemotactic factor/il-8. J Immunol 144:2205–2210.

Chapter 23

Models of Oral and Vaginal Candidiasis Based on *In Vitro* Reconstituted Human Epithelia for the Study of Host-Pathogen Interactions

Martin Schaller and Günther Weindl

Abstract

This protocol describes the setup, maintenance, and characteristics of models of oral and vaginal candidiasis based on well-established three-dimensional organotypic tissues of human oral and vaginal mucosa. Infection experiments are highly reproducible and can be used for the direct analysis of pathogen/epithelial cell interactions. Using the models, the several stages of infection by wild-type *Candida albicans* strains, the consequence of gene disruption of putative virulence factors in mutant cells, and the evaluation of the host immune response can be evaluated by histologic, biochemical, and molecular methods. As such, the models provide clear answers regarding protein and gene expression that are not complicated by nonepithelial factors. To study the impact of several host components, the mucosal infection models can be supplemented with immune cells, saliva, and probiotic bacteria, which might be relevant for host defense. It requires at least 3 days to be established and can be maintained thereafter for 2 to 4 days.

Key words: Reconstituted human epithelium, RHE, candidiasis, siRNA, mucosal infection, confocal laser microscopy, electron microscopy, *Candida albicans*.

1. Introduction

The mucosal epithelium has immense importance in host defense and immune surveillance, as it is the primary cell layer that initially encounters the majority of microorganisms. This specialized interaction will result in either a favorable coexistence between microbe and host, as in the case of commensal microbes, or in a breach of the mucosal barrier and subsequent cell injury, as in the case of microbial pathogens *(1)*. Barrier function alone is usually adequate to restrain commensal microbes but is often insufficient to protect against microbial pathogens. Accordingly, the

Steffen Rupp, Kai Sohn (eds.), *Host-Pathogen Interactions*, DOI: 10.1007/978-1-59745-204-5_23,

oral epithelium is able to secrete a variety of defense effector molecules and to orchestrate an immune inflammatory response to activate myeloid cells in the submucosal layers to clear the invading pathogens *(2, 3)*.

Models of mucosal candidiasis that closely parallel the *in vivo* situation and allow studies of relevant physiologic functions are highly desirable. In this protocol, we will focus on the use of well-established three-dimensional organotypic models of human oral and vaginal mucosa. The use of the SkinEthic models was pioneered to study *Candida albicans* infection and specifically the role of the secreted aspartyl proteinases (*SAP*) gene family in *C. albicans* pathogenicity *(4–6)*. In recent years, the RHE models have been widely used to analyze the expression patterns of many *C. albicans* genes and to evaluate the consequence of gene disruption on pathogenicity *(7–12)* and the epithelial cytokine pattern that may favor a chemotactic immune response and an environmental switch from an anti-inflammatory to a proinflammatory milieu *(13, 14)*. These observations support the hypothesis of an active host-fungus interaction at the epithelial surface, which comprises a dynamic adaptation of both the host and *C. albicans* during the transition from commensalism to parasitism.

Numerous methods have been successfully used for interpretation of the experimental results. The epithelial damage can be visualized by histologic analysis of the embedded reconstituted human epithelia (RHE) and quantified by the lactate dehydrogenase (LDH) activity in the culture medium. We used immunoelectron microscopy, confocal laser microscopy, and ELISA to measure protein expression and RT-PCR for gene expression studies.

The epithelial tissues are provided by SkinEthic Laboratories, which is a leading tissue production company specializing in the reconstitution of human epidermal and epithelial tissues for *in vitro* test applications for the pharmaceutical, chemical, academic, and consumer product industry (www.skinethic.com).

The epithelial cells are seeded on inert filter substrates that are lifted to the air-liquid interface in a humidified-air incubator. A fully defined nutrient medium feeds the basal cells through the filter substratum. After 5 days, a stratified epithelium has formed that closely resembles human epithelium *in vivo*. Because the oral and vaginal epithelial tissues are reconstituted in a physiologically natural environment and on a chemically defined medium, they express all natural major markers of the epithelial basement membrane and of epithelial differentiation and behave like human *in vivo* epithelium when treated with pharmacologically active but also irritating products. Absence of toxic effects immediately induces tissue repair reflecting the *in vivo* wound-healing process. Moreover, the totally defined and serum-free culture environment allows the detection of very small quantities of inflammatory mediators, cytokines, or growth factors secreted by the epithelium in response to topical application of test substances in a very reproducible way. Biological controls of these RHE prior to use include guaranteed absence of HIV integrated proviral DNA, hepatitis C viral DNA, cytomegalovirus DNA, mycoplasma, hepatitis B antigen HBs, and bacteria and fungi (www.skinethic.com).

In summary, the mucosal RHE model appear to closely mimic the *in vivo* situation and provides a very promising tool for the study of host-pathogen interactions and therapeutic effects of antimicrobial substances. They are highly beneficial in dissecting the different stages of fundamental fungus-cell interactions from attachment via invasion to tissue destruction. Defense relevant immune cells can be supplemented directly to the infected RHE and indirectly by application of the host cells on the basal side of the filter. This allows systematic dissection of no-contact host defense mechanisms in the early stage of infection via released soluble factors and additional protective effects by direct interaction of the immune cells with *C. albicans* infected epithelial tissue. The overall procedure is outlined in flowchart form (**Fig. 23.1**).

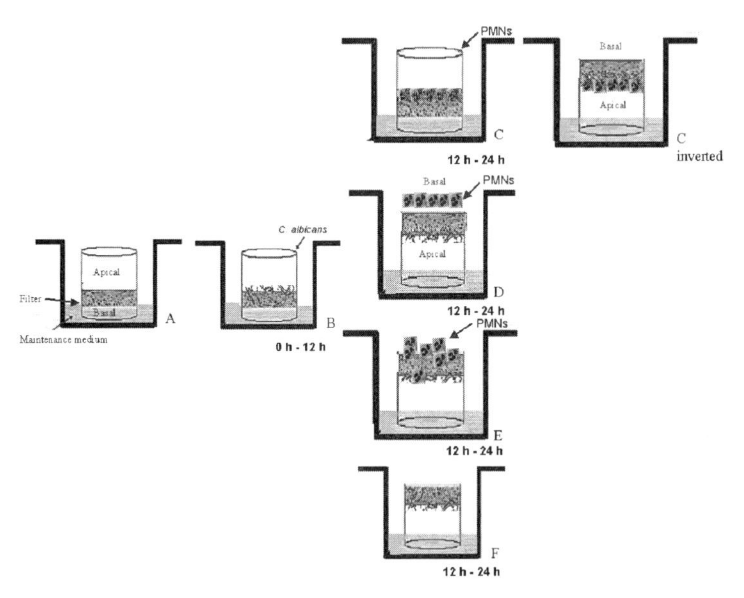

Fig. 23.1. Schematic diagram of the experimental design. (**A**) RHE on a microporous polycarbonate filter fed by a fully defined nutrient medium through the filter substratum. (**B**) Infection of RHE by *C. albicans.* (**C–F**) The model of oral candidiasis was supplemented with PMNs 6 or 12 h after infection in three different ways. (**C**) First, PMNs (10^6 cells in 50 μL PBS) were added directly to the apical epithelial layers of the preinfected model. Six hours after PMNs supplementation, the samples were inverted to mimic experimental conditions of the following samples. (**D**) Second, PMNs were added to the basal side of the polycarbonate filter. PMNs were not able to migrate through this microporous layer. (**E**) Third, the polycarbonate filters were perforated with a thin needle before addition of the PMNs to enable transepithelial migration. All samples were incubated for further 6 or 12 h after addition of the PMNs. Culture medium was applied to the basal side of the filter every 60 min to feed the cells. (**F**) Histologic sections and LDH analysis of the supplemented samples were compared with that of the nonsupplemented inverted model of oral candidiasis. (From Schaller, M., Boeld, U., Oberbauer, S., Hamm, G., Hube, B., and Korting, H. C. 2004. Polymorphonuclear leukocytes (PMNs) induce protective Th1-type cytokine epithelial responses in an in vitro model of oral candidosis. *Microbiology* **150**, 2807–2813. Reprinted with permission.)

2. Materials (*see* Note 1)

2.1. Establishment of the Models of Oral or Vaginal Candidiasis

2.1.1. Semisynchronization of Fungal Cells (Start 2 Days Before Planned Delivery of the RHE)

1. Counting chamber Neubauer (Carl Roth, Karlsruhe, Germany).
2. Incubation shaker, with temperature control (HT, Infors AG, Bottmingen, Switzerland).
3. Laminar Flow (Clean Air, Kendro, Hanau, Germany).
4. Phosphate-buffered saline, Dulbecco's PBS (1x), liquid (Invitrogen Gibco, Karlsruhe, Germany).
5. Sabouraud dextrose agar (Difco, Augsburg, Germany).
6. Sodium chloride (NaCl) solution, 0.9% (Sigma, Munich, Germany).
7. YPD (Yeast-extract-peptone-dextrose) broth (Difco).

2.1.2. Preparation, Preincubation of the RHE Tissue Culture (Start 1 Day Before Infection Will Be Established), and Infection of RHE

1. CO_2 incubator (MiniGalaxy A, Nunc, Wiesbaden, Germany).
2. Laminar Flow (Clean Air, Kendro).
3. SkinEthic Growth Medium (Small bottle) (SkinEthic, Nice, France).
4. SkinEthic Maintenance Medium (Small bottle) (SkinEthic).
5. SkinEthic Reconstituted Human Oral Epithelium, small: tissue surface 0.5 cm^2, age day 5, oral head and neck squamous cell carcinoma cell line TR146 (SkinEthic).
6. SkinEthic Reconstituted Human Vaginal Epithelium, small: tissue surface 0.5 cm^2, age day 5, human epidermoid carcinoma cells A431 (SkinEthic).
7. Tissue-culture plate 6-well, sterile with lid (Greiner Bio-one, Frickenhausen, Germany).
8. Tweezers, sterile.

2.1.3. Collection of the Culture Medium and Removal and Dissection of the Epithelia

1. Surgical disposable scalpels, sterile (Feather, Osaka, Japan; no. 11 and 21).

2.2. Use of the RHE as a Model to Study C. albicans Pathogenicity at Mucosal Surfaces and the Epithelial Immune Response Against C. albicans

Epithelial Cell Damage Assay

1. Hitachi 904 automatic analyzer (Roche Diagnostics, Grenzach-Wyhlen, Germany).
2. LDH IFCC liquid (Roche Diagnostics).

Histologic Studies

1. Distilled water.
2. Cacodylate buffer 0.1 M (Merck, Darmstadt, Germany).

2.2.1. Analysis of RHE Cell Damage

3. Ethanol 50%, 70%, 95% and 100% (Merck).
4. Glutaraldehyde 3.5% (Sigma, München, Germany) and formaldehyde 3.0% (Merck) in a 0.1 M cacodylate-buffered solution at pH 7.3 (Karnovsky solution).
5. Glycide ether (Serva, Heidelberg, Germany).
6. Osmium tetroxide 1% (Carl Roth, Karlsruhe, Germany), potassium ferrocyanide 1% (Carl Roth) solution in 0.1 M cacodylate buffer at pH 7.3 at room temperature.
7. Propylene oxide (Carl Roth).
8. Toluidine blue (Tebu, Frankfurt, Germany).
9. Ultracut Nova (Leica, Stuttgart, Germany).
10. Ultra-microtome (Ultracut, Reichert, Wien, Austria).
11. Uranyl acetate solution (0.05 g uranyl acetate in 2.5 mL H_2O) (Merck).

2.2.2. RNA Isolation and Quantitative RT-PCR

1. DNase I amplification grade (Invitrogen Gibco).
2. First Strand cDNA Synthesis Kit for RT-PCR (AMV) (Roche, Grenzach-Wyhlen, Germany).
3. Glass beads, acid-washed, 425 to 600 μm (30 to 40 U.S. sieve) (Sigma, Munich, Germany).
4. LightCycler (Roche).
5. LightCycler FastStart DNA MasterPLUS SYBR Green I (Roche).
6. peqGOLD RNAPure (Peqlab, Erlangen, Germany).

2.2.3. Confocal and Immunoelectron Microscopy

1. Confocal laser scanning microscope (Leica TCS SP, Leica, Stuttgart, Germany).
2. Donkey-anti-mouse-Cy5 (Dianova, Hamburg, Germany).
3. Donkey-anti-rabbit-Cy3 (Dianova).
4. PBS (Biochrom, Berlin, Germany).
5. PBS/10% donkey serum (Sigma, München, Germany).
6. PBS/BSA 0.1%/Tween 20 0.1%.
7. PLP (paraformaldehyde 2% [Merck] and lysine 0.125% [Sigma] in PBS).
8. RPMI 1640 medium (Invitrogen Gibco).
9. Silane-coated slides (Sigma).
10. Specific rabbit and mouse antibodies.
11. YOPRO 1:10 000 (Invitrogen Molecular Probes, Karlsruhe, Germany).
12. Zeiss 109 transmission electron microscope (Zeiss, Oberkochen, Germany).

13. Formvar-coated (Serva, Heidelberg, Germany) nickel grids (Stork, Eerbeek, Holland).
14. Freeze substitution apparatus (Bal-Tec, Witten, Germany).
15. Lowicryl K4M (Polysciences Ltd., Eppelheim, Germany).
16. PBS/10% donkey serum.
17. PBS/BSA 0.1%/Tween 20 0.1%.
18. PLP (paraformaldehyde 2% [Merck] and lysine 0,125% [Sigma] in PBS).
19. Primary specific antibody.
20. Secondary gold-conjugated antibody.
21. Uranyl acetate (Merck) and lead citrate (Merck).

2.2.4. ELISA

1. ELISA kits (Quantakine/Duo Set, R&D Systems, Wiesbaden, Germany)

2.3. Modifications of the Infection Protocol

2.3.1. Supplementation of the RHE by Host Cells

Supplementation of the RHE with PMNs and Transepithelial Migration Assay

1. Distilled water.
2. Fetal bovine serum, certified, heat-inactivated (Invitrogen Gibco).
3. Giemsa stain (Carl Roth).
4. Histopaque-1077 (Sigma).
5. Histopaque-1119 (Sigma).
6. May-Grunwald stain (Carl Roth).
7. RPMI 1640 medium (1x), liquid (Invitrogen Gibco).
8. Sodium chloride (NaCl) solution, 0.2%.
9. Sodium chloride (NaCl) solution, 1.6%.
10. Thin needle, sterile.

Supplementation with *Lactobacillus acidophilus*

1. *Lactobacillus acidophilus* (ATCC4356).

2.3.2. Modification of Host Gene and Protein Expression

Knockdown of RHE Gene Expression Using RNA Interference

1. HiPerFect transfection reagent (Qiagen, Hilden, Deutschland).
2. Negative Control siRNA Alexa Fluor 488 (Qiagen).

Inhibition of RHE Protein Expression Using Neutralizing Antibodies

1. Anti-human monoclonal antibodies (mAb).
2. Isotype controls.

3. Methods

3.1. Establishment of the Models of Oral or Vaginal Candidiasis

3.1.1. Semisynchronization of Fungal Cells (Start 2 Days Before Planned Delivery of the RHE)

1. Cultivate yeast cells for 24 h at 37°C on Sabouraud dextrose agar in an incubator.
2. Take a sample of the culture, suspend it in 5 mL 0.9% NaCl solution, and wash it 3 times. Use a Neubauer chamber to assess the number of *C. albicans* cells. Suspend a sample of approximately 2 × 105 cells in 10 mL Yeast-extract-peptone-dextrose medium. Culture the suspension for 16 h at 25°C with orbital shaking (150 U/min).
3. Take a suspension of 4×10^6 cells and incubate with shaking (150 U/min) in fresh 10 mL modified yeast-extract-peptone-dextrose medium for 24 h at 37°C. After washing 3 times with phosphate-buffered saline (PBS), adjust the final inoculum to the desired density (4×10^7/mL) with PBS solution (*see* **Note 2**).

3.1.2. Preparation, Preincubation of the RHE Tissue Culture (Start 1 Day Before Infection Will Be Established), and Infection of RHE

1. Upon arrival, remove the multiwell plate from the aluminum foil packing and strip off the white tape. Open the 24-well plate under a sterile airflow and remove the sterile filter paper (*see* **Note 3**).
2. With tweezers, carefully take out each 0.5 cm^2 insert containing the epithelial tissue, rapidly remove any remaining agarose that adheres to the outer sides of the insert by gentle blotting on the sterile filter paper, and immediately place in a 35-mm-diameter culture dish (6-well plate) previously filled with 1 mL of SkinEthic Maintenance Medium (at room temperature) (*see* **Note 4**).
3. Place the 35-mm-diameter culture dishes (6-well plates) in the incubator at 37°C, 5% CO_2, and saturated humidity for 24 h (*see* **Note 5**).
4. For infection of the 0.5 cm^2 RHE, use 2×10^6 *C. albicans* yeast cells in 50 μL PBS for each insert. Uninfected controls should contain 50 μL PBS alone (*see* **Note 6**).
5. Incubate infected and uninfected cultures at 37°C with 5% CO_2 at 100% humidity for the desired test period.
6. Use SkinEthic Maintenance Medium and change the medium after 24 h if you want to maintain the cultures for 48 h only. Use SkinEthic Growth Medium if you want to maintain the cultures for more than 48 h and change daily (1 mL of medium per tissue insert per day) (*see* **Note 7**).
7. RHE is widely used for comparing epithelial damage by *C. albicans* mutants with the corresponding wild-type infection strains. To investigate the consequence of gene

disruption of putative virulence factors, use a test period of 12 h and 24 h.

8. Do always include uninfected controls to ensure that the morphology of the RHE was normal before infection.

3.1.3. Collection of the Culture Medium and Removal and Dissection of the Epithelia

1. For analysis at the end of the experiments, carefully take out each 0.5 cm^2 insert, invert and place it on a culture dish.
2. Cut out the polycarbonate filter with the epithelial tissue from the basal side of the plastic insert with a sharp scalpel (no. 11).
3. Place the removed tissue (epithelial side up) on a culture dish and use a disposable scalpel (no. 21) to cut the RHE into several sections for separate analysis if necessary.
4. Collect the culture medium in order to determine epithelial cell damage by LDH assays and epithelial cytokine secretion by ELISA (*see* **Note 8**).

3.2. Use of the RHE as a Model to Study C. albicans Pathogenicity at Mucosal Surfaces and the Epithelial Immune Response Against C. albicans

The use of the SkinEthic RHE models was pioneered to study *C. albicans* infection and specifically the role of the secreted aspartyl proteinases (*SAP*) gene family in *C. albicans* pathogenicity *(4–6)*.

In recent years, the RHE models have been widely used to analyze the expression patterns of many *C. albicans* genes and to evaluate the consequence of gene disruption on pathogenicity *(7–12)* and the epithelial cytokine pattern that may favor a chemotactic immune response and an environmental switch from an anti-inflammatory to a proinflammatory milieu *(13, 14)*. These observations support the hypothesis of an active host-fungus interaction at the epithelial surface, which comprises a dynamic adaptation of both the host and *C. albicans* during the transition from commensalism to parasitism.

3.2.1. Analysis of RHE Cell Damage

Infection of the RHE with reference wild-type strain SC5314 *(15)* should induce signs of tissue damage characterized by edema, vacuolization, and detachment of keratinocytes by 24 h (**Fig. 23.2A**). A temporal progression of epithelial invasion and damage due to the time course of infection should be evident in time-course experiments by both microscopic analysis of histologic sections and LDH secretion into the culture medium.

Epithelial cell damage by fungal cells can be visualized by histologic investigation and analyzed by the release of LDH from epithelial cells into the surrounding medium. The LDH secretion levels into the cell culture medium are representative for the epithelial damage of the complete RHE whereas histologic sections are demonstrating only a distinct small part of the infected tissue. Therefore, histologic changes should be evaluated on the basis of 50 sections from five different sites for each RHE sample.

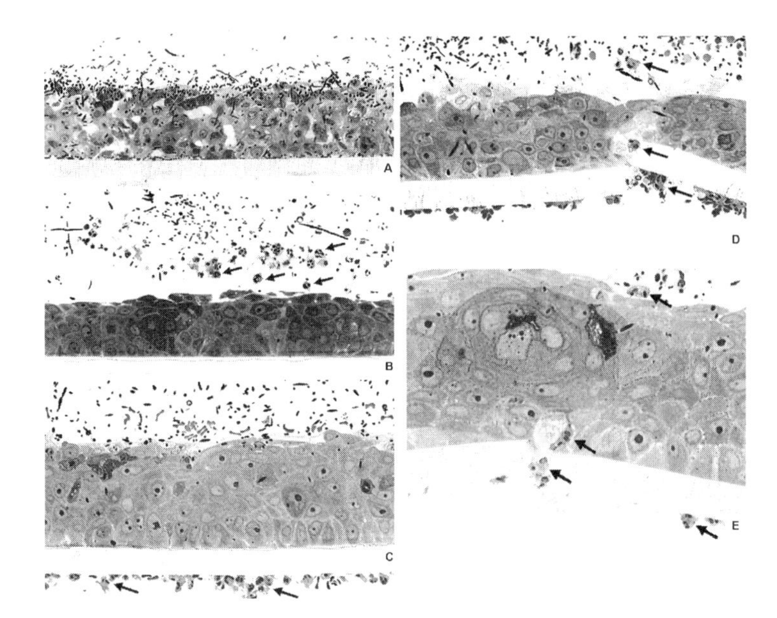

Fig. 23.2. Light micrographs of reconstituted human oral epithelium (RHE) 24 h after infection with *C. albicans* SC5314 in the absence and presence of PMNs. (**A**) *C. albicans* invasion of all epithelial layers by 24 h with extensive edema and vacuolization in the absence of PMNs. (**B**) Strongly reduced virulence phenotype of SC5314 when PMNs (arrows) were added to the apical epithelial layer 12 h after infection. (**C**) Firm attachment of PMNs (arrows) added to the basal side of the nonperforated filter 12 h after infection of an inverted culture resulting in a protective effect similar to (**B**). (**D**) Attenuated virulence phenotype of *C. albicans* was also seen when PMNs (arrows) were added to preinfected and inverted cultures after 12 h to the basal side of a perforated filter enabling transepithelial migration of the immune cells through the infected epithelium. (**E**) Chemoattraction and transepithelial migration of PMNs (arrows) through a pore of the polycarbonate layer of the RHE into epithelial layers and to the surface of the mucosa. (From Schaller, M., Boeld, U., Oberbauer, S., Hamm, G., Hube, B., and Korting, H. C. 2004. Polymorphonuclear leukocytes (PMNs) induce protective Th1-type cytokine epithelial responses in an in vitro model of oral candidosis. *Microbiology* **150**, 2807–2813. Reprinted with permission.)

Epithelial Cell Damage Assay

1. Collect the surrounding medium (*see* **Note 8**).
2. Analyze LDH activity spectrophotometrically and give it as U/l at 37°C.

Histological Studies (Semithin Sections)

1. Fix the specimen with 3.5% glutaraldehyde and 3.0% formaldehyde in a 0.1 M cacodylate-buffered Karnovsky solution at pH 7.3 for 24 h at 4°C.
2. Rinse twice with 0.1 M cacodylate buffer.
3. Postfix the specimen in osmium tetroxide 1%, potassium ferrocyanide 1.5% solution in 0.1 M cacodylate buffer at pH 7.3 at room temperature in a dark room.

4. Rinse in aqua dest. (3 × 5 min).
5. Contrast with uranyl acetate solution at 22°C in a dark room for 30 to 45 min.
6. Rinse again in distilled water (3 × 5 min).
7. Dehydration is achieved by transferring the material through an ascending ethanol series into absolute ethanol; for example, 50% (2 × 10 min) 70% (2 × 10 min) followed by 95% (2 × 5 min) and finally 100% 3 × 1 min).
8. After dehydration, specimens are infiltrated with the embedding medium by passing them through a sequence of solutions (propylene oxide [2 × 30 s], propylene oxide/glycide ether 1:1 [1 × 45 min], glycide ether [1 × 2 h]) until the dehydrating agent has been completely replaced by the final embedding medium.
9. Embed in plastic capsules for polymerization at 60°C for 2 days.
10. Cut the small blocks of tissue, using an ultra-microtome.
11. Study the semithin sections (1 μm) with a light microscope after staining with 1% toluidine blue.

3.2.2. RNA Isolation and Quantitative RT-PCR

1. Isolate total RNA from RHE samples using RNAPure according to the manufacturer's protocol (*see* **Note 9**).
2. Digest 1 μg total RNA with DNase and reverse transcribe with cDNA synthesis kit using random primers.
3. Perform quantitative real-time PCR in a LightCycler using 20 ng of cDNA and FastStart DNA Master PLUS SYBR Green I (*see* **Note 10**).
4. Normalize fold difference in gene expression to appropriate housekeeping genes (*see* **Note 11**).

3.2.3. Confocal and Immunoelectron Microscopy

Confocal Microscopy

1. For confocal microscopy, rinse the RHE in RPMI 1640 medium and then cryofix in liquid nitrogen.
2. Place 5-μm sections on silane-coated slides. Fix the sections in PLP (paraformaldehyde 2% and lysine 0,125% in PBS) for 2 min.
3. Incubate with PBS for 5 min, with PBS/BSA/Tween 20 for 10 min, and PBS containing 10% donkey serum for 30 min at room temperature.
4. Add the specific rabbit and mouse antibodies for 2 h.
5. Then, incubate the sections with PBS/BSA/Tween 20 for 30 min and with donkey-anti-rabbit-Cy3 and donkey-anti-mouse-Cy5 (both diluted 1:500) for 60 min.

6. Finally, wash with PBS/BSA/Tween 20 for 30 min.
7. Stain the nuclei with YOPRO.
8. Analyze the sections with a confocal laser scanning microscope at ×40 and ×250 magnification, respectively.

Immunoelectron Microscopy

1. For immunoelectron electron microscopy, fix RHE samples in PLP. (paraformaldehyde 2% and lysine 0.125% in PBS) for 4 h.
2. Freeze them rapidly by dipping into liquid propane, cooled to −185°C.
3. For freeze-substitution, fill up the samples in precooled (−90°C) uranylacetate 1% in methanol.
4. Transfer the samples in a freeze substitution apparatus.
5. Leave the samples for 90 h at −90°C, 8 h at −80°C, 8 h at −70°C and 8 h at −60°C.
6. Wash 2 times with methanol 100% at −40°C.
7. Embed the samples by passing them through a sequence of media starting with Lowicryl K4M /methanol 100% 1:1 at −40°C for 60 min followed by 2:1 Lowicryl K4M/ methanol 100% 2:1 at −40°C for 60 min and finally Lowicryl K4M at −40°C over night.
8. Polymerization for 48 h at −40°C with indirect ultraviolet light.
9. Polymerization for 48 h at room temperature with indirect ultraviolet light.
10. Mount ultrathin sections (50 nm) on formvar-coated nickel grids and incubate with PBS/BSA (PBS containing 10% goat serum) for 30 min.
11. Incubate with the specific rabbit antibody over night followed by PBS/BSA/Tween 20 for 30 min.
12. Then, incubate with secondary 10 nm gold-conjugated goat anti-rabbit antibody for 60 min.
13. Omit in control samples the primary Ab.
14. Counterstain grids with uranyl acetate and lead citrate and examine using a transmission electron microscope.

3.2.4. ELISA

1. Collect the culture medium at the conclusion of the RHE experiments (*see* **Note 8**).
2. Use commercially available ELISA kits according to manufacturer's instructions (*see* **Note 12**).

3.3. Modifications of the Infection Protocol

3.3.1. Supplementation of the RHE by Host Cells

A protective anti-*Candida* epithelial immune response is likely to contribute to the recruitment of polymorphonuclear cells (PMNs) and lymphocytes to the site of mucosal infection. To test this hypothesis, in oral RHE model, a PMN supplementation assay to study the effect of PMN cells during experimental oral candidiasis was established (*16*). Infection of RHE with *C. albicans* alone induced IL-1α, IL-1β, and TNF-α, with strong upregulation of GM-CSF, and IL-8, which was directly correlated with chemoattraction of PMNs to the site of infection. The addition of PMNs enhanced production of these cytokines. Notably, *C. albicans*–induced tissue damage was significantly reduced when PMNs migrated through a modified perforated basal polycarbonate filter (**Fig. 23.2D, E**) or when PMNs were applied to the apical epithelial surface (**Fig. 23.2B**). Interestingly, this protection of the epithelial tissue was also observed when PMNs were placed on the basal side of nonperforated filters, which prevents cell migration but allows free passage of soluble factors i.e. cytokines (**Fig. 23.2C**) *(16)*. Finally, the addition of lactobacilli to the vaginal model in the absence of PMNs also inhibited the *C. albicans* infection phenotype (unpublished results).

Supplementation of the RHE with PMNs and Transepithelial Migration Assay

1. Isolate PMNs from heparinized whole blood using Histopaque-1119 in combination with Histopaque-1077 according to the manufacturer's protocol.
2. Wash the PMNs recovered from the interface 3 times in PBS, remove residual erythrocytes by hypotonic lysis, and suspended the PMNs at a concentration of 4×10^7/mL in RPMI 1640 medium in the presence of 10% FCS (*see* **Note 13**).
3. Use Giemsa staining and light microscopy to ensure that a pure population of PMNs (>90% purity) with typical morphology has been isolated (*see* **Note 14**).
4. Assess the numbers of vital and nonvital leukocytes per sample using vital-staining by the trypan blue dye exclusion method. The viability should be ≥95% in all experiments.
5. Supplementation of the preinfected RHE with PMNs can be done in three ways (**Fig. 23.1C–E**).
   5.1. Add the PMNs (2×10^6 cells in 50 μL RPMI 1640/10% FCS) directly to the apical epithelial layers of the model 6 h or 12 h after infection with *C. albicans.*
   5.2. Invert the preinfected model 6 h or 12 h postinfection with *C. albicans* using two tweezers and add the PMNs to the basal side of the polycarbonate filter. In these samples, the filter prevents cell-cell contact

and PMN migration but allows soluble factors to pass through i.e. cytokines.

5.3. Perforate the polycarbonate filter with a thin needle before addition of the host cells to enable transepithelial migration of PMNs (**Fig. 23.1E**).

6. As a control, add PMNs to an uninfected RHE. Additional controls should include uninfected and infected samples that are inverted and supplemented with RPMI 1640 medium/10% FCS.
7. Invert and incubate all samples for a further 6 or 12 h after addition of the immune cells at 37°C with 5% CO_2 at 100% humidity (*see* **Note 15**).

Supplementation with *Lactobacillus acidophilus*

1. Supplementation of the vaginal RHE with probiotic bacteria can be done in different ways.

1.1. Inoculate the vaginal RHE with 2×10^6 *C. albicans* together with 10^7 *Lactobacillus acidophilus* cells in 50 μL PBS for 12 h and 24 h at 37°C with 5% CO_2 at 100% humidity in an incubator.

1.2. Add the lactobacilli (10^7 in 50 μL PBS) directly to the apical epithelial layers of the model 6 h or 12 h after infection with *C. albicans*. Incubate for further 6 h or 12 h.

1.3. Preincubate the uninfected RHE with 10^7 *Lactobacillus acidophilus* cells in 50 μL PBS for 2 h before infection with *C. albicans*. Incubate the samples for 12 and 24 h at 37°C with 5% CO_2 at 100% humidity in an incubator.

3.3.2. Modification of Host Gene Expression

Knockdown of RHE Gene Expression Using RNA Interference

1. Prepare transfection reagent and siRNA solutions according to the manufacturer's protocol (*see* **Note 16**).
2. Incubate the RHE with 50 μL of the transfection complexes (diluted in PBS) for an appropriate period of time at 37°C with 5% CO_2 at 100% humidity (*see* **Note 17**).
3. Remove the transfection solution and wash apical epithelial layers with 50 μL PBS.
4. Change the maintenance medium and start infection with *C. albicans* cells according to standard procedure.

Inhibition of RHE Protein Expression Using Neutralizing Antibodies

1. Block epithelial protein expression by using anti-human monoclonal antibodies (mAbs) and appropriate isotype controls (*see* **Note 18**).
2. Add the specific mAbs and the isotype controls (diluted in PBS), respectively, to RHE cultures 1 h at 37°C with 5%

CO_2 at 100% humidity prior to *C. albicans* infection (*see* **Note 7**).

3. Incubate the samples according to standard procedure.

4. Notes

1. All material should be sterile and maintained under sterile conditions. All solutions and materials coming into contact with RNA must be RNase-free, and proper techniques should be used accordingly. Furthermore, it is crucial that all material that comes in contact with the RHE samples is endotoxin-free, particularly in immunologic studies.
2. This procedure is useful to ensure that all *C. albicans* cells are in a similar growth phase when added to the RHE surface. This is important to get reproducible results.
3. It is very important to ask SkinEthic that the RHE and all culture media should be prepared without antibiotics and antimycotics.
4. Act quickly as the epithelial cultures dry out rapidly when not in contact with medium. Confirm that no air bubbles are formed underneath the insert. Make sure that the level of the medium outside of the RHE insert does not extend above the epithelial surface. If larger amounts of tissue are needed, RHE tissue with 4 cm^2 can be used and cultivated in 2 mL of Maintenance Medium. If smaller amounts of tissue are required, then specialized high-throughput 24-well (0.33 cm^2) or 96-well (0.1 cm^2) RHE tissue plates can be used and cultivated in 0.5 mL or 0.1 mL of Maintenance Medium, respectively.
5. SkinEthic Maintenance or Growth Medium should be stored at 4°C and in the dark. Use them at room temperature but do not preheat the media!
6. Make sure that the inocula are equally distributed on the epithelial surface.
7. For coincubation experiments with inhibitors, antimicrobial substances, or neutralizing antibodies, make sure that both the inoculum and the maintenance medium contain the desired concentration of the agent.
8. All samples should be collected under sterile conditions and centrifuged at 1500 to 2000 × *g* to remove any particulates. If samples are assayed on the day of collection, store on ice or at 4°C. Samples for determination of LDH can be stored at +2 to +8°C for a few days without significant loss of LDH activity. For ELISA, quick-freeze aliquots in liquid

nitrogen or a dry-ice/ethanol bath and store at −70°C (for up to 1 year) until assayed. Avoid repeated freeze-thaw cycles.

9. For simultaneous isolation of mammalian and fungal RNA, vortex RHE samples for 10 min with 0.5 mL glass beads in the presence of 1 mL RNAPure to break up fungal cell walls. Addition of glass beads can be omitted when only RNA from mammalian cells is investigated.

10. Prior to analysis, optimize PCR conditions for each primer pair. Confirm the formation of expected PCR products by melting curve analysis and agarose gel electrophoresis (2%). Dilute the corresponding DNA amplificate serially (6 logs) and use them to generate standard curves. Determine the threshold cycle (C_t) by use of the maximum-second-derivative function of the LightCycler software. Perform standard and experimental samples in triplicate and the average C_t reading used.

11. In gene expression studies, the selection of appropriate reference genes is critical. The expression level of these genes may vary among tissues or cells, and, more important, may undergo regulation in experimental treatments. For the infection studies in the oral RHE, we evaluated the expression stabilities of 10 different housekeeping genes in 20 samples according to the method by Vandesompele et al. *(17)*. The range of expression stability of the 10 investigated genes was (from most stable to least stable): YWHAZ (tyrosine 3-monooxygenase/tryptophan 5-monooxygenase activation protein, zeta polypeptide), G6PD (glucose-6-phosphate dehydrogenase), HMBS (hydroxymethyl-bilane synthase), POLR2A (polymerase (RNA) II (DNA directed) polypeptide A), TBP (TATA box binding protein), GAPDH (glyceraldehyde-3-phosphate dehydrogenase), B2M (beta-2-microglobulin), SDHA (succinate dehydrogenase complex, subunit A), UBC (ubiquitin C), ALDOA (aldolase A, fructose-bisphosphate).

12. When multiple cytokines from a small sample volume need to be quantified, other cytokine-detection systems may be used. We measured six different cytokines simultaneously using a commercial cytometric bead array (CBA, Pharmingen) *(17)*.

13. To remove residual red blood cells, subject cells to hypotonic lysis by resuspending the pellet in 5 mL cold 0.2% NaCl. After exactly 30 s, restore isotonicity by adding 5 mL ice-cold 1.6% NaCl and centrifuge. This procedure may be repeated once or twice until the cell pellet appears free of erythrocytes. The 30-s limit must be carefully observed as a more prolonged period of hypotonicity will result in neutrophil damage.

14. For Giemsa staining, smear cell preparation on standard microscope slides and air dry. Fix cells with 100% methanol for 1 to 2 min and stain with May-Grunwald solution for 4 min. Rinse with distilled water and add Giemsa stain for 4 min. After washing slides with distilled water, air dry the slides and examine stained slides under a light microscope.
15. Carefully place inverted samples in the middle of each well and ensure that the sample is not moving toward the edge of the well, which may cause loss of medium. Make sure that the Maintenance Medium will be added to the basal side of the inverted samples every 60 min to feed the keratinocytes/PMNs.
16. We have successfully used HiPerfect Transfection Reagent for a transfection period of 24 h and achieved 60% to 75% efficiency at mRNA level for siRNA concentrations of 10 nM (unpublished results). Using fluorescently labeled control siRNA, we observe that only the uppermost apical epithelial cell layers appear to be successfully transfected, whereas the remaining layers of the RHE show no fluorescence (**Fig. 23.3**). In addition, siRNA molecules do not appear to pass through the polycarbonate filter from the basal side (unpublished observations). However, in our experimental system, there may not be a requirement for a complete knockdown of gene expression throughout the RHE, as only the uppermost cell layers are directly interacting with the fungal cells.
17. Make sure that the transfection complexes are equally distributed on the epithelial surface. The amount of transfection

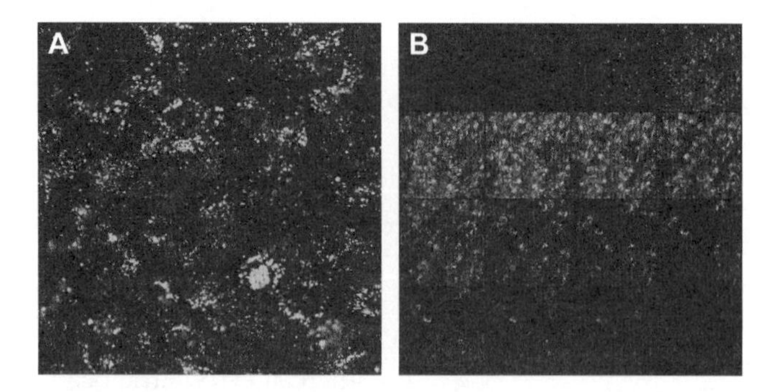

Fig. 23.3. Fluorescence microscopy of oral RHE (top view) 24 h after transfection with 10 nM Alexa Fluor 488–labeled nonsilencing siRNA using HiPerFect Transfection Reagent. Immediately after completion of the experiments, thin sections were made by hand with a sharp razor-blade and fixed in PLP for 2 min. The objective magnification used was 63× (oil immersion). (**A**) Cellular internalization of siRNA is observed in uppermost epithelial cells. (**B**) Z-stack image analysis of oral RHE demonstrates successful transfection within 15 μm of the apical surface, accounting for approximately 25% of the total epithelium (60 μm).

reagent and siRNA and the optimal incubation time for gene silencing analysis depends on the RHE model, the gene targeted, and the method of analysis. We recommend carrying out empirical testing by performing time-course experiments and varying amounts of transfection reagent and siRNA. Real-time RT-PCR and confocal fluorescence microscopy can be used to assess the successful transfection and knockdown of gene expression. To check for unwanted toxic effects of the transfection reagent, LDH release and histologic analysis may be conducted. Nonspecific off-target effects (e.g., TLR activation) may be determined by monitoring interferon responses.

18. An overview of mAbs suitable for blocking or neutralizing bioactivity is given in **Table 23.1**. In our experimental settings, the optimal concentration for the suggested monoclonal antibodies varies from 0.001 to 20 μg/mL. However, titration of antibodies is mandatory to obtain optimal results. Concentrations may not be applicable for antibodies from other manufacturers due to different bioactivity.

Table 23.1
Suggested anti-human monoclonal antibodies and isotype controls for blocking and neutralization of bioactivity.

Molecule	Concentration (μg/mL)	Isotype (mouse)	Antibody clone	Manufacturer	Cat. no.
GM-CSF	0.1–10	IgG_1	3209.1	R&D Systems	MAB215
IL-1α	0.01–1	IgG_1	4414	R&D Systems	MAB200
IL-1β	0.001–0.1	IgG_1	8516	R&D Systems	MAB201
IL-6	0.01–1	IgG_1	6708	R&D Systems	MAB206
IL-8/CXCL-8	0.1–1	IgG_1	6217	R&D Systems	MAB208
MCP-1/CCL2	1–10	IgG_1	24822	R&D Systems	MAB279
MIP-1β/CCL4	0.1–1	IgG_{2b}	24006	R&D Systems	MAB271
TNF-α	0.01–1	IgG_1	28401	R&D Systems	MAB610
TLR2	1–20	IgG_{2a}	TL2.1	eBioscience	16–9922
TLR4	1–20	IgG_{2a}	HTA125	eBioscience	16–9917
Isotype control		–	mouse IgG_1 (11711)	R&D Systems	MAB002
Isotype control		–	mouse IgG_{2a}	eBioscience	16–4724
Isotype control		–	mouse IgG_{2b}(20116)	R&D Systems	MAB004

Acknowledgments

We thank all the previous undergraduate and graduate students who have completed a project or a thesis in the laboratory and helped to build the foundations of these protocols. We thank Birgit Fehrenbacher, Renate Nordin, Helga Möller, and Hannelore Bischof, University of Tuebingen, for excellent technical assistance. M.S. and G.W. were supported by the Deutsche Forschungsgemeinschaft (Sch 897/1–3, Sch 897/3–1, Sch 897/3–2) and by NIH grant R21 DE015528–01.

References

1. Godaly, G., Bergsten, G., Hang, L., Fischer, H., Frendeus, B., Lundstedt, A. C., Samuelsson, M., Samuelsson, P., and Svanborg, C. (2001) Neutrophil recruitment, chemokine receptors, and resistance to mucosal infection. J Leukoc Biol **69,** 899–906.
2. Backhed, F., and Hornef, M. (2003) Toll-like receptor 4-mediated signaling by epithelial surfaces: necessity or threat? Microbes Infect **5,** 951–959.
3. Tjabringa, G. S., Vos, J. B., Olthuis, D., Ninaber, D. K., Rabe, K. F., Schalkwijk, J., Hiemstra, P. S., and Zeeuwen, P. L. (2005) Host defense effector molecules in mucosal secretions. FEMS Immunol Med Microbiol **45,** 151–158.
4. Schaller, M., Schafer, W., Korting, H. C., and Hube, B. (1998) Differential expression of secreted aspartyl proteinases in a model of human oral candidiasis and in patient samples from the oral cavity. Mol Microbiol **29,** 605–615.
5. Schaller, M., Korting, H. C., Schafer, W., Bastert, J., Chen, W., and Hube, B. (1999) Secreted aspartic proteinase (Sap) activity contributes to tissue damage in a model of human oral candidosis. Mol Microbiol **34,** 169–180.
6. Schaller, M., Bein, M., Korting, H. C., Baur, S., Hamm, G., Monod, M., Beinhauer, S., and Hube, B. (2003) The secreted aspartyl proteinases Sap1 and Sap2 cause tissue damage in an in vitro model of vaginal candidiasis based on reconstituted human vaginal epithelium. Infect Immun **71,** 3227–3234.
7. Heymann, P., Gerads, M., Schaller, M., Dromer, F., Winkelmann, G., and Ernst, J. F. (2002) The siderophore iron transporter of Candida albicans (Sit1p/Arn1p) mediates uptake of ferrichrome-type siderophores and is required for epithelial invasion. Infect Immun **70,** 5246–5255.
8. Zhao, X., Oh, S. H., Cheng, G., Green, C. B., Nuessen, J. A., Yeater, K., Leng, R. P., Brown, A. J., and Hoyer, L. L. (2004) ALS3 and ALS8 represent a single locus that encodes a Candida albicans adhesin; functional comparisons between Als3p and Als1p. Microbiology **150,** 2415–2428.
9. Zhao, X., Oh, S. H., Yeater, K. M., and Hoyer, L. L. (2005) Analysis of the Candida albicans Als2p and Als4p adhesins suggests the potential for compensatory function within the Als family. Microbiology **151,** 1619–1630.
10. Green, C. B., Cheng, G., Chandra, J., Mukherjee, P., Ghannoum, M. A., and Hoyer, L. L. (2004) RT-PCR detection of Candida albicans ALS gene expression in the reconstituted human epithelium (RHE) model of oral candidiasis and in model biofilms. Microbiology **150,** 267–275.
11. Li, D., Bernhardt, J., and Calderone, R. (2002) Temporal expression of the Candida albicans genes CHK1 and CSSK1, adherence, and morphogenesis in a model of reconstituted human esophageal epithelial candidiasis. Infect Immun **70,** 1558–1565.
12. Albrecht, A., Felk, A., Pichova, I., Naglik, J. R., Schaller, M., de Groot, P., Maccallum, D., Odds, F. C., Schafer, W., Klis, F., Monod, M., and Hube, B. (2006) Glycosylphosphatidylinositol-anchored proteases of Candida albicans target proteins necessary for both cellular processes and host-pathogen interactions. J Biol Chem **281,** 688–694.
13. Schaller, M., Mailhammer, R., Grassl, G., Sander, C. A., Hube, B., and Korting, H. C. (2002) Infection of human oral epithelia with Candida species induces cytokine

expression correlated to the degree of virulence. J Invest Dermatol **118,** 652–657.

14. Schaller, M., Korting, H. C., Borelli, C., Hamm, G., and Hube, B. (2005) Candida albicans-secreted aspartic proteinases modify the epithelial cytokine response in an in vitro model of vaginal candidiasis. Infect Immun **73,** 2758–2765.
15. Gillum, A. M., Tsay, E. Y., and Kirsch, D. R. (1984) Isolation of the Candida albicans gene for orotidine-5′-phosphate decarboxylase by complementation of S. cerevisiae ura3 and E. coli pyrF mutations. Mol Gen Genet **198**, 179–182.
16. Schaller, M., Boeld, U., Oberbauer, S., Hamm, G., Hube, B., and Korting, H. C. (2004) Polymorphonuclear leukocytes (PMNs) induce protective Th1-type cytokine epithelial responses in an in vitro model of oral candidosis. Microbiology **150,** 2807–2813.
17. Vandesompele, J., De Preter, K., Pattyn, F., Poppe, B., Van Roy, N., De Paepe, A., and Speleman, F. (2002) Accurate normalization of real-time quantitative RT-PCR data by geometric averaging of multiple internal control genes. Genome Biol **3,** research 0034.1–0034.12.

Chapter 24

Phagocytosis of *Candida albicans* by RNAi-Treated *Drosophila* S2 Cells

Shannon L. Stroschein-Stevenson, Edan Foley, Patrick H. O'Farrell, and Alexander D. Johnson

Abstract

Phagocytosis is a highly conserved aspect of innate immunity. *Drosophila melanogaster* has an innate immune system with many similarities to that of mammals and has been used to successfully model many aspects of innate immunity. The recent availability of Ribo Nucleic Acid interference (RNAi) libraries for *Drosophila* has made it possible to efficiently screen for genes important in aspects of innate immunity. We have screened an RNAi library representing 7216 fly genes conserved among metazoans to identify proteins required for the phagocytosis of the human fungal pathogen *Candida albicans*.

Key words: *Candida albicans, Drosophila*, S2 cells, cell culture, RNAi, double-stranded RNA, phagocytosis, immunofluorescence.

1. Introduction

Host phagocytic cells in the innate immune system play a critical role in defense against pathogens by engulfing them and subsequently activating other host defenses through cytokine production and antigen presentation *(1–3)*. The fruitfly *Drosophila melanogaster* has been well established as a model system for studying the interaction of conserved components of the innate immune system with human pathogens including *Listeria monocytogenes, Plasmodia, Mycobacterium marinum*, and *Candida albicans (4–8)*.

The main macrophage-like cells of *Drosophila* are the plasmatocytes, which phagocytose cell debris and invading microbes. The *Drosophila* S2 cell line is believed to be derived from plasmatocytes and, like plasmatocytes, phagocytoses pathogens *(9)*. Recently, Ribo

Steffen Rupp, Kai Sohn (eds.), *Host-Pathogen Interactions*, DOI: 10.1007/978-1-59745-204-5_24,

Nucleic Acid interference (RNAi) in S2 cells has been used to systematically study phagocytosis of *E. coli*, *Listeria monocytogenes*, *Mycobacterium fortuitum*, and *Candida albicans (10–13)*. In this paper, we describe the details of the methods of the latter study.

C. albicans is a common commensal fungal organism that in extremes of age, injury, antibiotic use, or a compromised immune response can predispose individuals to the development of mucosal or life-threatening systemic infections. *C. albicans* is now the fourth most common organism detected in systemic infections *(14)*, and mortality approaches 35% *(15)*. The availability of RNAi in *Drosophila* S2 cell phagocytes has identified proteins required for the recognition and phagocytosis of *C. albicans.*

2. Materials

2.1. Candida albicans Strains and Culture

1. *C. albicans* CAF2-1 strain (URA3/Δ*ura3*::λ*imm434*).
2. GFP-*C. albicans* (CAI-4 Δ*ura3*::λ*imm434*/Δ*ura3*::λ*imm434* expressing GFP under the control of the ADH1 promoter).
3. YEPD medium: 1% (w/v) yeast extract, 2% (w/v) peptone, 1% (w/v) dextrose.
4. Fetal bovine serum (FBS, Invitrogen, Carlsbad, CA).

2.2. S2 Cell Culture

1. S2 cells (Invitrogen).
2. Schneider's *Drosophila* Medium (Invitrogen) supplemented with 10% FBS and 100 units/mL penicillin and 100 μg/mL streptomycin (pen/strep, UCSF Cell Culture Facility).
3. Conditioned Schneider's Medium is prepared by culturing S2 cells for 3 to 4 days. Medium containing S2 cells is centrifuged to remove cells and medium is collected and stored at 4°C.

2.3. Preparation of dsRNA

1. RNAi library templates (Available from Open Biosystems, Huntsville, AL).
2. 10X PCR buffer: 100 mM Tris-HCl, pH 8.3, 500 mM KCl, 15 mM $MgCl_2$, 0.01% (w/v) gelatin. Store at −20°C. dNTPs: Prepare stock of 25 mM each dNTP by mixing an equal volume of 100 mM dATP, dTTP, dCTP, and dGTP (Invitrogen), Taq polymerase (Roche Diagnostics, Basel, Switzerland). Primer: TAATACGACTCACTATAGGGAGACCACGGGCGGGT (underlined part is anchor, non-underlined is T7 promoter).

3. T7 RNA polymerase kit: Contains T7 RNA polymerase, 5X reaction buffer, 100 mM DTT (Promega, Madison, WI). rNTPs: Prepare 25 mM stock by mixing an equal volume of 100 mM rATP, rUTP, rCTP, and rGTP (Roche Diagnostics). RNAse-free H_2O: Working in fume hood, add 3 mL diethyl pyrocarbonate (DEPC) to 2 L H_2O, mix well and let sit in fume hood overnight (Sigma-Aldrich, St. Louis, MO). Autoclave 15 minutes per liter.

2.4. RNAi Assay and Phagocytosis

1. Fluoroisothiocyanate (FITC isomer I, VWR, West Chester, PA) or tetramethyl rhodamine isothiocyanate (TRITC, Sigma-Aldrich). Stock solution is 5 mg/mL in DMSO. Make fresh immediately prior to use by dissolving FITC or TRITC powder in DMSO. Keep in dark.
2. Phosphate-buffered saline (PBS): Prepare 10X PBS stock: 1.37 M NaCl, 27 mM KCl, 43 mM Na_2HPO_4, 14 mM KH_2PO_4, adjust to pH 7.4, and autoclave. Dilute one part 10X PBS with nine parts water for PBS working stock.
3. 96-well Costar polystyrene cell culture plates (Fisher, Hampton, NH).
4. 96-well glass bottom plates (Greiner Bio-One, Kremsmuenster, Austria).
5. Concanavalin A (ConA) 0.5 mg/mL in H_2O (MB Biomedicals, Irvine, CA).

2.5. Immunofluorescence

1. Fix solution: 1% formaldehyde (Fisher) in PBS. Prepare fresh solution right before use.
2. Wash solution: PBS.
3. Block solution: 5% FBS in PBS.
4. Primary antibody: anti-*Candida* antibody (cat. no. B65411R, Biodesign International, Saco, ME).
5. Secondary antibody: goat anti-rabbit IgG conjugated with Cy3 (Jackson Immunoresearch Laboratories, West Grove, PA).
6. DAPI or Hoechst 33258 stain for DNA (DAPI will stain S2 and *Candida* nuclei under these conditions, Hoechst will only stain S2 cell DNA, both from Molecular Probes, (Eugene, OR).
7. Mounting media: Fluoromount-G (Southern Biotech, Birmingham, AL).
8. Microscope: Zeiss 200M immunofluorescence microscope capable of detecting and distinguishing FITC, Rhodamine, and DAPI stains (Carl Zeiss, Oberkochen, Germany).

3. Methods

The availability of large libraries of dsRNAs for *Drosophila* cells has enabled researchers to perform large-scale screens of many aspects of molecular, cellular, and developmental biology, including innate immunity. The libraries are in 96-well format to facilitate ease of amplification of both PCR products and dsRNAs. These are then easily added to *Drosophila* S2 cells also arrayed in a 96-well format. After 4 days of RNAi treatment, the cells can be assayed for phenotypes of interest. We describe here the assay for phagocytosis of the human fungal pathogen *C. albicans.*

To quickly and efficiently screen many wells of dsRNA-treated cells, it is important to be able to rapidly distinguish internalized versus external *C. albicans.* This can be accomplished by using FITC-labeled or GFP-expressing *C. albicans* (green) in the phagocytosis assay. After combining S2 cells and *C. albicans* and allowing time for phagocytosis, gentle fixation conditions are employed that do not permeabilize the S2 cells. Immunofluorescence then labels the *C. albicans* remaining external (red) from the S2 cells with a Cy3 conjugated secondary antibody that recognizes a primary antibody against the cell surface of *C. albicans* (**Fig. 24.1**). To obtain more quantitative results, the numbers of S2 cells phagocytosing *C. albicans* are counted after various times of phagocytosis (**Fig. 24.2**).

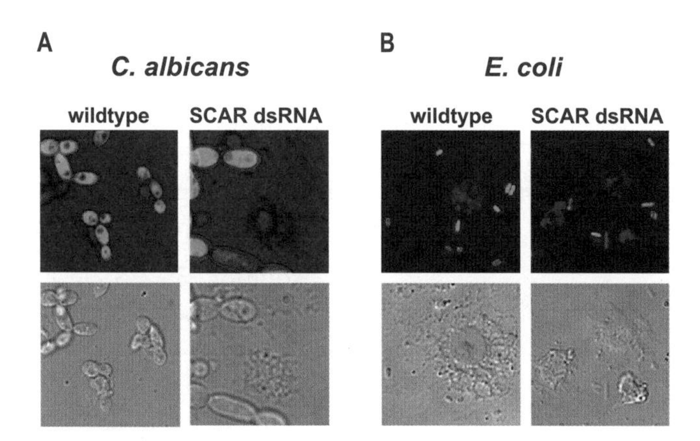

Fig. 24.1. Immunofluorescence of phagocytosis of pathogens. (**A**) GFP-expressing *C. albicans* (green) were co-incubated with S2 cells to allow phagocytosis. Cells were lightly fixed, and nonphagocytosed *C. albicans* were secondarily labeled with a rabbit anti–*C. albicans* antibody and Cy3-labeled anti-rabbit antibody (red). S2 cell DNA (blue) was labeled with Hoechst 33258. Left panels: wild-type S2 cells. Right panels: S2 cells treated with RNAi against SCAR (an actin regulator required for phagocytosis). (**B**) GFP-expressing *E. coli* (green) were co-incubated with S2 cells as in the *C. albicans* assay. A similar procedure was used to fix and stain nonphagocytosed *E. coli* using a goat anti–*E. coli* antibody and a Cy3-labed anti-goat antibody (red). S2 cells DNA (blue) was labeled with Hoechst 33258. Left panels: wild-type S2 cells. Right panels: S2 cells treated with RNAi against SCAR.

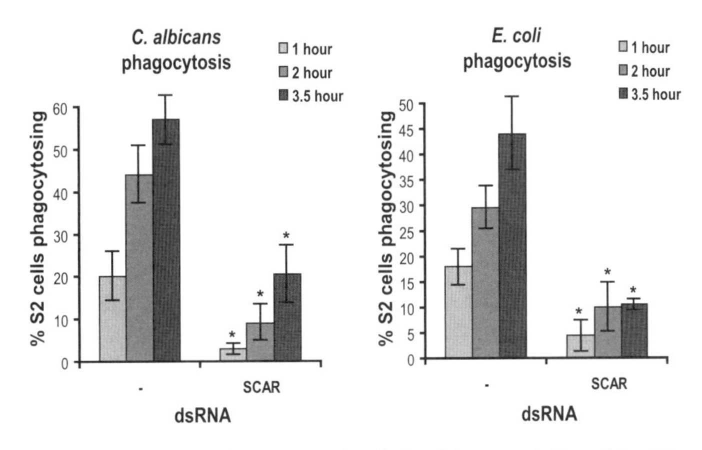

Fig. 24.2. Quantification of phagocytosis of *C. albicans* and *E. coli* by S2 cells. *C. albicans* or *E. coli* were co-incubated with wild-type or SCAR RNAi-treated S2 cells for various times, and the percentage of S2 cells that had phagocytosed one or more *C. albicans* or *E. coli* was quantified by counting 50 to 100 S2 cells. The maximum time shown is 3.5 h, as the levels of phagocytosis did not significantly increase after this time point. Results are the average of four experiments, and the error bars indicate the standard deviation. The results were evaluated for statistical significance using the *t*-test, assuming unequal variance. As indicated by the asterisks, the values for S2 cells treated with SCAR RNAi were statistically different from that of wild-type S2 cells with a confidence level $p < 0.01$.

3.1. *Candida albicans* Strains and Culture

1. Stocks of the *C. albicans* strains are stored at −80°C. To freeze cells, a toothpick is used to scrape a generous portion of *C. albicans* from a fresh plate. The cells are swirled in 750 μL YEPD in a cryotube to resuspend the yeast. Then, 750 μL of sterile 50% glycerol is added and mixed well. The cells are snap frozen in liquid nitrogen and transferred to the −80°C freezer. To start a fresh plate, a toothpick is used to scrape a small amount of cells from the frozen tube, streaked onto a YEPD agar plate, and grown at 30°C for 1 day. This plate can be stored at room temperature for up to 2 weeks (*see* **Note 1**).
2. To start an overnight liquid culture, a toothpick is used to pick a small amount of *C. albicans* from a patch on the plate that is then swirled in YEPD media. This is grown with shaking overnight at 30°C. Before phagocytosis assays, the *C. albicans* is diluted 1:40 with fresh YEPD and grown an additional 3 h before counting with a hemocytometer. As a guideline, we have found the OD_{600} to be approximately 0.2 to 0.5 after dilution and OD_{600} ~1.0 after growth for 3 h.
3. To induce hyphal growth, FBS is added to a final concentration of 10% to the newly diluted *C. albicans*. The cells are grown for 3 h at 37°C with shaking.

3.2. S2 Cell Culture

1. The general methods for freezing, thawing, and culturing of S2 cells are followed from the Invitrogen product materials available for downloading from the Invitrogen website. In brief, cells are passaged every 3 to 4 days by pipetting to resuspend the cells and diluted 1:5 to 1:10 in a new flask with new Schneider's medium containing 10% FBS and pen/strep. In general, cells are maintained at a concentration of 1×10^6 to 1×10^7 per mL. New cells are thawed from a frozen stock no later than every 4 weeks (*see* **Note 2**).
2. To count S2 cells, cells were resuspended by gently pipetting up and down. Then, 10 μL of cells were placed on a hemocytometer and counted.

3.3. Preparation of dsRNA

1. The template is diluted 1:200 and used in the following 50 μL PCR reaction to produce DNA template for making dsRNA: Template 1 μL, primer (10 μM) 2 μL, PCR buffer (10X) 5 μL, dNTPs (25 mM each) 0.4 μL, Taq 0.5 μL, H_2O 41.1 μL. Cycling conditions are as follows:
   - Step 1: 94°C (2 min).
   - Step 2: 94°C (30 s).
   - Step 3: 42°C (45 s).
   - Step 4: 72°C (60 s).
   - Step 5: Loop to **step 2** four times.
   - Step 6: 94°C (30 s).
   - Step 7: 60°C (45 s).
   - Step 8: 72°C (60 s).
   - Step 9: Loop to **step 6** 29 times.
   - Step 10: 72°C (10 min).
   - Step 11: 4°C (hold).
2. The Promega T7 RNA synthesis kit is used to prepare dsRNA. In brief, for a 10 μL reaction, combine T7 buffer (5X) 2 μL, DTT (100 mM) 1 μL, rNTP (25 mM each) 2 μL, DNA PCR template 4 μL, T7 RNA polymerase 1 μL. Incubate reaction at 37°C for 4 h.
3. Heat reactions to 65°C to 70°C for 30 min and slowly cool to room temperature to anneal dsRNA.
4. Quantify dsRNA and dilute reactions with H_2O to a final concentration of 1 mg/mL dsRNA. DNA and RNA are kept at −80°C for long term storage.

3.4. Phagocytosis Assays

3.4.1. RNAi Assay

1. Add 75 μL of fresh Schneider's media to 96-well plates. Add 2 μL dsRNA (1 mg/mL) to each well and shake well. The plates can be stored at 25°C for 2 to 3 h if needed before adding S2 cells.
2. Count S2 cells and add 50,000 cells per well in 75 μL of conditioned media. Mix well and seal the plate edges

with Parafilm to minimize evaporation. Incubate 4 days at 25°C.

3. For the phagocytosis assay, GFP expressing *C. albicans* may be used or alternatively, the *C. albicans* can be labeled with FITC (or TRITC). An overnight culture grown in YEPD is diluted 1:40 and grown an additional 3 h. Cells are centrifuged and resuspended to a final concentration of 1×10^7 cells/mL. Add FITC or TRITC to a final concentration of 100 μg/mL. (20 μL per 1 mL cells). Incubate at room temperature for 20 min. It is critical to keep FITC and TRITC in the dark as light will bleach the fluors. Wash 3 times with 1 mL PBS. Resuspend in PBS and count the cells for use in phagocytosis assays (*see* **Note 3**).

3.4.2. High-Throughput Phagocytosis Assays

1. After cells have been treated with dsRNA for 4 days, choose 10 wells and resuspend the S2 cells by pipetting up and down. Take 10 μL from each well, pool and count with a hemocytometer to determine an average cell density (*see* **Note 4**).
2. Add 1×10^5 S2 cells to a new 96-well polystyrene tissue culture dish with a total of 150 μL Schneider's media per well (*see* **Note 5**).
3. Immediately add 2×10^5 FITC-labeled or GFP-expressing *C. albicans* and shake to mix well. Incubate the plates for 2 h at 25°C to allow phagocytosis to occur (*see* **Note 6**).
4. Transfer S2 and *C. albicans* cells to Concanavalin A (ConA)-coated 96-well glass-bottom microplates and incubate for 1 h. Resuspend the S2 and *C. albicans* cells by pipetting up and down vigorously 5 to 10 times while minimizing air bubbles. The glass-bottom plates are ConA coated by adding 50 μL of the ConA solution the day before and letting sit at room temperature overnight. Remove the ConA solution before adding the S2 and *C. albicans* cells.
5. After co-incubating the cells, fix and process as in **Section 3.5** (*see* **Notes** 7 and **8**).

3.4.3. Quantitative Phagocytosis Assays

1. These instructions assume a time-course with three time points including 1 h, 2 h, and 3.5 h. Any time points may be chosen although at times longer than 3.5 h, overcrowding may occur and the amounts of *C. albicans* added should be reduced accordingly.
2. Count and plate 1×10^5 S2 cells in a 96-well polystyrene tissue culture dish in a total of 150 μL Schneider's media.
3. Add 5×10^5 GFP-expressing or FITC-labeled *C. albicans* and mix well. Try to add the *C. albicans* as soon as possible after counting and disturbing the S2 cells (*see* **Note 6**).
4. Incubate the plate at 25°C for various times (30 min, 1 h, and 2.5 h).

5. Transfer S2 and *C. albicans* cells to ConA-coated 96-well glass-bottom microplates as in **Section 3.4.2** and incubate for various times. (The samples that were incubated for 30 min in the step above should be incubated an additional 30 min for a total time of 1 h. The samples that were incubated for 1 h or 2.5 h above should be incubated an additional 1 h for total incubation times of 2 h and 3.5 h).
6. After co-incubating the cells, fix and process as in **Section 3.5** (*see* **Notes** 7 and **8**).

3.5. Immunofluorescence Staining

1. Gently remove media from each well. It is critical to gently tilt the plate to the side and pipette off the media without disturbing the cells because S2 cells do not adhere strongly. Let the wells air-dry for 2 min (*see* **Note 9**).
2. Fix cells for 5 min by gently adding 100 μL 1% formaldehyde in PBS. The formaldehyde should be freshly made. The gentle fixing does not permeabilize the S2 cells allowing the differentiation of phagocytosed versus nonphagocytosed *C. albicans*. All solutions are gently added to the side of the well and allowed to settle onto the samples in order to not wash away the cells (*see* **Note 10**).
3. All incubations should be done in the dark to preserve the FITC and Cy3 fluorescence.
4. Wash the samples one time with 100 μL PBS by removing the formaldehyde (discard in hazardous waste container) and adding PBS. Let sit at room temperature for 5 min.
5. The cells are blocked with 100 μL 5% FBS in PBS for 1 h at room temperature or overnight at 4°C. While blocking the samples, the primary antibody is also preblocked to decrease background. The primary anti-*Candida* antibody is diluted 1:1000 in 5% FBS in PBS and stored at 4°C during the block step.
6. The block solution is removed and 100 μL of the preblocked primary antibody is added and incubated overnight at 4°C. When leaving the plates at 4°C overnight, the plates should be wrapped in Parafilm to prevent evaporation of the antibody solution.
7. The samples are washed 2 times with 100 μL PBS for 5 min.
8. The goat anti-rabbit secondary antibody (1:1000 dilution in 5% FBS/PBS) is added and plates are incubated for 2 h at room temperature.
9. The samples are again washed 2 times with 100 μL PBS for 5 min.
10. DAPI or Hoechst stain (1:2000 dilution) diluted in PBS is added and cells are stained until desired levels (10 to 45 min).

11. Media is removed and 45 to 50 μL fluoromount mounting media is added. The plates can be stored at 4°C for up to a couple of months but the fluorescence does fade and best results are obtained by finishing analysis of the plates within 2 weeks.
12. To rapidly screen through many plates of S2 cells phagocytosing *C. albicans*, check each well for the presence of phagocytosed (green only) *C. albicans*. These wells are phagocytosing normally. An RNAi that disrupts phagocytosis will lead to the Cy3 staining of most or all *C. albicans* (external *C. albicans*). Examples of wild-type S2 cells and cells RNAi-treated for SCAR (an actin regulator required for phagocytosis) which disrupts phagocytosis are shown in **Fig. 24.1**. Wells that are not phagocytosing need to be checked for the presence of S2 cells as some dsRNAs are lethal to the cells. Lethal dsRNAs are excluded from further study. dsRNAs that disrupt phagocytosis in the first large screen are followed up by more detailed time courses of phagocytosis. An example of phagocytosis of *E. coli* by these cells is also shown in **Fig. 24.1**.
13. After completing the immunofluorescence, 100 S2 cells in each well are counted and the number of S2 cells phagocytosing one or more *C. albicans* are noted. This was the percentage of phagocytosis (**Fig. 24.2**). The differences between conditions can be statistically analyzed with a *t*-test (*see* **Note 11**).

4. Notes

1. It is important that *C. albicans* be restreaked from a frozen stock every 2 weeks as chromosome loss, gains, and rearrangements can occur after extended growth on plates. Additionally, only a very small amount of *C. albicans* cells are needed to start an overnight culture in YEPD.
2. New aliquots of S2 cells are thawed and used every 4 weeks at the maximum. We have found that as the S2 cells age in culture, the efficiency of phagocytosis decreases to a point of almost no phagocytosis of *C. albicans*. We usually begin a new culture when the old culture is 3 weeks old to allow a 1-week overlap for the new culture to recover and start growing normally. In general, the phagocytosis varies somewhat from day to day. If the levels of wild-type S2 cells phagocytosing *C. albicans* were low, we did not trust the experiment for that day and thawed new cells.
3. Other pathogens can be similarly stained for use in phagocytosis assays. We have found that the same methods can be

used to stain *S. cerevisiae* and *S. aureus* with FITC. *E. coli* is not stained by FITC but can be similarly stained with FM4–64 from Molecular Probes. Approximately 5 μL of a saturated overnight culture of *E. coli* and 30 μL of a saturated overnight culture of *S. aureus* were added per well of 96-well plates for phagocytosis using these pathogens. Appropriate levels of pathogen to add can be empirically determined by using a dilution series of pathogen.

4. RNAi-treated S2 cells can also be used in other techniques such as Western blotting and QPCR. The appropriate number of cells are treated with dsRNA for 4 days and the cells harvested and assayed. For examples of their use in Western blotting, see Stroschein-Stevenson et al. and Foley et al. *(13, 16)*.
5. Because both *C. albicans* and S2 cells will adhere fairly well to the ConA-treated plates, incubation in these ConA-treated plates for the entire phagocytosis will result in the two cell types settling on the bottom and phagocytosis will be very inefficient. The preincubation in polystyrene plates allows movement and interaction between S2 and *C. albicans* cells enabling phagocytosis to occur. Transfer to the glass-bottom plates then allows completion of the phagocytosis, fixation, immunofluoresence, and screening on an immunofluoresence microscope.
6. We have found that it is important to add *C. albicans* to the S2 cells as quickly as possible after resuspending and adding the S2 cells to new media in 96-well plates. After settling down in the plate, the S2 cells phagocytose less efficiently.
7. To complete the phagocytosis assays with *E. coli* or *S. aureus*, we used primary antibodies against the cell surface of these bacteria purchased from Biodesign International (cat. no. B47711G and B65881R).
8. The phagocytosis protocol can be adapted to other cell types such as the mouse macrophage–derived RAW264.7. Treatment with RNAi would of course have to be specific for these cells. Similar to S2 cells, the other cell types can be plated in 96-well plates and *C. albicans* added to allow phagocytosis to occur. The same immunofluorescence protocol can be used (unpublished observation).
9. S2 cells are not very adherent and can be resuspended by simply pipetting up and down. If care is not taken during the immunofluorescence washes, significant numbers of S2 cells will be washed away. Solutions should be removed and added by tilting the plates gently to the side and removing media from the bottom corner and adding to the side of the plate.
10. The fixation of the S2 cells is very gently done with a relatively low concentration of formaldehyde and no detergent.

This is critical for the antibody staining to differentiate between internalized and external *C. albicans*. Detergent or harsher fixation conditions permeabilize the S2 cell membrane resulting in the staining of all *C. albicans*.

11. After counting the numbers of S2 cells phagocytosing *C. albicans* or other pathogen, the differences can be statistically analyzed by a *t*-test assuming unequal variances. Microsoft Excel has a user-friendly program for this *t*-test.

Acknowledgments

The dsRNA library used in this screen was produced by Ben Eaton, Edan Foley, Nico Stuurman, Graeme Davis, Patrick O'Farrell, and Ron Vale at the University of California – San Francisco. We are grateful to Matt Lohse and Soo-Jung Lee for comments on the manuscript. This work was supported in part by grants from the National Institutes of Health (NIH) to A.D.J. (RO1 AI49187) and P.H.O. (RO1 AI60102) and a Jane Coffin Childs postdoctoral research grant to S.L.S.

References

1. Elrod-Erickson, M., Mishra, S., and Schneider, D. (2000) Interactions between the cellular and humoral immune responses in *Drosophila*. *Curr Biol* **10,** 781–784.
2. Hoffmann, J. A., Kafatos, F. C., Janeway, C. A., and Ezekowitz, R. A. (1999) Phylogenetic perspectives in innate immunity. *Science* **284,** 1313–1318.
3. Romani, L. (2004) Immunity to fungal infections. *Nat Rev Immunol* **4,** 1–23.
4. Alarco, A. M., Marcil, A., Chen, J., Suter, B., Thomas, D., and Whiteway, M. (2004) Immune-deficient *Drosophila melanogaster*: a model for the innate immune response to human fungal pathogens. *J Immunol* **172,** 5622–5628.
5. Cheng, L. W., and Portnoy, D. A. (2003) *Drosophila* S2 cells: an alternative infection model for *Listeria monocytogenes*. *Cell Microbiol* **5,** 875–885.
6. Dionne, M. S., Ghori, N., and Schneider, D. S. (2003) *Drosophila melanogaster* is a genetically tractable model host for *Mycobacterium marinum*. *Infect Immun* **71,** 3540–3550.
7. Mansfield, B. E., Dionne, M. S., Schneider, D. S., and Freitag, N. E. (2003) Exploration of host-pathogen interactions using *Listeria monocytogenes* and *Drosophila melanogaster*. *Cell Microbiol* **5,** 901–911.
8. Schneider, D., and Shahabuddin, M. (2000) Malaria parasite development in a *Drosophila* model. *Science* **288,** 2376–2379.
9. Lavine, M. D., and Strand, M. R. (2002) Insect hemocytes and their role in immunity. *Insect Biochem Mol Biol* **32,** 1295–1309.
10. Agaisse, H., Burrack, L. S., Philips, J. A., Rubin, E. J., Perrimon, N., and Higgins, D. E. (2005) Genome-wide RNAi screen for host factors required for intracellular bacterial infection. *Science* **309,** 1248–1251.
11. Philips, J. A., Rubin, E. J., and Perrimon, N. (2005) *Drosophila* RNAi screen reveals CD36 family member required for mycobacterial infection. *Science* **309,** 1251–1253.
12. Ramet, M., Manfruelli, P., Pearson, A., Mathey-Prevot, B., and Ezekowitz, R. A. (2002) Functional genomic analysis of phagocytosis and identification of a *Drosophila* receptor for *E. coli*. *Nature* **416,** 644–648.

13. Stroschein-Stevenson, S. L., Foley, E., O'Farrell, P. H., and Johnson, A. D. (2006) Identification of *Drosophila* gene products required for phagocytosis of *Candida albicans*. *PLoS Biol* **4,** e4.

14. Edmond, M. B., Wallace, S. E., McClish, D. K., Pfaller, M. A., Jones, R. N., and Wenzel, R. P. (1999) Nosocomial bloodstream infections in United States hospitals: a three-year analysis. *Clin Infect Dis* **29,** 239–244.

15. Calderone, R. (Ed.) (2002) Host recognition by *Candida* species, ASM Press, ASM Press, Washington, DC.

16. Foley, E., and O'Farrell, P. H. (2004) Functional dissection of an innate immune response by a genome-wide RNAi screen. *PLoS Biol* **2,** E203.

Chapter 25

Oral Mucosal Cell Response to *Candida albicans* in Transgenic Mice Expressing HIV-1

Louis de Repentigny, Daniel Lewandowski, Francine Aumont, Zaher Hanna, and Paul Jolicoeur

Abstract

Controlled studies on the immunopathogenesis of mucosal candidiasis in HIV infection have been hampered by the lack of a relevant animal model. We have previously reported that oral *Candida* infection in CD4C/HIV transgenic mice expressing gene products of HIV-1 in immune cells and developing an AIDS-like disease closely mimics oropharyngeal candidiasis in human HIV infection. The role of defective dendritic cells and CD4+ T cells in impaired induction of protective immunity and in the phenotype of chronic oral carriage of *C. albicans* can now be investigated under controlled conditions in these transgenic mice.

Key words: *Candida albicans*, candidiasis, animal models, transgenic mice, immune response, pathogenesis.

1. Introduction

Oropharyngeal candidiasis (OPC) is the most frequent opportunistic fungal infection among HIV-infected patients *(1)* and remains a significant cause of morbidity despite the dramatic ability of antiretroviral therapy to reconstitute immunity *(2)*. The critical immunologic defects that cause the onset and maintenance of mucosal candidasis in patients with HIV infection have not been identified *(3)*. The availability of CD4C/HIV transgenic (Tg) mice expressing HIV-1 in immune cells and developing an AIDS-like disease has provided the opportunity to devise a novel model of mucosal candidiasis that closely mimics the clinical and pathologic features of candidal infection in human HIV-1 infection *(4)*. These transgenic mice allow, for the first time, a precise cause-and-effect

Steffen Rupp, Kai Sohn (eds.), *Host-Pathogen Interactions*, DOI: 10.1007/978-1-59745-204-5_25,

analysis of the immunopathogenesis of mucosal candidiasis in HIV infection under controlled conditions in a small laboratory animal *(5, 6)*.

2. Materials

2.1. Tg Mice Expressing HIV-1

1. The CD4C/HIV^{MutA} Tg mice, which express *rev*, *env*, and *nef* of HIV-1, have been described elsewhere *(7)*. The CD4C/HIV^{MutA} construct harbors the mouse CD4 enhancer and human CD4 promoter elements to drive the expression of HIV-1 genes in CD4+ CD8+ and CD4+ CD8− thymocytes, in peripheral CD4+ T cells, and in macrophages and dendritic cells (DCs). Selective expression of the *nef* gene is required and sufficient to elicit an AIDS-like disease in these Tg mice, characterized by failure to thrive, wasting, severe atrophy, and fibrosis of lymphoid organs, loss of CD4+ T cells, interstitial pneumonitis, and focal segmental glomerulosclerosis associated with tubulointerstitial nephritis and tubular microcystic dilatation *(7)*.
2. Founder mouse F21388 is bred on the C3H/HeN Hsd background. Animals from this line express moderate levels of the transgene, with about 50% survival at 3 months *(7)*.

2.2. Animal Model of Mucosal Candidiasis

1. *C. albicans* strain LAM-1 *(8)* is maintained as a suspension in 65% glycerol, 10 mM Tris (pH 7.5), 10 mM $MgCl_2$, at −70°C.
2. Sabouraud dextrose broth (BD Difco, Sparks, MD) and rotary agitator.
3. Solution for anesthesia: 1 mg/mL of xylazine and 15 mg/mL of ketamine, in 0.01 M phosphate-buffered saline (PBS), pH 7.4.
4. Sterile calcium alginate tipped applicator Calgiswabs (Puritan, Guilford, ME).
5. Ringer's citrate buffer: 10 mM NaCl, 0.35 mM KCl, 0.2 mM $CaCl_2$, 0.15 mM $NaHCO_3$, 34 mM sodium citrate.
6. Sabouraud dextrose agar (BD BBL, Sparks, MD).

2.3. Flow Cytometry Analysis of Immune Cell Populations

1. Cleaning buffer solution: 20 mM Tris-HCL (pH 7.5), 20 mM NaCl, 40 mM EDTA, and 1 mM dithiothreitol (DTT).
2. Hanks' balanced salt solution (HBSS; Invitrogen Life Technologies, Grand Island, NY).
3. Complete tissue culture medium: RPMI 1640 (Invitrogen Life Technologies) supplemented with 10% of heat-inactivated

fetal bovine serum (FBS) (Invitrogen Life Technologies), 20 mM HEPES buffer, 2 mM L-glutamine, 5×10^{-5} M β-mercaptoethanol, 100 U/mL penicillin, 100 μg/mL streptomycin, 0.25 μg/mL amphotericin B, and 50 μg/mL of gentamicin.

4. Nylon mesh (pore size, 80 μm; Millipore, Bedfo MA).
5. FACS Lysing Solution (BD Biosciences, San Jose, CA).
6. Collagenase type IV (cat. no. C5138; Sigma-Aldrich, St. Louis, MO). A 1% stock solution in RPMI 1640 is kept at −20°C.
7. Anti-mouse anti-CD45–PE (30-F11), anti-CD11b–PerCP (MI/70), anti-I-A^K (MHC class II alloantigen)–FITC (11–5.2), anti-CD11c–APC (HL3), anti-CD3–APC (145–2C11), anti-CD4–PErCP (RM4–5) monoclonal antibodies, and their respective isotype controls (hamster IgG1, λ; rat IgG2b, k; mouse IgG2b, k; mouse IgG2a, k; rat IgG1; rat IgG2b) (BD Biosciences).
8. FACSCalibur flow cytometer (BD Biosciences) equipped with CellQuest software.

2.4. Flow Cytometry Determination of Intracellular Cytokines

2.4.1. CD4+ T Cells

1. Biotin-conjugated antibody specific for CD4 (RM4–5; BD Biosciences).
2. Streptavidin Captivate ferrofluid particles (Molecular Probes, Eugene, OR).
3. Anti-CD3 (145–2C11), anti-CD28 (37.51), and anti-CD69 (H1.2F3) monoclonal antibodies (BD Biosciences).
4. Supplemented RPMI 1640 (*see* **Section 2.3 point 3**) containing 5 μg/mL anti-CD3 antibody.
5. Ionomycin (500 ng/mL) and phorbol 12-myristate 13-acetate (PMA; 5 ng/mL) (Sigma-Aldrich).
6. Monensin (GolgiStop; BD Biosciences) and brefeldin A (GolgiPlug; BD Biosciences).
7. Cytofix/Cytoperm kit (BD Biosciences).
8. Anti-mouse IFN-γ-PE (XMG1.2), anti-IL-2-FITC (JES6–5H4), anti-IL-4-APC (11B11), anti-IL-10-FITC (JES5–16E3), anti-TNF-α-PE (MP6-XT22), and their respective isotype control antibodies (rat IgG1 or rat IgG2b) (BD Biosciences).
9. FACSCalibur flow cytometer (BD Biosciences) equipped with CellQuest software.

2.4.2. DCs

1. 5-mL syringe equipped with a 25-gauge needle.
2. Nylon mesh (pore size, 80 μm) (Millipore).
3. RBC lysing buffer: In 800 mL distilled water, dissolve 8.3 g NH_4Cl, 1.0 g $KHC0_3$, 1.8 mL of 5% EDTA. Filter

(0.22 μm) and add distilled water to a final volume of 1000 mL.

4. HBSS (Invitrogen Life Technologies).
5. Supplemented RPMI 1640 (*see* **Section 2.3 point 3**) containing GM-CSF (1000 U/mL) and IL-4 (500 U/mL) (Cedarlane Laboratories, Hornby, Ontario).
6. LPS (Sigma-Aldrich).
7. Live *C. albicans* blastoconidia (*see* **Section 3.2 point 1**).
8. Brefeldin A.
9. Amphotericin B (2.5 μg/mL; cat. no. 15240–096, Invitrogen).
10. Anti-mouse-IL-12-PE (C15.6) and isotype control antibody (rat IgG1) (BD Biosciences).
11. FACSCalibur flow cytometer (BD Biosciences) equipped with CellQuest software.

2.5. Quantitative RT-PCR Analysis of Cytokines in Oral Mucosal Tissue

1. RNA later (Qiagen, Valencia, CA).
2. RNeasy Mini kit (Qiagen).
3. Omni TH-115 homogenizer (Omni International, Marietta, GA).
4. RiboGreen RNA Quantitative Reagent and kit (Molecular Probes).
5. QuantiTect Reverse Transcription kit (Qiagen).
6. QuantiTect SYBR Green PCR kit (Qiagen).
7. Primers: IL-4, I.D. no 10946584a1 (Primer Bank); IFN-γ, forward: TCA AGT GGC ATA GAT GTG GAA GAA, reverse: TGG CTC TGC AGG ATT TTC ATG; β-actin, forward: AGA GGG AAA TCG TGC GTG AC, reverse: CAA TAG TGA TGA CCT GGC CGT.
8. SmartCycler real-time thermal cycler (Cepheid, Sunnyvale, CA).

3. Methods

3.1. Tg Mice Expressing HIV-1

1. Male and female specific pathogen-free CD4C/HIV^{MutA} Tg mice and non-Tg littermates are housed in sterile microisolators with a 12-h light-dark cycle, in sterilized individual cages equipped with air filter hoods. The animals are supplied with sterile water and are fed with sterile mouse chow.
2. Male and female mice are kept in separate cages after weaning.

3. Tg mice and non–Tg mice are housed in separate cages during experimental infection with *C. albicans.*
4. Breeding of Tg mice and experimental infection are done in separate rooms, with restricted access. Mice are bred as heterozygotes for the transgene.
5. All personnel coming into contact with the Tg mice are required to wash their hands and to wear protective clothing (mask, gloves, cap, gown, shoe covers).

3.2. Animal Model of Mucosal Candidiasis

1. *C. albicans* is grown to late-log-phase in Sabouraud dextrose broth for 18 h at 30°C with rotary agitation.
2. Yeast cells are washed twice in sterile PBS, counted in a hemocytometer, and deposited at 10^8 cells/1.5 mL Eppendorf tube. The tubes are centrifuged and the supernatant is discarded. A separate tube is prepared for each mouse to be inoculated with *C. albicans.*
3. Mice are anesthetized with 50 μL/10 g of body weight of prepared ketamine/xylazine solution, equivalent to 75 mg/kg of ketamine and 5 mg/kg of xylazine, by intraperitoneal injection. The mice remain anesthetized for approximately 20 min.
4. The inoculum of 10^8 pelleted *C. albicans* blastoconidia is recovered by rotating a Calgiswab at the bottom of an individual tube. A single Calgiswab is used to inoculate a single mouse.
5. The oral cavity of the anesthetized mouse is carefully opened, and the tongue, cheeks, and palate are repeatedly rubbed in a nontraumatic, gentle way with the Calgiswab over a 10-s period. Anesthesia eliminates potential discomfort of oral inoculation and favors adherence of *Candida* to the oral epithelium by preventing immediate swallowing of the inoculum.
6. Longitudinal quantitation of *C. albicans* in the oral cavity of individual mice is done daily for the first 7 days after oral inoculation and then twice weekly throughout the observation period. *C. albicans* is collected by thoroughly rubbing oral surfaces with a Calgiswab. This procedure is performed without anesthesia and produces minimal discomfort to the animals. The Calgiswabs used for sampling are dissolved in 2-mL volumes of Ringer's citrate buffer, and 200 μL is plated on Sabouraud dextrose agar. Plates are incubated for 24 h at 37°C, and the total colony forming units recovered per oral cavity per mouse is calculated.
7. Oral infection with *C. albicans* further augments the morbidity and hastens premature death in these Tg mice in

comparison with uninfected controls. Accordingly, the health status of these mice is monitored daily. Criteria for euthanasia are applied as recommended by the Canadian Council on Animal Care (1998). Euthanasia of animals attaining these defined end points is performed by inhalation of CO_2.

3.3. Flow Cytometry Analysis of Immune Cell Populations

1. Groups of 6–10 CD4C/HIV^{MutA} and non–Tg littermates (45 to 55 days old) are orally infected or not with 10^8 *C. albicans* blastoconidia and assessed at 7, 45, or 70 days postinfection.
2. Independent experiments are conducted by pooling oral mucosal cells from all mice within each group.
3. Heparinized blood is collected by cardiac puncture under anesthesia, and the mice are exsanguinated by perfusion with PBS. The latter maneuver greatly facilitates dissection of the oral mucosa and analysis of immune cell populations by flow cytometry (*see* **Note 1**).
4. Spleens are removed, mechanically disrupted by pressing through a nylon mesh, and deposited in 25-mm-diameter dishes containing 2 mL of HBSS. Cell suspensions are washed twice in HBSS, resuspended in complete tissue culture medium, and filtered through a nylon mesh to obtain a homogeneous suspension. Remaining red blood cells are removed with FACS lysing solution.
5. The cheeks and the hard and soft palate are dissected free of the underlying muscle layer and washed for 5 min in cleaning buffer solution. Tissues are washed in HBSS, cut longitudinally, and minced into 1 mm^2 fragments in complete medium. Minced tissues are digested by incubating with collagenase type IV (final concentration, 0.25%) in complete medium at 37°C for 30 min with gentle agitation, replacing the medium, and incubating for a further 30 min. Tissue debris are excluded by twice filtering cell suspensions through an 80-μm nylon mesh.
6. Cell suspensions are resuspended in complete medium, and splenocytes are adjusted to 1×10^6 cells/mL. Cell viability, determined by trypan blue exclusion, should be >90%.
7. Cells suspensions are incubated with 1 μg of conjugated monoclonal antibodies for 30 min at 4°C. Cells are twice washed in cold PBS.
8. Spleen cells are used as a control for comparison of flow cytometry profiles with the oral mucosal cell population. Electronic compensation is performed to specifically quantitate each fluorochrome in multicolor analysis. Data are acquired on 10,000 events by gating on CD45+ cells and expressed according to the various combinations of antibodies (**Fig. 25.1**).

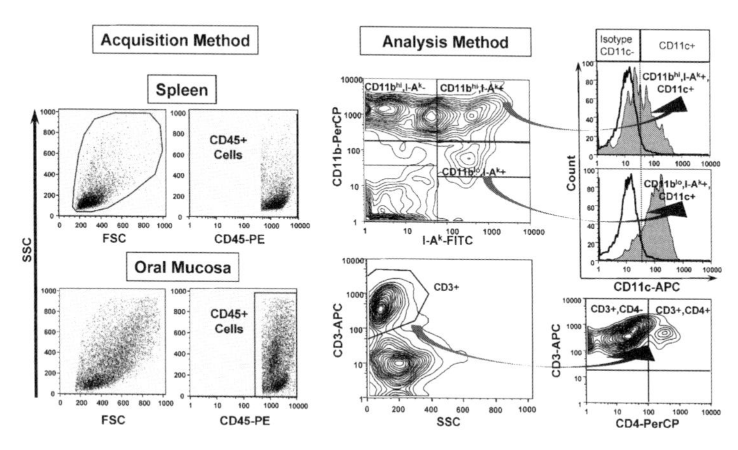

Fig. 25.1. Flow cytometry analysis of oral mucosal cell populations in CD4C/HIV^{MutA} Tg mice. Acquisition was gated on CD45+ cells from spleen and oral mucosa, and multicolor analysis was conducted to identify and quantify specific cell populations. Immature and mature DCs were identified as CD11b^{high}, I-A^{k}+, CD11c+ and CD11b^{low}, I-A^{k}+, CD11c+, respectively, whereas CD4+ T cells were CD3+, CD4+. (Reproduced from Lewandowski D, Marquis M, Aumont F, et al. 2006. Altered CD4+ T cell phenotype and function determine the susceptibility to mucosal candidiasis in transgenic mice expressing HIV-1. *J Immunol* 177(1):479–491; copyright 2006 The American Association of Immunologists, Inc.)

3.4. Flow Cytometry Determination of Intracellular Cytokines

3.4.1. CD4+ T Cells

1. Cervical lymph nodes (CLNs) are carefully removed and processed as for spleens (*see* **Section 3.3 point 4**), with the exception that removal of remaining red blood cells with FACS lysing solution is omitted.
2. Positive selection of CD4+ T cells from CLN cell suspensions is done using biotin-conjugated antibody specific for CD4 and streptavidin Captivate ferrofluid particles, according to the manufacturer's instructions (Captivate Ferrofluid Conjugates and Related Products, Molecular Probes, Eugene, OR), yielding >95% CD4+ T cells (*see* **Note 2**).
3. Selected CD4+ T cells (10^5/100 μL) are cultured for 72 h on anti-CD3–coated (5 μg/mL) and anti-CD28–coated (5 μg/mL) 96-well tissue culture plates in supplemented RPMI 1640 containing 5 μg/mL anti-CD3 antibody.
4. The cells are restimulated for 4 h with ionomycin (500 ng/mL) and PMA (5 ng/mL) in presence of monensin or brefeldin A, and surface stained with anti-CD69 antibody.
5. After washing, intracellular staining is performed with the Cytofix/Cytoperm kit and each of the anti-cytokine antibodies, and their respective isotype control antibodies, according to the manufacturer's instructions.
6. Flow cytometry is performed by gating on CD69+ cells, to determine the percentage of CD4+ T cells expressing each of the cytokines.

3.4.2. DCs

1. Bone marrow–derived DCs (BMDCs) are generated using a modification of methods previously described *(9;10)*.
2. Femurs are dissected free of surrounding tissues using small scissors, rinsed for 10 s in a Petri dish containing 70% ethanol, and transferred to a Petri dish containing PBS.
3. The head of the femur is sectioned and the bone marrow is recovered by flushing the femoral marrow cavity with 2 to 3 mL of PBS contained in a 5-mL syringe equipped with a 25-gauge needle.
4. The cell suspension is dispersed by pumping the syringe back and forth, filtered through a 80-μM nylon mesh, and centrifuged for 10 min at 300 × *g*. The supernatant is removed and discarded.
5. Red blood cells are lysed by adding 1 mL of RBC lysing buffer for 5 min, and the cell suspension is washed twice with HBSS.
6. BM cells are cultured in 2 mL of supplemented RPMI 1640 containing GM-CSF (1000 U/mL) and IL-4 (500 U/mL), in 6-well tissue culture plates.
7. At days 2 and 3, nonadherent cells are removed and fresh medium supplemented with GM-CSF and IL-4 is added.
8. Nonadherent cells are harvested at day 7 and determined to be DCs by morphology and flow cytometry (>90% CD11b+, CD11c+, I-A^k+).
9. 10^6 BMDCs are stimulated with LPS (100 ng/mL or 1 μg/mL) or pulsed with live *C. albicans* blastoconidia at different DC to *Candida* ratios (2:1, 1:1, 1:5) for 2 h.
10. Brefeldin A and 2.5 μg/mL of amphotericin B (to prevent *Candida* overgrowth; *see* **Note 3**) are added, and incubation is continued for a further 16 h.
11. Intracellular staining is performed with anti-IL-12-PE.
12. The percentage of CD11c+ DCs expressing IL-12 is determined by flow cytometry.

3.5. Quantitative RT-PCR Analysis of Cytokines in Oral Mucosal Tissue

1. Cheeks are collected (*see* **Section 3.3 point 5**) into RNAlater and stored at −20°C.
2. Total RNA is extracted using the RNeasy Mini kit with tissue disruption using an Omni Homogenizer and DNase digestion as recommended by the manufacturer (RNeasy Mini Handbook; Qiagen).
3. The extracted RNA is quantitated using the RiboGreen RNA Quantitation Reagent kit and its integrity is confirmed by agarose gel electrophoresis.
4. RNA (600 ng) is reverse-transcribed using the QuantiTect Reverse Transcription kit, according to the manufacturer's

instructions (QuantiTect Reverse Transcription Handbook; Qiagen).

5. The entire cDNA thus obtained is amplified in duplicate by real-time PCR in a SmartCycler instrument using the QuantiTect SYBR Green PCR kit and recommended protocol (QuantiTect SYBR Green PCR Handbook; Qiagen). Individual assays are conducted with specific primer pairs for IFN-γ, IL-4, and β-actin as a control housekeeping gene.
6. Cytokine mRNA expression is normalized against the expression of β-actin, and relative quantitation is performed using the comparative threshold (Ct) cycle number method (User Bulletin #2, ABI Prism 7700 Sequence Detection System, Applied Biosystems; http://hcgs.unh.edu/protocol/realtime/UserBulletin2.pdf).
7. Comparative threshold of the target genes, normalized ΔCt (Ct $_{\text{cytokine}}$ – Ct $_{\text{actin}}$), $\Delta\Delta$Ct (average ΔCt – average ΔCt $_{\text{non-Tg}}$), and relative level ($2^{-\Delta\Delta \text{Ct}}$) are calculated in Excel (*see* **Note 4**).

4. Notes

1. Exsanguination by perfusion with PBS prevents contamination of oral mucosal cell suspensions with circulating leukocytes, as shown by the very low percentage of red blood cells (<1%) in these cell suspensions.
2. No difference in the generated data is observed by using the Dynal Mouse CD4 Negative Isolation kit (Dynal Biotech, Oslo, Norway).
3. Amphotericin B is reported to not modify cytokine production by DCs (*11*).
4. Control experiments demonstrated that expression of β-actin in cheek mucosa is not altered by HIV-1 transgene expression or candidal infection.

Acknowledgments

The authors would like to thank Serge Sénéchal for skilled assistance with flow cytometry and Claire St-Onge for manuscript preparation. These studies are supported by a grant from the Canadian Institutes of Health Research HIV/AIDS Research Program (HOP-41544).

References

1. Samaranayake LP. (1992) Oral mycoses in HIV infection. Oral Surg Oral Med Oral Pathol; 73(2):171–180.
2. Martins MD, Lozano-Chiu M, Rex JH. (1998) Declining rates of oropharyngeal candidiasis and carriage of Candida albicans associated with trends toward reduced rates of carriage of fluconazole-resistant C. albicans in human immunodeficiency virus-infected patients. Clin Infect Dis; 27(5):1291–1294.
3. de Repentigny L, Lewandowski D, Jolicoeur P. (2004) Immunopathogenesis of oropharyngeal candidiasis in human immunodeficiency virus infection. Clin Microbiol Rev; 17(4):729–759.
4. de Repentigny L, Aumont F, Ripeau JS, et al. (2002) Mucosal candidiasis in transgenic mice expressing human immunodeficiency virus type 1. J Infect Dis; 185(8):1103–1114.
5. Marquis M, Lewandowski D, Dugas V, et al. (2006) CD8+ T cells but not polymorphonuclear leukocytes are required to limit chronic oral carriage of Candida albicans in transgenic mice expressing human immunodeficiency virus type 1. Infect Immun; 74(4):2382–2391.
6. Lewandowski D, Marquis M, Aumont F, et al. (2006) Altered CD4+ T cell phenotype and function determine the susceptibility to mucosal candidiasis in transgenic mice expressing HIV-1. J Immunol; 177(1):479–491.
7. Hanna Z, Kay DG, Rebai N, Guimond A, Jothy S, Jolicoeur P. (1998) Nef harbors a major determinant of pathogenicity for an AIDS-like disease induced by HIV-1 in transgenic mice. Cell; 95:163–175.
8. Lacasse M, Fortier C, Trudel L, Collet AJ, Deslauriers N. (1990) Experimental oral candidosis in the mouse: microbiologic and histologic aspects. J Oral Pathol Med; 19(3):136–141.
9. Inaba K, Inaba M, Romani N, et al. (1992) Generation of large numbers of dendritic cells from mouse bone marrow cultures supplemented with granulocyte/macrophage colony-stimulating factor. J Exp Med; 176(6):1693–1702.
10. Poudrier J, Weng X, Kay DG, Hanna Z, Jolicoeur P. (2003) The AIDS-like disease of CD4C/human immunodeficiency virus transgenic mice is associated with accumulation of immature CD11bHi dendritic cells. J Virol; 77(21):11733–11744.
11. d'Ostiani CF, Del Sero G, Bacci A, et al. (2000) Dendritic cells discriminate between yeasts and hyphae of the fungus Candida albicans. Implications for initiation of T helper cell immunity in vitro and in vivo. J Exp Med; 191(10):1661–1674.

Chapter 26

Proteomic Profiling of Serologic Response to *Candida albicans* During Host-Commensal and Host-Pathogen Interactions

Aida Pitarch, César Nombela, and Concha Gil

Abstract

Candida albicans is a commensal inhabitant of the normal human microflora that can become pathogenic and invade almost all body sites and organs in response to both host-mediated and fungus-mediated mechanisms. Serologic responses to *C. albicans* that underlie its dichotomist relationship with the host (host-commensal and host-pathogen interactions) display a high degree of heterogeneity, resulting in distinct serum anti-*Candida* antibody signatures (molecular fingerprints of anti-*Candida* antibodies in serum) that can be used to discriminate commensal colonization from invasive disease. We describe the typical proteomic strategy to globally and integratively profile these host antibody responses and determine serum antibody signatures. This approach is based on the combination of classic immunoproteomics or serologic proteome analysis (two-dimensional electrophoresis followed by quantitative Western blotting and mass spectrometry) with data mining procedures. This global proteomic stratagem is a useful tool not only for obtaining an overview of different anti-*Candida* antibodies that are being elicited during the host-fungus interaction and, consequently, of the complex *C. albicans* immunome (the subset of the *C. albicans* proteome targeted by the immune system), but also for evaluating how this pathogen organism interacts with its host to trigger infection. In contrast with genomics and transcriptomics, this proteomic technology has the potential to detect antigenicity associated with posttranslational modification, subcellular localization, and other functional aspects that can be relevant in the host immune response. Furthermore, this strategy to define molecular fingerprints of serum anti-*Candida* antibodies may hopefully bring to light potential candidates for diagnosis, prognosis, risk stratification, clinical follow-up, therapeutic monitoring, and/or immunotherapy of candidiasis, especially of its life-threatening systemic forms.

Key words: *Candida albicans*, immunome, systemic candidiasis, antibody response, proteomics, immunoproteomics, SERPA, two-dimensional gel electrophoresis, mass spectrometry.

Steffen Rupp, Kai Sohn (eds.), *Host-Pathogen Interactions*, DOI: 10.1007/978-1-59745-204-5_26,

1. Introduction

Candida albicans is a harmless member of the microflora on the mucosal surfaces (the oral cavity, gastrointestinal tract, and vaginal canal) of most healthy individuals and warm-blooded animals. However, in response to mechanisms mediated both by the host (predisposing factors that involve changes in the host microflora, natural barrier, and/or immune system) and by the fungus (virulence factors that include morphologic and phenotypic switching, adhesion to host structures, and/or secretion of hydrolytic enzymes), this endogenous organism (acquired at or soon after birth) ceases to be an apparently benign commensal and evolves into a deadly opportunistic pathogen that is poised to penetrate into deeper tissues of the host, enter its bloodstream, and invade and grow within its tissues and organs (**Fig. 26.1**) *(1–4)*. The damage (candidiasis) inflicted on the host by this invasive "Mr. Hyde" can range from superficial mucosal lesions (oropharyngeal, esophageal,

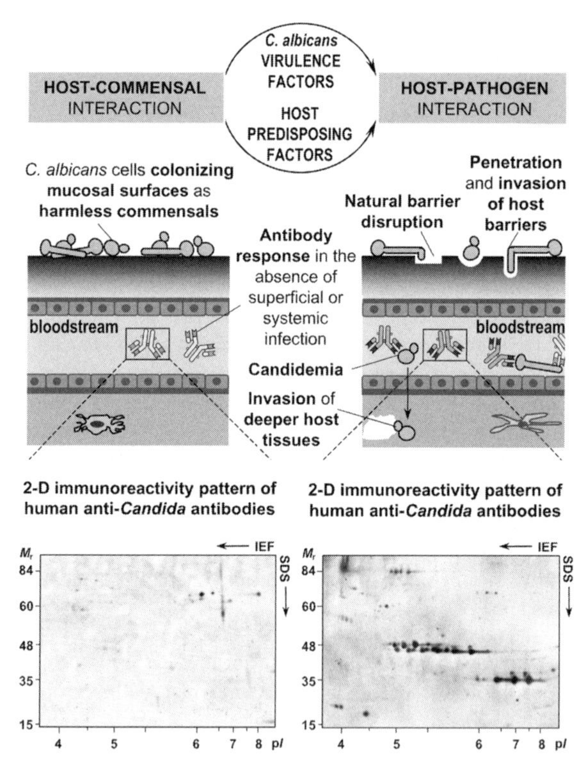

Fig. 26.1. Representative proteomic profiles of serologic response to *C. albicans* during host-commensal and host-pathogen interactions. Two-dimensional (2-D) immunoreactivity patterns of serum human antibodies to *C. albicans* proteins during commensalism (*left*) and pathogenicity (*right*) are shown. The development of *C. albicans* infection is determined by the nature of the host-fungus interactions. In response to both host-mediated and fungus-mediated mechanisms, this endogenous organism (on the mucosal surfaces) can switch from the commensal to the pathogenic state and cause host damage (candidiasis). *See* **Section 1** for further information.

and vulvovaginal candidiasis) to life-threatening systemic forms of infection (candidemia and localized deep-seated and disseminated candidiasis), relying on the underlying host dysfunction *(5, 6)*.

A distinctive hallmark of the natural course of the host-commensal interaction is undoubtedly the presence of circulating antibodies to *C. albicans* secreted, cell wall and cytoplasmic components (of carbohydrate and/or protein nature) into bloodstream in most individuals *(7–10)*. Clues as to this ubiquity of anti-*Candida* antibodies in human sera may stem from continuous exposure to these antigens during commensal colonization of *C. albicans* (starting early during infancy), which supposedly appear to be able to elicit specific serum antibody responses in the absence of superficial or systemic infection. Be that as it may, serologic responses to *C. albicans* that lie behind these two states (commensal colonization and invasive disease) have however proved to be heterogeneous and distinct (**Fig. 26.1**) *(7, 8, 10)*. Accordingly, human sera intriguingly contain anti-*Candida* antibody signatures (molecular fingerprints of anti-*Candida* antibodies in serum) reflective of host-commensal and host-pathogen interactions that may potentially be used to assess health and disease conditions *(8)*.

These serum antibody signatures can easily be defined by proteomic profiling of the serologic response of the host to *C. albicans* (i.e., by serum anti–*C. albicans* antibody expression profiling) using high-throughput technologies such as traditional immunoproteomics or serologic proteome analysis (SERPA) *(7, 11–13)* in conjunction with data mining procedures *(8, 14–16)*. The combination of classic immunoproteomics (i.e., two-dimensional electrophoresis (2-DE) followed by quantitative Western blotting and mass spectrometry (MS) analysis with computational and statistical methods is a useful and straightforward approach (**Fig. 26.2**) to:

1. Obtain a global and integrated view of different serum anti-*Candida* antibodies that are being produced during the host-fungus interaction (i.e., of molecular fingerprints of host antibodies in serum) and, consequently, characterize the *C. albicans* immunome (the subset of the proteome that acts as a target for the immune system *(8, 17, 18)* (*see* **Chapter 15**).
2. Directly study the host-fungus interaction and evaluate how this double-edged organism (commensal and opportunistic pathogen) interacts with its host to produce infection. The characterization of antigens that are being expressed during the different states may also be crucial to shed some light on the underlying mechanisms that enable *C. albicans* to switch from a commensal to a pathogen lifestyle (or in other words, from the harmless "Dr. Jekyll" to the lethal "Mr. Hyde").
3. Uncover antibodies directed against posttranslationally modified *C. albicans* antigens *(7)*. Unlike genomics and

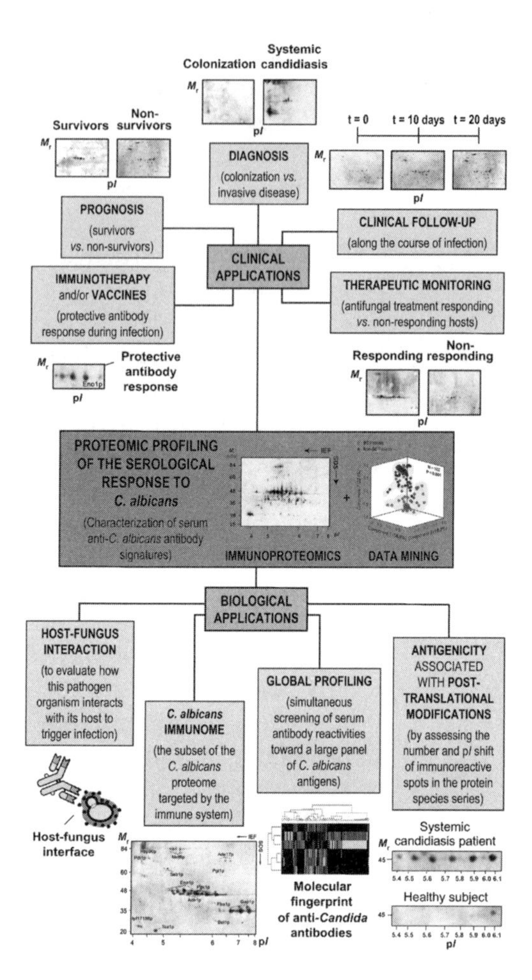

Fig. 26.2. Main biological and clinical applications of proteomic profiling of serologic response to *C. albicans* by using classic immunoproteomics or SERPA and data mining procedures.

transcriptomics, classic immunoproteomics enables the detection of antigenicity associated with posttranslational modification, subcellular localization, and other functional aspects that may play a key role in the host immune response.

4. Complement and extend traditional one-by-one approaches, which reduce a complex system, such as serum, to single antigen-antibody interactions. In fact, the present global profiling technique allows the simultaneous screening of serum antibody reactivities toward a large panel of *C. albicans* antigens, many of which have not been characterized using these one-by-one methods *(7)*.
5. Establish the spectrum of circulating anti-*Candida* antibody specificities during mucosal surface colonization and invasive disease, and identify potential infection-specific markers (antibodies and/or their related antigens) that may

be useful in screening, diagnosis, and risk stratification of candidiasis, especially of the systemic or invasive forms of infection *(7, 8, 19–21)*.

6. Evaluate changes in the global serum anti-*Candida* antibody-expression profiles between survivors and nonsurvivors during the host-pathogen interaction, and discover serologic markers with prognostic significance for predicting clinical outcomes in these *Candida* infected hosts *(7, 8, 19)*.
7. Define the global reactivity patterns of host anti-*Candida* antibodies produced along the course of *Candida* infection, and identify serum anti-*Candida* antibody signatures that may be used to follow the progression of disease during the host-pathogen interaction in the infected hosts *(7, 19)*.
8. Compare global serum anti-*Candida* antibody-reactivity profiles between antifungal treatment responding and nonresponding hosts during the host-pathogen relationship, and uncover molecular fingerprints of serum anti-*Candida* antibodies that may have utility in monitoring therapeutic response and evaluating the efficacy of antifungal treatment regimens *(7)*.
9. Identify *C. albicans* antigens with the best potential for eliciting a protective antibody response during infection. It may provide a rationale for the design of novel vaccine-based or immunotherapy-based strategies to prevent and control candidiasis, particularly its systemic forms *(8, 19, 22, 23)*.

In this chapter, we describe the classic immunoproteomic approach for anti-*Candida* antibody-reactivity profiling *(7, 8, 24)*. Serum antibody signatures directed against *C. albicans* housekeeping enzymes (including diverse chaperones, heat shock proteins, all highly conserved glycolytic enzymes, fermentative proteins, other metabolic enzymes, elongation factors, ribosomal proteins, porins and redox enzymes, among others) and cell wall proteins (β-1,3-glucosyltransferase or glucan 1,3-beta-glucosidase, and glycolytic enzymes) have successfully been characterized using this strategy *(7, 8, 25, 26)*. However, global gel-free profiling procedures, such as serologic expression cloning (SEREX, based on cDNA expression library screening with serum antibodies) *(27, 28)*, *in vivo* induced antigen technology (IVIAT, a variation of colony immunoscreening) *(29)*, multiple affinity protein profiling (MAPPing, based on two-dimensional (2-D) immunoaffinity chromatography and MS) *(12, 30)*, phage-epitope microarray analysis *(31)*, protein microarray/chip technology associated with surface-enhanced laser desorption/ionization (SELDI) time-of-flight MS *(32)*, to name but a few, could also alternatively be applied to define and analyze molecular fingerprints of serum anti-*Candida* antibodies elicited during host-commensal and host-pathogen interactions. Remarkably, the current global proteomic profiling approach can be extended, with the appropriate adjustments, to other infectious diseases, cancers, allergies, or autoimmune disorders (**Fig. 26.2**).

2. Materials

All solutions and buffers must be prepared with chemical reagents of the highest purity (electrophoresis grade or better) and ultrapure water (double-distilled, deionized water with a resistivity >18MΩ/cm, as provided by Nanopure or Milli-Q systems [Millipore, Bedford, MA]) in clean glassware to avoid impurities that may interfere with the analysis (*see* **Note 1**). Growth media and solutions should be sterilized by autoclaving before use when working under sterile conditions.

2.1. Preparation of C. albicans Protein Sample

2.1.1. Cell Extracts

1. Yeast-Peptone-D-glucose (YPD) plates: 1% (w/v) yeast extract (Difco Laboratories, Detroit, MI), 2% (w/v) peptone (Difco), 2% (w/v) D-glucose, 2% (w/v) agar (Difco).
2. YPD medium: 1% (w/v) yeast extract, 2% (w/v) peptone, 2% (w/v) D-glucose.
3. Lysis buffer: 50 mM Tris-HCl, pH 7.5, 1 mM EDTA, 150 mM NaCl, 1 mM dithiothreitol (DTT), 0.5 mM phenylmethylsulfonyl fluoride (PMSF; Fluka, Chelmsford, MA), and 5 μg/mL each of antipain, leupeptin, and pepstatin (Sigma, St. Louis, MO) (*see* **Notes 2** and **3**).
4. 0.40- to 0.60-mm chilled, acid-washed glass beads (Sartorius, Goettingen, Germany; *see* **Note 4**).
5. Bead mill homogenizer: Fast-Prep cell breaker (Q-Biogene, Carlsbad, CA).

2.1.2. Protoplast Lysates

1. YPD plates and YPD medium (*see* **Section 2.1.1**).
2. Pretreatment buffer: 10 mM Tris-HCl, pH 9.0, 5 mM EDTA, 1% (v/v) β-mercaptoethanol (*see* **Note 2**).
3. 1 M sorbitol solution: Dissolve 182.17 g of sorbitol in sufficient water to yield a final volume of 1 L.
4. Glusulase (Du Pont; NEN Life Science Products, Boston, MA).
5. Lysis buffer (*see* **Section 2.1.1**).

2.2. Separation of C. albicans Proteins by 2-DE

2.2.1. Isoelectric Focusing (IEF)

1. Rehydration buffer: 7 M urea (Bio-Rad, Hercules, CA), 2 M thiourea (Fluka), 2% (w/v) 3-[(3-cholamidopropyl)dimethylamino]-1-propanesulfonate (CHAPS; Sigma), 65 mM dithioerythritol (DTE; Merck, Darmstadt, Germany), 0.5% (v/v) immobilized pH gradient (IPG) buffer pH 3–10 (GE Healthcare Limited, Buckinghamshire, UK), and a trace of bromophenol blue (*see* **Notes 2, 5** to 7).
2. Immobiline pH 3 to 10 nonlinear (NL) gradient DryStrips (18 cm long, GE Healthcare Limited) (*see* **Notes** 7 and **8**).
3. IPGphor IEF system (GE Healthcare Limited) (*see* **Note 7**).

4. Ceramic strip holders (GE Healthcare Limited).
5. IPG cover fluid (GE Healthcare Limited).
6. Electrode pads (GE Healthcare Limited).
7. Strip holder cleaning solution (GE Healthcare Limited).
8. Internal 2-DE standards (molecular weight and isoelectric point marker proteins; Bio-Rad) (*see* **Note 7**).

2.2.2. Sodium Dodecyl Sulfate–Polyacrylamide Gel Electrophoresis (SDS-PAGE)

1. Equilibration buffer 1: 50 mM Tris-HCl, pH 6.8, 6 M urea, 30% (v/v) glycerol, 2% (w/v) SDS, and 2% (w/v) DTE (Merck) (*see* **Notes 2, 6**, and **9**).
2. Equilibration buffer 2: 50 mM Tris-HCl, pH 6.8, 6 M urea, 30% (v/v) glycerol, 2% (w/v) SDS, 2.5% (w/v) iodoacetamide (Sigma), and a trace of bromophenol blue (*see* **Notes 2, 6**, and **9**).
3. Homogeneous SDS-polyacrylamide gels (10%T/1.6%C, piperazine diacrylamide (PDA); 1.5 mm-thick; *see* **Notes 2** and **10**).
4. SDS-PAGE running buffer (Laemmli buffer): 25 mM Tris-base (do not adjust pH), 192 mM glycine, 0.1% (w/v) SDS.
5. 0.5% (w/v) agarose in SDS-PAGE running buffer or in Milli-Q grade water (Millipore) (*see* **Note 11**).
6. Protean II gel running tank (Bio-Rad) (*see* **Note 7**).
7. Refrigerated thermostatic circulator unit.

2.3. Visualization of the Total C. albicans Protein Pattern by 2-DE Gel Staining

1. Glass or polyethylene containers with tight-fitting lid of appropriate size for gel. For SYPRO Ruby staining, use dark polypropylene or polyvinyl chloride (PVC) dishes, but not glass containers (*see* **Note 12**).
2. Reciprocal ("ping-pong") shaking platform.
3. Vacuum aspiration apparatus.

2.3.1. Colloidal Coomassie Blue Staining

1. Fixative solution: 50% (v/v) methanol and 2% (v/v) phosphoric acid.
2. Wash solution: Milli-Q grade water.
3. Pre-treatment solution: 17% (w/v) ammonium sulfate, 3% (v/v) phosphoric acid, 33% (v/v) methanol, and 0.066% (w/v) Coomassie Brilliant Blue G-250 (Bio-Rad). Dissolve 85 g ammonium sulfate in 15 mL phosphoric acid 85% and 330 mL Milli-Q grade water. Add 165 mL methanol and stir for 1 h.
4. Colloidal Coomasie blue solution: Add a solution containing 330 mg Coomasie Blue G-250 in 5 mL methanol to the pre-treatment solution (*see* **Note 13**).
5. Destaining solution: Milli-Q grade water.

2.3.2. Silver Staining (Compatible with Mass Spectrometry)

1. Fixative solution: 50% (v/v) methanol and 5% (v/v) acetic acid.
2. Wash solution 1: 50% (v/v) methanol.
3. Wash solution 2: Milli-Q grade water.
4. Sensitizing solution: 0.02% (w/v) sodium thiosulfate.
5. Silver nitrate solution: 0.1% (w/v) silver nitrate (*see* **Note 2**).
6. Developing solution: 2% (w/v) sodium carbonate in 0.04% (v/v) formalin (35% formaldehyde) (*see* **Notes 2** and **14**). Chill the solution at 4°C.
7. Stopping solution: 5% acetic acid.
8. Storing solution: 1% acetic acid.

2.3.3. SYPRO Ruby Protein Gel Staining

1. Fixative solution: 10% (v/v) methanol and 7% (v/v) acetic acid.
2. Staining solution: SYPRO Ruby protein gel stain (Bio-Rad).
3. Wash solution: 10% (v/v) methanol and 7% (v/v) acetic acid.

2.4. Detection of Global Immunoreactivity Profiles of Serum Antibodies to C. albicans Proteins by 2-D Western Blotting

2.4.1. Electrophoretic Blotting (Electroblotting) of 2-DE Gels onto Nitrocellulose Membranes

1. Nitrocellulose membrane (HyBond-ECL, GE Healthcare Limited).
2. 3MM Chr chromatography filter paper (Whatman, Maidstone, UK).
3. Fiber pads (Bio-Rad) (*see* **Note** 7).
4. Plastic gel holder cassette (Bio-Rad) (*see* **Note** 7).
5. Trans-blot electrophoretic transfer cell (Bio-Rad) (*see* **Note** 7).
6. Refrigerated thermostatic circulator unit.
7. Transfer buffer: 48 mM Tris-base (do not adjust pH), 39 mM glycine, 0.037% (w/v) SDS, and 20% (v/v) methanol.

2.4.2. Visualization of the Total C. albicans Protein Pattern on 2-D Blot: SYPRO Ruby Protein Blot Staining

1. Fixative solution: 10% (v/v) methanol and 7% (v/v) acetic acid.
2. Staining solution: SYPRO Ruby protein blot stain (Bio-Rad).
3. Wash solution: Milli-Q grade water.
4. Reciprocal ("ping-pong") shaking platform.
5. Dark polypropylene or polyvinyl chloride (PVC) dishes with tight-fitting lid of appropriate size for gel. Glass containers are not recommended (*see* **Note 12**).

2.4.3. 2-D Immunoblotting: Immunodetection of Global Reactivity Profiles of Host Antibodies to C. albicans Antigens

2.4.3.1. Antigen-Antibody Binding

1. Tris-buffered saline (TBS) solution: 20 mM Tris-HCl, pH 7.5, and 50 mM NaCl.
2. Blocking solution: 5% (w/v) nonfat dry milk (without calcium) in TBS solution.
3. Wash (Tween-TBS, TTBS) solution: 0.1% (v/v) Tween-20 (Fluka) in TBS solution.
4. Antibody solution: 1% (w/v) nonfat dry milk (without calcium) in TTBS solution.

5. Serum specimens from patients with systemic candidiasis (SC) and superficial candidiasis, and controls (non-SC patients and healthy subjects), among others. Store at −80°C in small aliquots.
6. Horseradish peroxidase (HRP)-labeled anti-human IgG antibody (GE Healthcare Limited).
7. Reciprocal ("ping-pong") shaking platform.
8. Plastic container.
9. Heat-sealable plastic bags.

2.4.3.2. Enhanced Chemiluminescent (ECL) Detection

1. ECL detection system (GE Healthcare Limited).
2. Acetate sheet protectors.
3. X-ray film cassette (Hypercassette; GE Healthcare Limited).
4. High performance films (Hyper-film ECL; GE Healthcare Limited).
5. Photographic developing reagent kit (Kodak, Rochester, NY).

2.4.4. Stripping and Reprobing of 2-D Blots

1. Stripping buffer: 130 mM glycine-HCl pH 2.2, 1% (w/v) SDS, and 0.05% (v/v) NP-40.
2. Wash (TTBS) buffer: 0.1% (v/v) Tween-20 in TBS.
3. Reciprocal ("ping-pong") shaking platform.
4. Plastic container.

2.5. Data Acquisition and Analysis of the Densitometric Profiles of Antibody Reactivities to C. albicans Proteins

1. Imaging densitometer (GS-800; Bio-Rad) (*see* **Note 7**).
2. Epi illuminated laser-scanning instrument (Molecular Imager FX; Bio-Rad) (*see* **Note 7**).
3. Quantity One software (Bio-Rad) (*see* **Note 7**).
4. ImageMaster 2D Platinum software (GE Healthcare Limited) (*see* **Note 7**).

2.6. Identification of C. albicans Immunoreactive Proteins by MS (see Chapter 15)

1. Peptide sample preparation: The reader is directed to **Chapter 15** for buffers and reagents.
2. Matrix-assisted laser desorption/ionization time-of-flight (MALDI-TOF) mass spectrometer (Voyager-DE STR; PerSeptive Biosystems, Framingham, MA).
3. Mascot search engine: http://www.matrixscience.com.

3. Methods

The typical proteomic strategy for studying the serologic response to *C. albicans* during its dichotomist relationship with the host (i.e., during host-commensal and host-pathogen interactions) includes the following steps: (i) isolation and solubilization of the

C. albicans proteome or subproteome of interest (used as a source of *C. albicans* antigens), (ii) simultaneous separation of its protein constituents by 2-DE, (iii) visualization of its 2-D pattern by specific staining methods, (iv) detection of global immunoreactivity profiles of host antibodies to its corresponding immunogenic proteins by 2-D Western blotting using serum specimens (screened individually) from *Candida* colonized and infected hosts, (v) acquisition and comparison of the densitometric patterns of antibody reactivities in the different host groups by mono- and multi-parametric analyses, and (vi) identification of immunorelevant *C. albicans* proteins (and indirectly of their related host antibodies) by MS. A flowchart of this immunoproteomic procedure is illustrated in **Fig. 26.3**. The protocols presented below basically correspond with proteomic profiling of serologic response of the host to *C. albicans* (**steps i** to **v**), whereas those associated with MS-identification of the *C. albicans* immunome (**step vi**) are described in detail in **Chapter 15**. Although these methods are suitable for this polymorphic fungus *(7, 8)*, it may be necessary to adjust them when using other infectious agents.

3.1. Preparation of C. albicans Protein Sample

Sample preparation is the key to high-quality 2-DE results. This procedure must be optimized empirically for each protein sample type with the goal of yielding a sample in which all its proteins are fully solubilized, disaggregated, denatured, and reduced (*see* **Note 5**). The general strategy involves (i) preparation of cells, (ii) cell lysis, and (iii) protein solubilization (*see* **Note 15**). Cell disintegration can be performed using several techniques (*see* **Note 16**), the most common methods for *C. albicans* cells and, overall, for yeasts and fungi being:

1. Mechanical disruption procedures, such as (i) shaking or stirring of the cell culture with glass beads either in a bead mill homogenizer or in a vortex mixer, and (ii) grinding of freeze-dried mycelia in a mortar and pestle, among others.
2. Enzymatic digestion of their cell walls to generate protoplasts, which are then lysed by an osmotic shock to release intracellular proteins (protoplast lysates).

The methods given here outline two protocols to isolate and solubilize *C. albicans* cytoplasmic proteins (using both techniques) *(24)*, as many of these have been shown to be important targets of the host antibody response to *Candida* infection *(7)*. Although the choice between both procedures will depend on the specific application, that of protoplast lysates (*see* **Section 3.1.2**) leads to an easier and more sensitive immunodetection system because of the higher expression of some immunogenic proteins under these conditions *(24)*. Remarkably, given that the cell wall subproteome is often responsible for initial host-fungus interactions and contains the major antigens and host recognition molecules *(9, 33, 34)*, we also strongly recommend the use of this protein sample type for analyzing the serological response to *C. albicans (8)*. Detailed protocols for its isolation and solubilization have recently been described (*see* Refs. *35* and *36*), so will, therefore, not be addressed in this chapter.

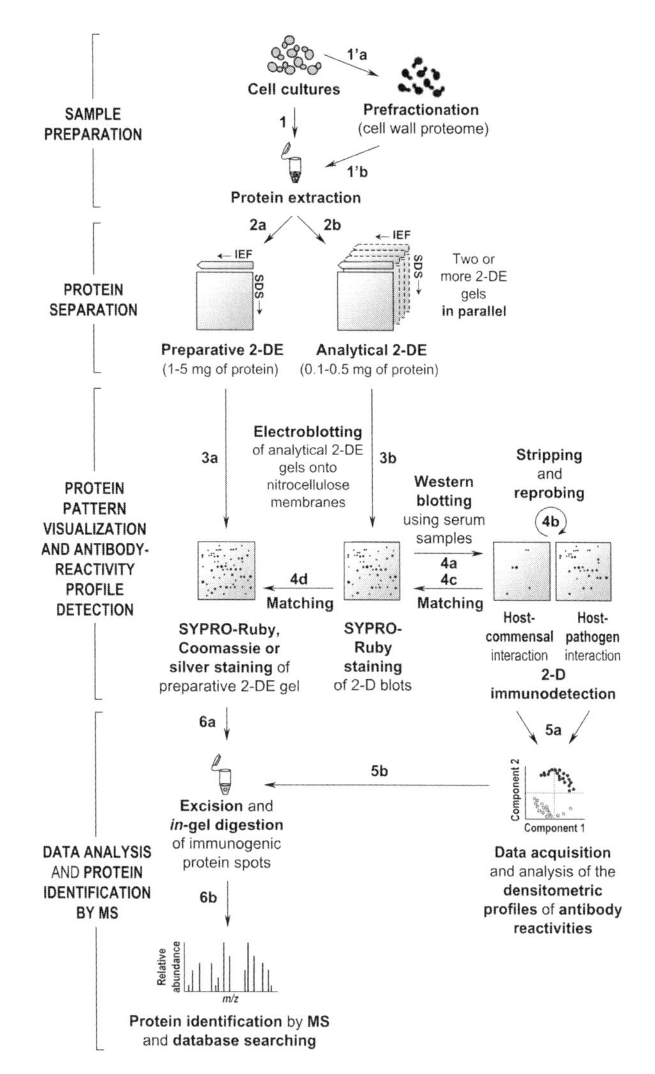

Fig. 26.3. Flowchart of the typical strategy for analyzing the serologic response to *C. albicans* (molecular fingerprinting of anti-*Candida* antibodies in serum) during host-commensal and host-pathogen interactions by using classic immunoproteomics or SERPA and data mining procedures. Serum samples from colonized and infected hosts (humans or warm-blooded animals) are screened individually by 2-D Western blotting for circulating anti-*Candida* antibodies to 2-DE–separated *C. albicans* proteins (used as a source of antigens). Immunoreactive *C. albicans* proteins (and indirectly their related antibodies) are then identified by MS analysis. Specific serum antibody signatures can be defined using different data mining procedures. *See* **Section 3** for further details.

3.1.1. Cell Extracts

1. Grow *C. albicans* cells on a YPD plate (stock maintenance medium) at 30°C for 2 days. Use a single colony to inoculate 50 mL of YPD (or selective) medium in a 250-mL flask (*see* **Note 17**), and grow overnight at 30°C in a shaking incubator (200 rpm).
2. Use this 50-mL preculture to inoculate a 250-mL flask containing 50 mL fresh YPD medium, and grow at 30°C with

vigorous rotary shaking (200 rpm) until the culture reaches log phase growth (*see* **Note 18**).

3. Harvest the yeast cells by centrifugation at 4500 × *g* for 5 min and discard the supernatant. Resuspend the cell pellet in 50 mL water, and centrifuge 5 min at 4500 × *g*. Discard the supernatant.
4. Resuspend the cell pellet in 50 mL ice-cold lysis buffer, and centrifuge 5 min at 4500 × *g*. Decant the supernatant.
5. Resuspend the cells in 3 volumes of ice-cold lysis buffer, and add 3 to 4 volumes of 0.5-mm acid-washed glass beads (*see* **Note 19**). Lyse the cells using a bead mill homogenizer (e.g., the Fast-Prep cell breaker; level of 5.5) or a vortex mixer for 15 to 20 s, and incubate on ice for 1 min (*see* **Note 15**). Repeat this step until complete cell breakage, as determined with a phase-contrast microscope.
6. Separate the cell extract from glass beads and cell debris by centrifugation at 4°C and 4500 × *g* for 15 min (*see* **Notes 15** and **20**). Transfer the supernatant to another tube, and centrifuge again.
7. Collect the clarified supernatant carefully, quantify the protein content by standard protein determination methods, and store at −80°C for later analysis (or alternatively use immediately for following IEF; *see* **Section 3.2.1**).

3.1.2. Protoplast Lysates (see ***Note 21****)*

1. Prepare *C. albicans* cells as described in **steps 1** to **3** of **Section 3.1.1**.
2. Gently resuspend the cells in 5 mL of pretreatment buffer to a density of 1×10^9 to 2×10^9 cells/mL, and incubate at 28°C with gentle rotary shaking (80 rpm) for 30 min. Centrifuge 10 min at 600 × *g*, and discard the supernatant.
3. Gently resuspend the cell pellet in 50 mL of a 1 M sorbitol solution. Centrifuge 10 min at 600 × *g*, and discard the supernatant.
4. Gently resuspend cell pellet in a 1 M sorbitol solution to a density of 5×10^8 cells/mL, and add 30 μL/mL ice-cold Glusulase (*see* **Note 22**). Incubate cells with very gentle shaking (80 rpm) at 28°C until obtaining more than 90% to 95% protoplasts (~45 min to 1 h). Monitor the degree of protoplast formation with a phase-contrast microscope (by counting spherical cells and observing cell lysis in hypotonic solution). Centrifuge 10 min at 600 × *g*, and decant the supernatant carefully.
5. Gently wash the protoplast pellet with 50 mL of a 1 M sorbitol solution. Centrifuge 15 min at 600 × *g* and decant the supernatant carefully. Repeat this step two more times to eliminate any trace of Glusulase (*see* **Note 22**).

6. Resuspend the protoplast pellet in 3 volumes of ice-cold lysis buffer (*see* **Note 19**). Lyse the protoplasts by vortexing for 15 to 20 s, and incubate on ice for 1 min (*see* **Note 15**). Repeat this step as indicated in **step 5** of **Section 3.1.1**.
7. Centrifuge the protoplast lysate at 4°C and 4500 × *g* for 15 min (*see* **Note 15**). Transfer the supernatant to another tube, and centrifuge again.
8. Process the clarified supernatant as described in **step** 7 of **Section 3.1.1**.

3.2. Separation of C. albicans Proteins by 2-DE

2-DE remains the method of choice for simultaneously separating the complex antigenic composition of *C. albicans* at a given time point *(25, 37–39)*. In this high-resolution technique, denatured and reduced proteins are resolved on the basis of (i) their isoelectric point (p*I*, specific pH value at which their net charge is zero) by isoelectric focusing (IEF) within a pH gradient (*see* **Note 23**) in the first dimension, and (ii) their molecular weight by electrophoresis on SDS-polyacrylamide gels (SDS-PAGE) in the second dimension.

The protein sample should be run in parallel on two or more 2-DE gels (**Fig. 26.3**). One gel (preparative 2-DE gel loaded with 1 to 5 mg of protein) is stained with Coomassie blue, silver, or SYPRO Ruby, as appropriate, with the purpose of visualizing the whole *C. albicans* protein pattern (*see* **Section 3.3**), and then used for protein spot excision (according to immunoblotting results) and MS identification (*see* **Section 3.6** and **Chapter 15**). The other gels (analytical 2-DE gels loaded with 0.1 to 0.5 mg of protein) are used for electroblotting (*see* **Section 3.4**). The resulting 2-D blots are stained with SYPRO Ruby with the aims of verifying protein transfer and of spot matching with the previous preparative 2-DE gel and ensuing immunostained 2-D blots (*see* **Note 24**). These 2-D blots are then probed up to four times with different serum specimens by Western blotting to obtain 2-D immunoreactivity patterns of host antibodies to *C. albicans* proteins (*see* **Section 3.4**), which are quantified by densitometric analysis and compared in the different host groups by multivariate data analysis (*see* **Section 3.5**).

3.2.1. IEF (First-Dimension Separation)

1. Mix the *C. albicans* protein sample (about 0.1 to 0.5 mg of protein for analytical 2-DE gels, and 1 to 5 mg of protein for preparative 2-DE gels) with the rehydration buffer in a final volume of 340 μL per strip (*see* **Note 25**). Incubate for 30 min at room temperature with continuous shaking to enable full denaturation and solubilization of the proteins in the sample (*see* **Note 5**). Centrifuge for 15 min at 12,000 × *g* to pellet insoluble material.
2. Pipette the supernatant into each strip holder (*see* **Note 26**) as a streak between its two electrodes without generating any bubbles. Use one holder per strip.

3. Remove the protective cover film from each IPG strip with forceps (with rounded, nonserrated tips), starting at the anodic (arrow pointed) end. Carefully place each IPG strip into a holder, with its gel-side down and its anodic (+) end directed toward the pointed end of the strip holder, in such a way as to avoid trapping air bubbles under the IPG strip. Remove them by lifting the IPG strip up again with forceps. Check that each end of the gel contacts a holder electrode.
4. Position the strip holder(s) on the cooled electrode contact areas (gold section; *see* **Note 27**) of the IPGphor IEF system, which must be placed on a level surface.
5. Add 3 to 4 mL IPG cover fluid to completely overlay each strip (starting on its two ends; *see* **Note 28**). Place the plastic coverlid on each strip holder without making bubbles. Close the safety lid of the IPGphor IEF apparatus.
6. Perform rehydration and IEF separation at 15°C and 50 μA per strip using the following running conditions:
   (a) For analytical runs: (i) passive rehydration: 0 V for 16 h (*see* **Note 29**), and (ii) IEF: 500 V for 1 h, 500 to 2000 V for 1 h, and 8000 V for 5.5 h.
   (b) For preparative runs: (i) active rehydration: 30 V for 13 h (*see* **Note 29**), and (ii) IEF: 500 V for 1 h, 1000 V for 1 h, 2000 V for 1 h, 2000 to 5000 V for 3 h, and 8000 V for 11 h.
7. After IEF, remove excess oil from each IPG strip on a filter paper, and proceed to IPG strip equilibration immediately (*see* **Section 3.2.2**) or store the IPG strip(s) at −80°C with its/their gel-side up in individual 25 × 200 mm screw-cap tubes or culture tubes capped with flexible paraffin film for future SDS-PAGE analysis.
8. Carefully clean the strip holders immediately after use (*see* **Note 26**).

3.2.2. SDS-PAGE (Second-Dimension Separation)

1. Before second-dimension separation, place each IPG strip into an individual Petri dish (or in a screw-cap tube) on a shaker. Add equilibration buffer 1 and incubate for 15 min. Transfer each strip into another Petri dish, add equilibration buffer 2, and incubate again for 15 min (*see* **Fig. 26.4** and **Note 30**). Rinse each strip carefully with water or running buffer, and remove excess running buffer on filter paper, not allowing the paper to touch the gel.
2. Place each equilibrated strip on the edge of the SDS-polyacrylamide gel (10% T; 1.6% C; *see* **Note 10**) sideways. Check that no air burbles are trapped between the IPG strip and the SDS-PAGE gel. Seal each IPG strip with melted agarose (*see* **Note 11**) without introducing air bubbles, and wait until this is polymerized (*see* **Note 31**).

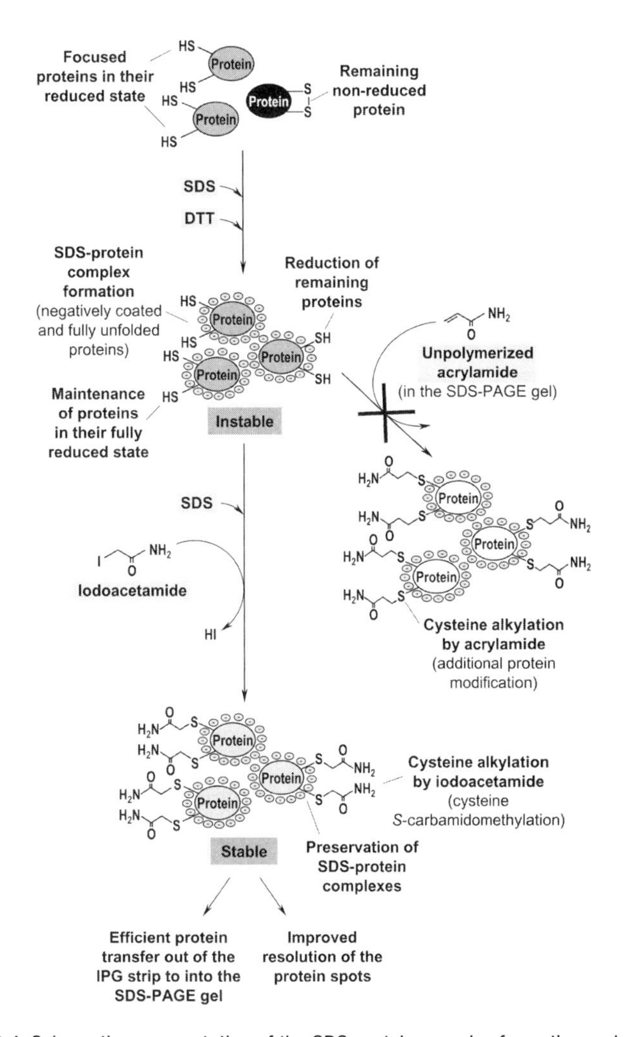

Fig. 26.4. Schematic representation of the SDS-protein complex formation and protein reduction/alkylation reactions performed on the IPG strip equilibration. After IEF and before protein transfer from the IPG strips to the SDS-PAGE gel, these should be equilibrated in a solution containing SDS and a reductant agent (DTT) and then in another solution containing SDS and an alkylating agent (iodoacetamide). These steps are carried out to (i) unfold and coat the proteins with negative charges only (by saturation of the IPG gel with SDS, leading to SDS-protein complexes), (ii) preserve protein reduction (by the DTT addition), (iii) prevent cysteine alkylation by unpolymerized acrylamide in the polyacrylamide gel (with the iodoacetamide addition), (iv) minimize electroendosmotic effects that reduce protein transfer (by adding urea and glycerol), and (v) resolubilize proteins. This results in an efficient transfer of proteins out of the IPG strip to into the SDS-PAGE gel and improved resolution of the protein spots (*see* **Note 30**).

3. Load the gel cassette(s) into the running tank containing an appropriate volume of Laemmli buffer (following the vertical system) and perform SDS-PAGE at 15°C using a cooling unit and at 40 mA per gel until the bromophenol blue front(s) reach(s) the bottom of each gel (for ~5 to 6 h; *see* **Notes 10** and **30**).

4. Remove 2-DE gel(s) from the electrophoresis unit, and proceed to gel staining (for preparative 2-DE gels; *see* **Section 3.3**) or electroblotting (for analytical 2-DE gels; *see* **Section 3.4.1**).
5. Rinse the gel running tank in water to remove any residual running buffer.

3.3. Visualization of the Total C. albicans Protein Pattern by 2-DE Gel Staining

The ideal protein gel staining technique should meet the following requirements: wide linear quantification range, high sensitivity, good protein-to-protein consistency, minimal background, quickness, simplicity, low toxicity, environmentally friendly, reasonable cost, and compatibility with MS. However, there is currently no method available that combines all these properties or is universal for all proteins. Each stain interacts with proteins in a different way *(40, 41)*. In practice, the most used techniques are Coomassie blue, silver, and fluorescence staining, which each have advantages and disadvantages. The methods presented below describe three protocols for detection of the whole *C. albicans* protein pattern, which do not interfere with subsequent MS analysis.

3.3.1. Colloidal Coomassie Blue Staining

Coomassie blue staining is 50- to 100-fold less sensitive than silver staining but is simpler and more quantitative (with a linear quantification range of ~2 orders of magnitude) and has better protein-to-protein consistency. Colloidal staining methods, such as the following protocol, are usually preferred, as these are 5- to 10-fold more sensitive than conventional (classic alcohol/acetic acid) Coomassie blue methods.

1. Place the preparative 2-DE gel in a plastic or glass container and incubate in fixative solution for 3 h to overnight with gentle agitation on a platform shaker (*see* **Note 32**).
2. Discard fixative solution, and rinse the gel with three changes of water (10 min each).
3. Incubate the gel in pretreatment solution for 1h.
4. Incubate the gel in colloidal Coomassie blue solution overnight (~18 to 24 h) (*see* **Note 13**).
5. Remove colloidal solution, and wash the gel repeatedly with water until the desired contrast is reached (*see* **Note 33**).
6. Proceed to densitometric scanning of the stained gel (*see* **Section 3.5.1**) and/or protein spot excision for further MS analysis (*see* **Section 3.6**).

3.3.2. MS-Compatible Silver Staining

Despite its high sensitivity (below 1 ng), silver staining exhibits a very narrow linear quantification range (~1 order of magnitude) and greater protein-to-protein variation. Furthermore, its reproducibility can significantly be influenced by small differences, for example, in reaction times and temperatures, because it is a complex and multistep procedure. Conventional silver staining methods can be made MS compatible by using milder chemical conditions that minimize protein modifications during the staining process and/or interference with trypsin digestion; that is, by excluding (i) glutardialdehyde, a protein cross-linking agent (*see* **Note 34**), from the sensitizing solution, and (ii) formaldehyde from the silver nitrate solution *(42)*. However, this leads to a reduction in sensitivity (~20%, compared with conventional, nonmodified methods). The present MS-compatible protocol is taken from a modified silver-staining version reported by Shevchenko et al. *(42)*.

1. Place the preparative 2-DE gel in a glass or plastic container, and incubate in fixative solution for 30 min on a platform shaker with gentle agitation (*see* **Notes 32** and **35**).
2. Discard fixative solution, and rinse the gel first with wash solution 1 for 15 min and then with water for another 15 min.
3. Incubate the gel in sensitizing solution for 1 min.
4. Remove the solution, and rinse the gel with two changes of water, each time for 1 min.
5. Incubate the gel in prechilled silver nitrate solution for 20 min at 4°C.
6. Discard the solution, and rinse the gel with two changes of water (1 min each).
7. Add developing solution and agitate until protein spots appear as desired (~0.5 to 5 min) (*see* **Note 36**). Replace the solution when it turns yellow.
8. Quickly remove the solution after the gel is fully developed (*see* **Note 36**), and quench development by incubating for 5 min in stop solution.
9. Keep the silver-stained gel in storing solution or in water at 4°C.
10. Proceed to densitometric scanning of the stained gel (*see* **Section 3.5.1**) and/or protein spot excision for subsequent MS analysis (*see* **Section 3.6**).

3.3.3. SYPRO Ruby Protein Gel Staining

SYPRO Ruby protein gel stain is a commercially available fluorescence dye that is virtually as sensitive as MS-compatible silver staining methods but has a wider linear quantification range (~3 orders of magnitude) than colorimetric stains (such as Coomassie blue and silver stains). The protocol outlined here is adapted from the procedure described by the supplier (Bio-Rad) of this fluorophore, which noncovalently binds to proteins through similar mechanisms to Coomassie blue stains.

1. Place the preparative 2-DE gel in a plastic container (*see* **Note 12**), and incubate in fixative solution for 30 min to 1 h on a platform shaker with continuous, gentle shaking (*see* **Note 32**).

2. Remove fixative solution, and incubate the gel with 330 mL SYPRO Ruby protein gel stain (~10 times the volume of the gel) for 3 h to overnight. Protect the gel from light from this step forward (*see* **Note 37**).
3. Discard staining solution, and rinse the gel with wash solution for 30 to 60 min to reduce background fluorescence and increase sensitivity.
4. Wash the gel with two changes of water (10 min each) and proceed to gel imaging (*see* **Note 38** and **Section 3.5.1**) and/or protein spot excision for further MS analysis (*see* **Note 39** and **Section 3.6**).

3.4. Detection of Global Immunoreactivity Profiles of Serum Antibodies to C. albicans Proteins by 2-D Western Blotting

2-D Western blotting (or immunoblotting), using 2-DE–separated *C. albicans* proteins as a source of antigens and sera from *Candida* colonized and infected hosts (humans or warm-blooded animals) as a source of antibodies, is a powerful tool for obtaining a comprehensive and integrated view of global immunoreactivity profiles of host serum antibodies (referred to as serum antibody signatures) to a broad range of *C. albicans* antigens in health and disease, or in other words, of serologic response to *C. albicans* during host-commensal and host-pathogen interactions, respectively (**Fig. 26.1**) *(7, 8, 19, 25)*. Furthermore, antigenicity associated with posttranslational modification can also be detected with this technique (as a result of the high resolving power of 2-DE) by evaluating the number and p*I* shift of immunoreactive spots in the protein species series (**Fig. 26.2**). Overall, the anti-*Candida* antibody response appears to be unevenly directed to the different protein species of *C. albicans* antigens *(7, 25)*.

The classic procedure implies the following steps: (i) electroblotting of proteins from the 2-DE gel onto the surface of a chemically inert membrane, (ii) staining of the total protein pattern on the electroblotted membrane, (iii) immunodetection of reactive proteins that induce a specific serum antibody response, and (iv) reprobing of 2-D blots with other serum samples. During all the steps of this process, it is essential to carefully handle the membranes with forceps (with rounded, nonserrated tips) at their edges so as not to damage their surface and/or create potential staining artifacts.

3.4.1. Electroblotting of 2-DE Gels onto Nitrocellulose Membranes

2-DE-separated proteins are transferred to the surface of an immobilizing membrane using an electric field. This electroblotting procedure can be performed using a tank (wet, vertical) transfer system, in which the gel is submerged in a large volume of buffer, or using a semidry (horizontal) transfer system, where only a small volume of transfer buffer is used. Although the semidry method is faster and cheaper, this may lead to incomplete binding and lower yields. The protocol described below is based on tank blotting, which bypasses these drawbacks (by diluting SDS from the gel more efficiently) and, consequently, enables better recoveries of proteins, particularly of those with high molecular weights and/or in small quantities (low-abundance protein spots) *(43)*.

1. Place the analytical 2-DE gel(s) in a plastic or glass container, and briefly rinse with water. Equilibrate the gel in transfer buffer for 10 to 15 min to remove excess SDS.
2. Cut two pieces of filter paper and one piece of nitrocellulose membrane (*see* **Note 40**) per gel to the dimensions of the 2-DE gel. Soak the nitrocellulose membrane in the transfer tank or in a plastic or glass dish containing transfer buffer, and equilibrate it for 15 to 20 min. Wet filter papers and two fiber pads per gel in transfer buffer.
3. Assemble the blot sandwich for transfer as indicated in **Fig. 26.5A,** ensuring that no air bubbles are trapped in the resulting sandwich (i.e., both between membrane and gel and between filter paper and gel (*see* **Note 41**)).
4. Place the plastic gel holder cassette (containing the blot sandwich) into the transfer tank with transfer buffer such that the nitrocellulose membrane is placed on the anode side of the gel (**Fig. 26.5A**).
5. Perform electroblotting at 250 mA with a constant current for 3 h (for 1.5-mm gels), or alternatively at 50 mA with a constant current overnight, at 4°C using a refrigerated thermostatic circulator unit.
6. Remove the nitrocellulose membrane (*see* **Note 42**) from the blotting apparatus, and proceed to blot staining (*see* **Section 3.4.2**) or, directly, to immunodetection (*see* **Note 24** and **Section 3.4.3**).

3.4.2. Visualization of the Total C. albicans Protein Pattern on 2-D Blot: SYPRO Ruby Protein Blot Staining

Proteins adsorbed on the nitrocellulose membrane (2-D blot) can be stained with SYPRO Ruby protein blot stain (a commercially available, fluorescent, sensitive, and permanent protein stain) before immunodetection because their epitopes are not blocked but are freely available for their related antibodies. This procedure is carried out to assess the efficiency of transfer and define the total *C. albicans* protein pattern on the blot for subsequent 2-D immunoaffinity identification, thus eliminating the need for duplicate 2-D blots or 2-DE gels (run in parallel and stained with the appropriate method; *see* **Note 24**). The current protocol is taken from that reported by the manufacturer (Bio-Rad). Note that the protocols to stain gels and blots with SYPRO Ruby are different (*see* **Section 3.3.3**), although both stains should reveal identical protein patterns.

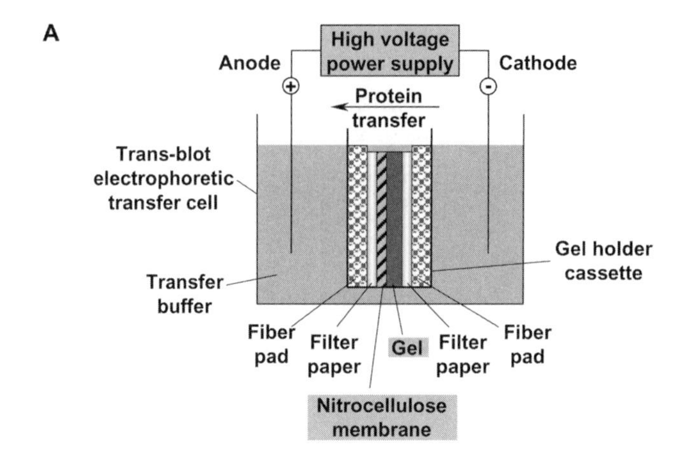

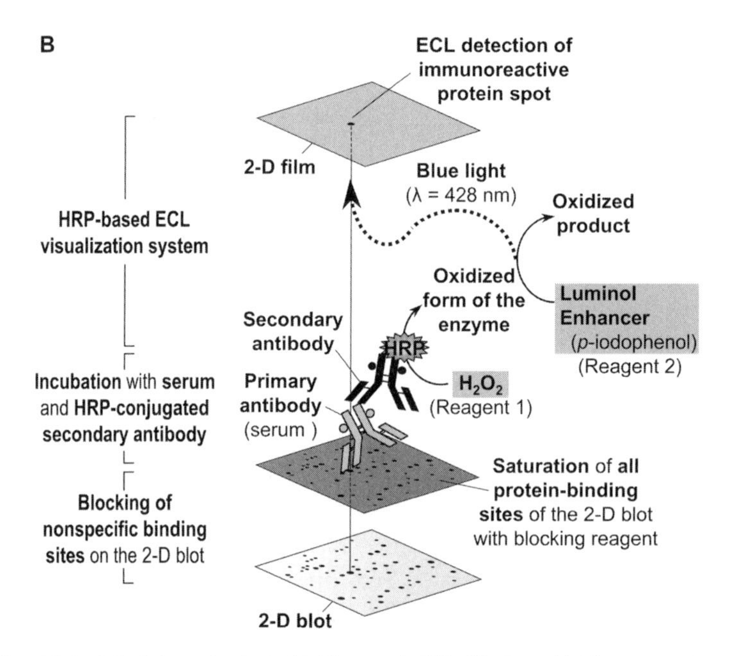

Fig. 26.5. Principles of electroblotting and ECL Western blotting procedures. (**A**) Diagram of the electroblotting procedure using a tank transfer system. The nitrocellulose membrane must be placed on the anode side of the (SDS-polyacrylamide) gel to allow the transfer of negatively charged proteins from the 2-DE gel onto the membrane. Accordingly, the plastic gel holder cassette (containing the blot sandwich) is oriented in the transfer tank (containing transfer buffer) as follows: anode (+), fiber pad, filter paper, nitrocellulose membrane, gel, filter paper, fiber pad, and cathode (−). If this orientation is wrong, then proteins from the gel will be lost into the buffer rather than electroblotted onto the nitrocellulose membrane. (**B**) Schematic illustration of ECL Western blotting. After blocking of nonspecific binding sites on the membrane and subsequent incubation with serum and HRP-conjugated secondary antibody, ECL substrate reacts with HRP to give off light, and a blue-light sensitive autoradiography film is then exposed to emitted light (*see* **Note 49**). The reagents used in this HRP-based ECL visualization system are luminol/H_2O_2/*p*-iodophenol. The oxidized luminol substrate (in an excited state) emits blue light. The *p*-iodophenol is a chemical enhancer that increases light output.

1. Place the 2-D blot in a plastic container (*see* **Note 12**), and incubate it face down in fixative solution for 15 min on a platform shaker with gentle agitation (*see* **Note 32**).
2. Discard fixative solution, and rinse the blot with four changes of water (5 min each).
3. Incubate the blot with SYPRO Ruby protein blot stain for 15 min.
4. Remove staining solution, and wash the blot with 4 to 6 changes of water (1 min each) to reduce background fluorescence.
5. Proceed to blot imaging (*see* **Section 3.5.1**) and then to immunodetection (*see* **Section 3.4.3**).

3.4.3. 2-D Immunoblotting: Immunodetection of Global Reactivity Profiles of Host Antibodies to C. albicans Antigens

At this point, the immobilized proteins are probed with serum samples (containing host antibodies) to detect their global reactivity patterns toward *C. albicans* antigens. The immunoaffinity identification of 2-D *C. albicans* antigen recognition profiles involves (i) saturation of all protein-binding sites of 2-D blot with a nonreactive protein or detergent, (ii) application of the primary antibody (serum, or antibody directed against the antigen) and then of the secondary antibody (enzyme-conjugated antibody directed against the primary antibody), and (iii) ECL detection of the antigen/primary-antibody/secondary-antibody/enzyme complex bound to the nitrocellulose membrane (**Fig. 26.5B**). To obtain consistent and reproducible results, it is crucial to carry out this protocol with each serum specimen on at least two different 2-D blots, one of which should be used for the first time with each serum sample (*see* **Note 43** and **Section 3.4.4** for reprobing 2-D blots).

3.4.3.1. Antigen-Antibody Binding

1. Place the 2-D blot in a clean plastic container with 100 mL blocking solution, and incubate for 1 to 2 h at room temperature with constant agitation on a platform shaker (*see* **Notes 32** and **44**) to block nonspecific binding sites.
2. Rinse the membrane 3 times with 200 mL TTBS solution (5 to 10 min each).
3. Prepare primary antibody solution by diluting serum sample 1:100 (or at the appropriate dilution; *see* **Note 45**) in antibody solution. Place the blot in heat-sealable plastic bag with 10 mL primary antibody solution and seal bag (*see* **Note 46**). Incubate for 1 h at room temperature.
4. Remove the membrane from the plastic bag and place in a plastic container with 200 mL TTBS solution. Rinse the blot with 3 to 4 changes of TTBS solution (15 to 20 min each time).

5. Prepare secondary antibody solution by diluting HRP-labeled antihuman IgG antibody 1:1000 (or at the suitable dilution; *see* **Note 45**) in the antibody solution (*see* **Note 47**). Place the blot in heat-sealable plastic bag with 10 mL secondary antibody solution and seal bag (*see* **Note 46**). Incubate for 2 h at room temperature.
6. Remove the membrane from the bag and rinse as in **step 4**. Eventually, wash the blot with TBS solution for 10–15 min to remove excess Tween-20.

3.4.3.2. ECL Detection

1. Mix equal volumes of each detection reagent from the ECL kit just prior to use to maximize its effectiveness (*see* **Note 48**).
2. Lay the membrane face up on an acetate sheet protector. Pipette the mixed detection reagent on the membrane, covering it completely (*see* **Note 48**), and incubate for 1 min at room temperature. Drain off excess detection reagent. Place another acetate sheet protector on the membrane without making air bubbles.
3. Place the membrane/acetate sandwich face up in an x-ray film cassette.
4. Perform all procedures from this point on *in a dark room* under safe light conditions. Cut away a corner from a high-performance film (autoradiography film; *see* **Note 49**) to define its orientation (*see* **Note 42**), and superimpose it on the membrane. Close the cassette.
5. Expose the film for the appropriate time (*see* **Note 49**).
6. Immediately develop the film with photographic developing reagents.
7. Proceed to film imaging and quantification of immunoreactivities by densitometry (*see* **Section 3.5.1**).
8. Remove the membrane from the film cassette, and proceed to stripping and reprobing with another serum sample (*see* **Section 3.4.4**).

3.4.4. Stripping and Reprobing of 2-D Blots

After ECL detection, primary and secondary antibodies can be removed from the membrane to perform sequential reprobing of 2-D blots with other serum specimens (*see* **Note 50**). This allows the confirmation of immunoblotting results (using sera previously tested on other 2-D blots) as well as reduction in costs for 2-DE and electroblotting runs. Each blot can be stripped up to three times with minimal loss of antigen (*see* **Note 43**).

1. Place the 2-D blot in a plastic container with stripping buffer, and incubate for 1 to 2 h at room temperature with constant agitation on a platform shaker (*see* **Note 32**) to dehybridize primary and secondary antibodies from the membrane (*see* **Note 51**).

2. Rinse the blot 3 times with 200 mL TTBS solution (15 min each).
3. Block the membrane and repeat immunodetection (using another serum sample) as described in **Section 3.4.3**.

3.5. Data Acquisition and Analysis of the Densitometric Profiles of Antibody Reactivities to *C. albicans* Proteins

Evaluation of the global profiles of antibody reactivities (host antibody signatures), especially in complex mixtures such as serum, toward a wide array of antigens in the different groups of host-fungus interactions (i.e., in health and disease) is unfeasible by manual comparison (with the eye). For this reason, digital data should be obtained from the 2-D gel, blot, and film images generated in the previous steps of the current immunoproteomic strategy (*see* **Sections 3.3, 3.4.2,** and **3.4.3,** respectively) to enable reliable data quantification, comparison, and interpretation.

Data analysis basically comprises two stages: preprocessing and postprocessing (**Fig. 26.6**). The first phase, data preprocessing, is essential to ensure that all data can be compared together. This includes digital image acquisition, background subtraction, intensity normalization, quantification, and spot alignment and matching using known landmarks, among others. Once data have been preprocessed appropriately, data mining is subsequently performed. This second phase, data postprocessing, is of prime importance to analyze and integrate high-throughput proteomic data as well as to discriminate between healthy and diseased populations (i.e., between host-commensal and host-pathogen interactions). This stage mainly involves unsupervised and supervised learning techniques.

3.5.1. Data Preprocessing: Imaging, Quantification, and Matching of Densitometric Patterns of Antibody Reactivities

1. Scan the colloidal Coomassie blue- and silver-stained gels, and ECL-developed films (*see* **Sections 3.3.1, 3.3.2**, and **3.4.3**, respectively) with an imaging densitometer, and digitize 2-D images with the Quantity One software (*see* **Note 52**).
2. Scan the SYPRO Ruby-stained gels and blots (*see* **Section 3.3.3** and **3.4.2**, respectively) with an epi illuminated laser-scanning instrument (*see* **Note 53**). Use the laser/filter combinations 532/550LP and 532/640DF35 for imaging gels and blots, respectively. Digitize 2-D images with the Quantity One software (*see* **Note 52**). Rotate left-right the digital images from the 2-D blots (*see* **Note 54**).
3. Process the different 2-D images using the ImageMaster 2D Platinum software. Perform detection (after spot filtering), background subtraction, normalization to the loading control, volumetric (integrated optical density of the related spot area) quantification, matching (using as landmarks the spots from SYPRO Ruby-stained blots; *see*

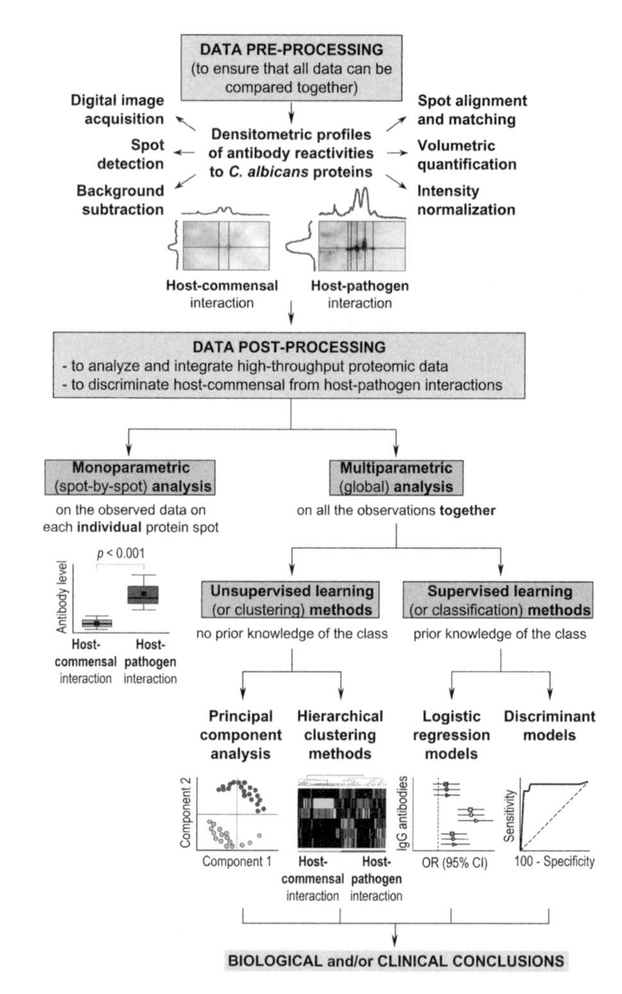

Fig. 26.6. Typical data processing from immunoreactivity profiles of anti–*C. albicans* antibodies during mucosal surface colonization and invasive disease. Data analysis basically consists of two phases: preprocessing and postprocessing. Data mining procedures reported previously to examine serum anti-*Candida* antibody signatures during host-commensal and host-pathogen interactions *(8)* are shown in this diagram.

Note 55), and editing of M_r and p*I* values (after gel calibration with internal 2-DE standards) of all the spots from *C. albicans* immunoreactive proteins (eliciting host antibody response).

3.5.2. Data Postprocessing: Data Mining

Proteomic profiling of serologic response of the host to *C. albicans* is a high-throughput method that generates data of a multidimensional nature. Sophisticated bioinformatic tools (**Fig. 26.6**) are therefore needed to identify specific antibody proteomic signatures for *Candida* colonized and infected host groups and to integrate the clinical parameters of the proteomic data obtained for diagnosis, risk stratification, prognosis, therapeutic monitoring, clinical follow-up, and/or immunotherapy of *Candida* infections *(8)* (*see* **Note 56**).

Although antibody reactivities can be compared between the different host groups using monoparametric analyses, these do not take into account the relationships among the proteomic data (*see* **Note 57**). Accordingly, multiparametric techniques are commonly used for these proteomic studies, which can be:

1. Unsupervised learning or clustering methods, in which there is no prior knowledge of the class (*i.e.*, commensalism or pathogenicity, healthy or disease, survivors or nonsurvivors, etc.) of the samples. These include self-organized maps, principal component analysis, and hierarchical and *k*-means clustering methods, among others. These techniques enable the individual samples to be distributed as far from each other as possible in a multidimensional space on the basis of various metric distances (*see* **Note 58**).
2. Supervised learning or classification methods, where there is prior knowledge of the class of the samples. These comprise support vector machines, decision trees, neural networks, and logistic and discriminant models.

3.6. Identification of *C. albicans* Immunoreactive Proteins by MS

Despite the wide variety of protein identification methods, MS is currently the technique of choice for characterizing immunorelevant *C. albicans* proteins detected by immunoblotting (and indirectly of their related host antibodies) and, overall, any *C. albicans* protein *(25, 26, 37)*. Although a brief protocol is outlined below, further information and practical steps on the MS-based identification of the *C. albicans* immunome are comprehensively detailed in **Chapter 15**, to which the reader should be directed.

1. Select and excise the immunoreactive protein spot of interest from the Coomassie-, silver-, or SYPRO Ruby-stained, preparative 2-DE gel (*see* **Section 3.3**), and transfer it into a low-binding microcentrifuge tube. Cut it into small fragments (~1-mm^3 cubes).
2. Destain and wash the gel particles.
3. Reduce, alkylate, and dry the destained gel fragments.
4. Rehydrate them with trypsin (protease), and perform the tryptic digestion at 37°C overnight.
5. Extract the resulting tryptic peptides from the gel particles by sonication.
6. Desalt and concentrate the tryptic digest.
7. Proceed to mass measurement of the peptide sample on a MALDI-TOF mass spectrometer, as recommended by the manufacturer.
8. Use the pattern of measured peptide masses to search for protein identity in appropriate annotated protein databases *(44, 45)* using the Mascot search engine.

4. Notes

1. The glassware should be rinsed with ultrapure water, or even with methanol and ultrapure water (to prevent contamination by ionic detergents and polymers), before use. In addition, it is essential to wear disposable latex gloves for all procedures handling 2-DE and MS material or equipment to avoid contamination by human epidermal proteins (keratins). The gloves should be washed on the outside surface with ultrapure water (while they are being worn) to reduce interference by powder or other contaminants.
2. PMSF, urea, thiourea, β-mercaptoethanol, DTT, DTE, iodoacetamide, acrylamide, PDA, ammonium persulfate, TEMED, silver nitrate, and formalin (35% formaldehyde), among others, should be handled with caution because they are *very hazardous*. Weigh or measure them in a fume hood, and wear suitable protective clothing including lab coat, gloves, goggles, and a mask. Dispose of the remains ecologically. Polymerize the remains of acrylamide and PDA with an excess of ammonium persulfate.
3. Mix all protease inhibitors just before use. PMSF is soluble in isopropanol, ethanol, methanol, and 1,2-propanediol but unstable in aqueous solution. Antipain is soluble in water, methanol, and dimethylsulfoxide (DMSO). Leupeptine is also soluble in water. Pepstatin can be solubilized in methanol and ethanol but is insoluble in water. All of them are added to reduce potential proteolytic processes. PMSF and leupeptine inhibit serine proteases (e.g., trypsin, chymotrypsin, and thrombin) and thiolproteases (e.g., papain). Antipain mainly inhibits papain and trypsin. Pepstatin inhibits acid proteases (e.g., pepsin, chymosin, cathepsin D, and renin).
4. The size of the glass beads is essential to attain an efficient cell homogenization. Optimal bead size for yeast cells and mycelia is 0.5 mm and for spores 0.1 mm.
5. The lysis or rehydration buffer for IEF must be specifically formulated to (i) convert all proteins into single conformations, (ii) prevent protein aggregates and complexes, (iii) ensure the complete sample solubilization, (iv) get and keep hydrophobic proteins in solution, (v) break disulfide and hydrogen bonds, (vi) maintain all proteins in their fully reduced state and avoid different oxidation steps of the proteins, and (vii) deactivate proteases, among others *(46)*. *See* **Fig. 26.7** for further details on the role(s) of each component of the standard lysis buffer. Select IPG buffers or carrier-ampholytes with the same pH interval as the IPG strip to be rehydrated.
6. Never heat any solutions containing urea above 30°C to dissolve it, because this would lead to protein carbamylation

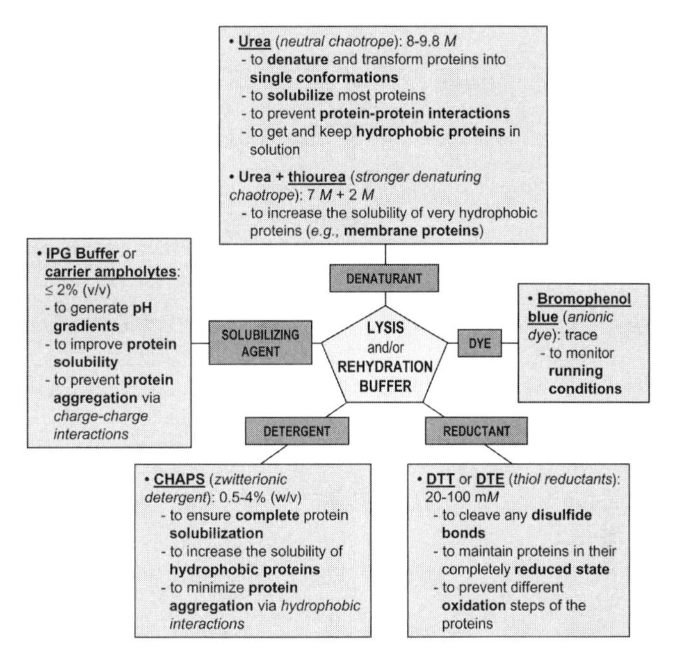

Fig. 26.7. Role(s) of each component of the lysis and/or rehydration buffer. IPG denotes immobilized pH gradient; CHAPS, 3-[(3-cholamidopropyl)dimethylamino]-1-propanesulfonate; DTT, dithiothreitol; and DTE, dithio-erythreitol. *See* **Notes 5** and **25** for further information.

(mediated by isocyanate, a urea degradation product) and, therefore, artifactual spots in the ensuing 2-DE patterns. DTE may be substituted for DTT. The reductant agent (DTE or DTT) should be added just before use.

7. The IPG buffers, IPG dry strips, internal 2-DE standards, IEF system and solutions, SDS-PAGE apparatus, Trans-blot electrophoretic transfer units, imaging densitometers or scanning instruments, and bioinformatic programs for 2-D image acquisition and analysis can be available from several suppliers, such as GE Healthcare Limited (previously Amersham Biosciences, which initially developed and commercialized IPG strips), Bio-Rad, and/or Genomic Solutions. Therefore, 2-DE, Western blotting, and bioinformatic analyses can be carried out suitably on any available system following the manufacturer's instructions *(47, 48)*.
8. Although we recommend the use of pH 3.0 to 10.0 NL IPG strips (of 18 cm length) because their sigmoidal gradient provides a better resolution between pH 5.0 and pH 7.0 (corresponding with the widest range of *C. albicans* proteins on a 2-DE gel; *see* **Fig. 26.8** and Ref. *24*), other commercially available pH ranges—medium (e.g., pH 4.0 to 7.0) and narrow (e.g., pH 5.0 to 6.0)—or even lengths (e.g., 7 cm for cost-effective screening and optimization of the sample preparation, or 24 cm for maximal resolution and loading capacity) of the

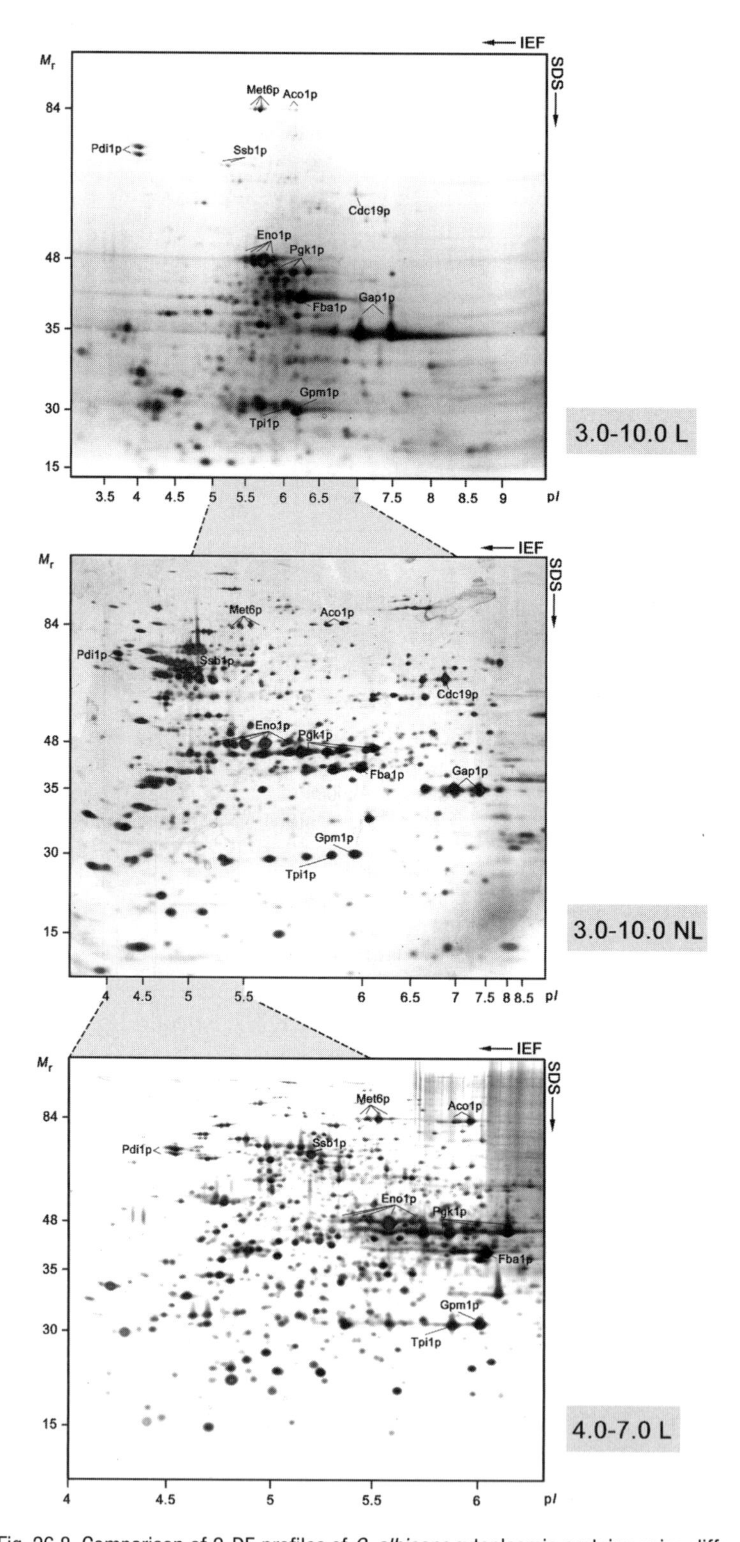

Fig. 26.8. Comparison of 2-DE profiles of *C. albicans* cytoplasmic proteins using different pH gradients. Silver-stained 2-DE gels of soluble *C. albicans* cytoplasmic proteins using pH 3.0 to 10.0 linear (*top panel*), pH 3.0 to 10.0 nonlinear (*middle panel*), and pH 4.0 to 7.0 linear (*bottom panel*) IPG strips are shown. The sigmoidal gradient of the pH 3.0 to 10.0 nonlinear IPG strips is especially useful to increase protein resolution between pH 5.0 and pH 7.0 and distribute the proteins more uniformly over the gel (*see* **Note 8**). *Labeled spots* represent *C. albicans* immunogenic proteins that elicit serum antibody responses in patients with systemic candidiasis *(7)*.

IPG strips may be more appropriate for other applications on the basis of the specific objectives of the experiment.

9. It is advisable to dissolve first urea (below 30°C; *see* **Note 6**) and then add the other components, otherwise a long time would be required to completely dissolve urea and SDS together. Alkylation of the cysteins may be improved using more basic resolving gel buffer (i.e., with pH of 8.8 rather than 6.8) *(46)*.
10. The pore size of the gel can be controlled by the total acrylamide concentration (T) and the degree of cross-linking (C). T (%) = $[(a + b) \times 100] / V$, and C (%) = $b \times 100/(a + b)$, where a is the mass of acrylamide in g, b the mass of cross-linking reagent (piperazine diacrylamide, PDA) in g, and V the volume in mL. It is convenient to use PDA as a cross-linker because it improves protein resolution and minimizes *N*-terminal protein blockage and silver-staining background *(49)*. The proposed values (10%T/1.6%C, PDA) can be modified to optimize resolution of proteins in a higher- or lower-molecular-weight range. Running time for the second-dimension separation may vary on the basis of different acrylamide concentrations, degrees of cross-linking, and samples. The 1.5-mm-thick gels allow greater protein load capacity. It is convenient to polymerize preparative SDS-PAGE gels overnight to minimize the amount of remaining unreacted acrylamide and therefore prevent alkylation of thiol groups present in proteins (by unpolymerized acrylamide in the SDS-PAGE gel; *see* **Fig. 26.4**).
11. Melt the agarose in a Bunsen burner, heating stirrer or microwave oven. Thereafter, cool the agarose solution to 50°C to 60°C (in a water bath) in order to avoid potential carbamylation of some proteins (due to the presence of urea) (*see* **Note 6**) after sealing the IPG strips with agarose.
12. It is essential to use plastic containers for SYPRO Ruby staining because these absorb a minimal amount of fluorescent dye.
13. Prepare colloidal solution immediately before staining the gel. Do not filter it, as the colloidal dye particles are retained on the filter. The role of ammonium sulfate is to increase the strength of hydrophobic interactions between proteins and dye.
14. Do not keep formaldehyde at 4°C to prevent its polymerization and deposition.
15. The sample preparation procedure should be as simple as possible to circumvent potential protein losses and modifications and, therefore, wrong conclusions in proteome analysis. Because proteolytic enzymes may be released upon cell disruption, protease inhibitors should be added to bypass putative protein degradation and/or modification (*see* **Note 3**). In addition, it is extremely important that the temperature of the cell suspension remains at 4°C during cell breakage to prevent

undesirable proteolytic activity, heat inactivation of proteins, protein denaturation, and/or overexpression of heat shock proteins, to name but a few. Use chilled solutions for cell disruption. The extraction should be carried out using non-ionic buffers to avoid the presence of salts and other ionic contaminants in the sample that may result in streaky 2-DE patterns. Overall, lipids, nucleic acids, salts, and solid material should be removed without selective protein loss.

16. After cell homogenization, the different cellular compartments or structures can be separated by several techniques, such as differential centrifugation, sucrose density gradients, sequential extraction and/or free-flow-electrophoresis (according to differences in their mass, density, relative solubility and/or charge, respectively), and/or immunoprecipitation (using organelle-specific antibodies) *(37)*.
17. Liquid cultures should be grown in a flask that is at least 4 to 5 times larger than the culture volume.
18. The incubation time and cell density of inoculum should be adjusted on the basis of the yeast or fungal strain. It is crucial that the *C. albicans* culture is in early-log (OD_{600nm} = 0.5 to 1) and mid-log (OD_{600nm} = 4) phase growth when preparing cytoplasmic extracts and protoplast lysates, because the cells are easier to disrupt using glass beads and more susceptible to wall lytic enzymes, respectively, than those close to or in stationary phase growth *(24)*.
19. The wet weight (in grams) of the cell pellet is virtually equal to the packed cell volume (in milliliters). Accordingly, add approximately 300 μL of ice-cold lysis buffer for each 100 mg of wet cell pellet and resuspend.
20. The cell extract can be separated from glass beads by drilling multiple small holes of a diameter less than 0.4 to 0.6 mm (with a flamed needle) in the tip of the centrifuge tube (containing cell extracts, cell debris, and glass beads) and subsequently placing it into another centrifuge tube (whose cap has been removed previously). After centrifuging these tubes, the supernatant should be collected carefully, without disturbing the pellet. This recommended procedure leads to an increase in the recovery of cell homogenate and, therefore, of protein material. The glass beads (with a diameter of 0.4 to 0.6 mm) are retained in the perforated centrifuge tube. *See* **Fig. 26.9** for further details. For large-scale experiments, the reader is directed to a recent review of methods *(35)*. To reuse the glass beads, rinse them by soaking in concentrated nitric acid for 1 h, and then flush through with water. Dry them in a baking oven, cool, and keep at 4°C.
21. All centrifugations for protoplast isolation should be carried out with the brakes off. Given the extreme fragility of the protoplasts, it is essential to handle these gently. The supernatants

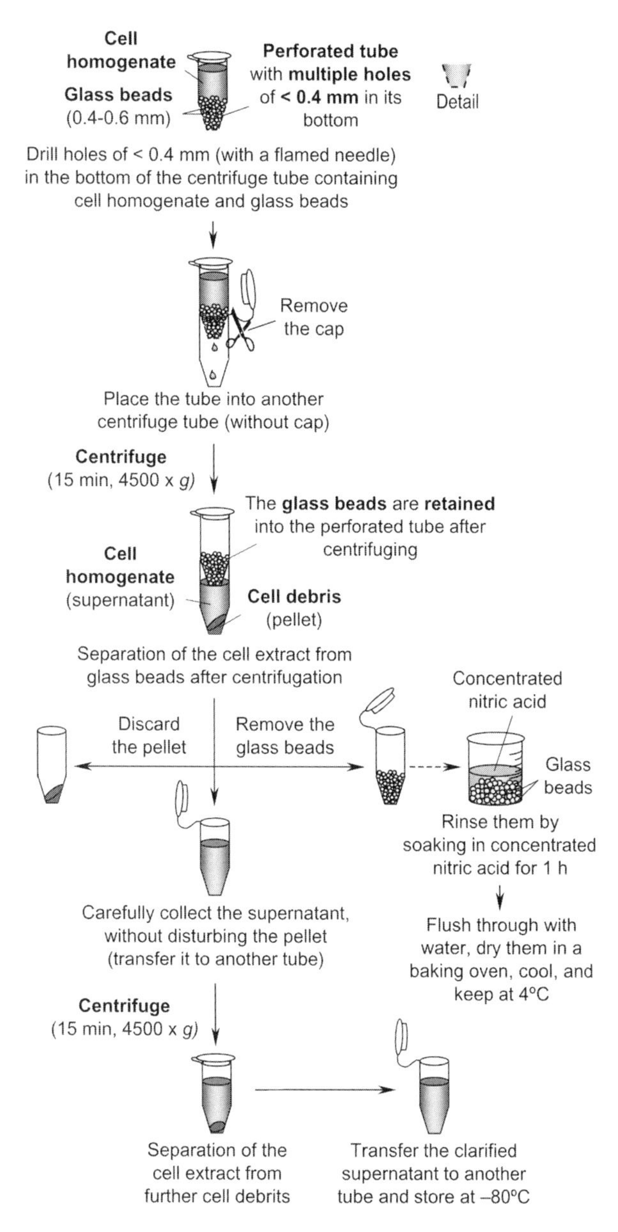

Fig. 26.9. Proposed strategy for separating the *C. albicans* cell extract from glass beads. A "strainer" for separating the cell homogenate from glass beads is created by making multiple holes of a diameter less than 0.4 to 0.6 mm in the bottom of the 1.5-mL centrifuge tube (which contains the cell homogenate and glass beads) with a flamed needle. *See* **Note 20** for further details.

should be decanted carefully to avoid material loss, as the protoplast pellet is less compact than the cell pellets. Use a small volume to resuspend the protoplasts (by gently swirling liquid across the surface of the pellet) and then add more solution until the correct final volume is reached. This proposed procedure is to facilitate their resuspension, because this is difficult and sticky.

22. Glusulase is a preparation of the intestinal juice of the Roman garden snail *Helix pomatia* and is composed of a mixture of lytic enzymes (β-glucuronidase, sulfatase, and a cellulase). This has proved to be useful for entirely digesting yeast cell walls (i.e., for obtaining protoplasts) in nearly all yeast species, including *C. albicans (50)*. It is important to use sterilized forceps to take the lid off the Glusulase bottle, and gently mix Glusulase with cell suspension (without shaking it). The protoplast cells should be washed several times before they are lysed in order to eliminate any trace of Glusulase and, therefore, any enzymatic activity that may modify the extracted proteins (protoplast lysates).
23. Isoelectric focusing (IEF) is carried out under the influence of an electric field (requiring the application of high voltages) and in a pH gradient, which can be established in a gel by two different ways *(46)*:
    (a) Carrier ampholyte-generated pH gradients, in which amphoteric buffers are in free solution (mobile carrier ampholytes). This original method for IEF leads to continuous, time- and sample-dependent, unstable, and drift pH gradients over a long IEF time.
    (b) Immobilized pH gradients (IPGs), in which acrylamide derivatives containing acidic and basic buffering groups (the so-called immobilines) are covalently incorporated into the polyacrylamide gel matrix. The development of this alternative procedure for IEF has resulted in substantial improvements in resolution, reproducibility, and protein loading capacity.
24. Optionally, one analytical 2-DE gel (run in parallel) can be silver-stained *(51, 52)* and used, instead of SYPRO Ruby-stained 2-D blots, for subsequent spot matching with the preparative 2-DE gel and following immunostained 2-D blots *(7)*.
25. Both the volume of rehydration solution containing the protein sample and the amount of protein loaded onto an IPG strip rely on the strip length chosen for the experiment. In practice, up to 125, 340, and 450 μL total volume may be loaded onto a 7-cm, 18-cm, and 24-cm strip, respectively, for optimal protein resolution *(46)*. Internal 2-DE standards can be loaded with and without protein sample to allow isoelectric point and molecular weight values of separated proteins to be estimated *(24)*. Protein sample can be applied to the IPG strip by two different methods (**Fig. 26.10**):
    (a) Rehydration loading. The sample (optionally in lysis buffer, *see below*) is loaded during IPG strip rehydration by including it in the rehydration buffer. This method is easy-to-perform and is usually recommended for large sample volumes (>100 μL) and high protein amounts (>1 mg). Furthermore, this system enhances the entry of

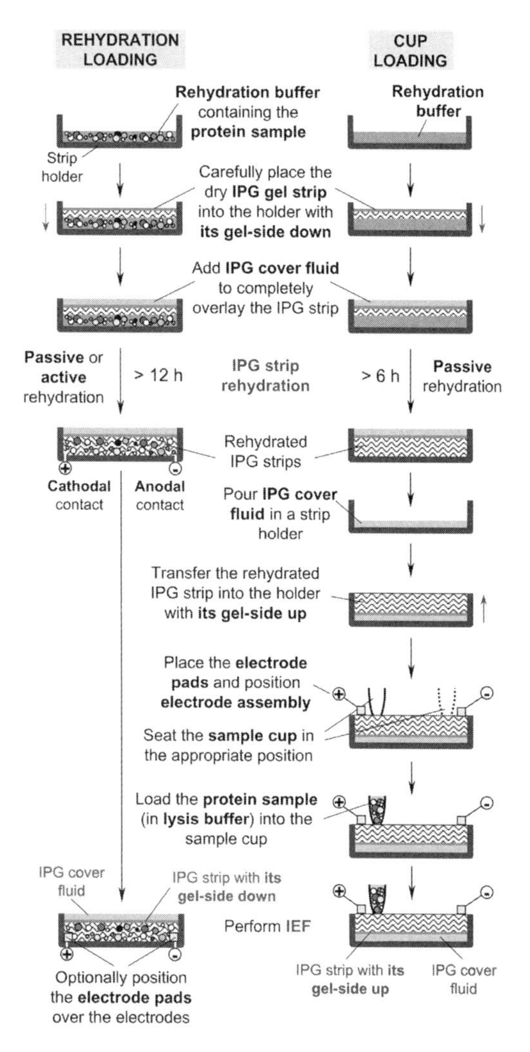

Fig. 26.10. Illustrative comparison of the two methods of protein sample application into the IPG strip. Protein sample can (i) be loaded during IPG strip rehydration by including it in the rehydration buffer (rehydration loading) or (ii) be applied to the pre-rehydrated IPG strip via a sample cup at a defined pH of the gradient (cup loading) (*see* **Note 25**).

high-molecular-weight proteins into the strip by applying a voltage during rehydration (active rehydration).

(b) Cup loading. The sample in lysis buffer (*see below*) is applied to the pre-rehydrated IPG strip (with rehydration buffer) through a sample cup at a defined pH of the gradient. This procedure (cathodic or anodic cup loading) is preferred for IEF in very acidic narrow or basic pH gradients, respectively. However, it can lead to precipitation at the sample application point.

Lysis buffer should be a solution similar in composition to the rehydration solution. The typical composition of lysis buffer for *C. albicans* proteins is: 7 M urea, 2 M thiourea,

4% (w/v) CHAPS, 40 mM Tris base, 65 mM DTE, 0.5% (v/v) IPG buffer pH 3–10, and a trace of bromophenol blue (*see* **Note 5** and **Fig. 26.7**) *(24)*.

26. The strip holders must be clean and entirely dry before use. It is crucial to clean them carefully after each IEF run to remove residual protein and avoid drying the protein solution in them. These should be soaked overnight in a solution of 2% to 5% strip holder cleaning solution (a neutral pH detergent supplied by GE Healthcare Limited; *see* **Note 7**) in water, and then meticulously brushed with a toothbrush using undiluted cleaning solution. After that, rinse them with water and air-dry them. Do not use new strip holders without cleaning them prior to the first run.
27. The electric field is applied at the IPG strip via two platinum contacts (located in the strip holder at fixed distances), which are placed on the cooled electrode contact areas of the IPGphor system with the arrow pointed end on the anodal (+) contact area. This contact area is large to adapt the different lengths of strip holders.
28. The IPG cover fluid (paraffin oil) serves to avoid evaporation during rehydration and IEF and, consequently, IPG strip drying, urea crystallization, and oxygen and dioxide uptake *(46)*.
29. After rehydration (but before IEF), place 5-mm-wide, damp (but not wet) electrode pads between the rehydrated IPG strip (the gel ends) and each holder electrode to adsorb excess free water on the gel surface, salts, and proteins whose pIs are outside the pH range of the IPG strip.
30. Prior to transfer of the IPG strip to the SDS-PAGE gel, the strips have to be incubated with equilibration buffer to maintain the focused proteins in their fully reduced state and transform them into SDS-protein complexes. This enables efficient transfer of proteins, completely unfolded and coated with negative charges only, to the SDS-PAGE gel and produces a better resolution of the protein spots. Iodoacetamide is added to (i) minimize point streaking due to excess DTT in silver-stained patterns and (ii) avoid cysteine alkylation by unpolymerized acrylamide in the SDS-PAGE gel (**Fig. 26.4**). *See* **Fig. 26.11** for further details on the role(s) of each component of the equilibration solution.
31. The combination of the IPG strip with agarose eliminates the need to use a stacking gel in the second dimension *(49)*.
32. It is important to prepare all reagents immediately before use, because this results in a lower background, more reliable stain, and reduction in any contamination of the 2-DE gel or 2-D membrane. Furthermore, all reagents should be removed

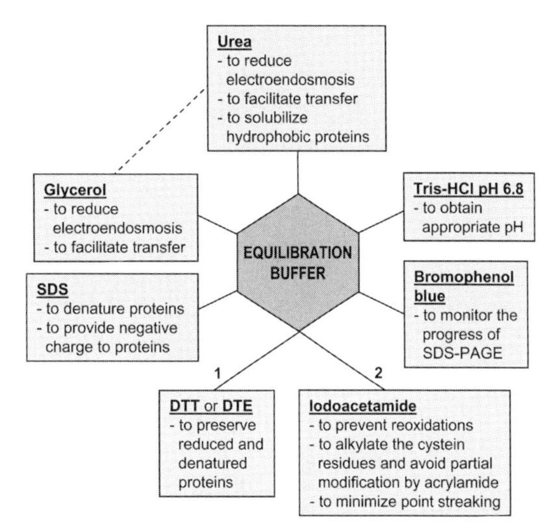

Fig. 26.11. Role(s) of each component of the equilibration solutions for IPG strips. Electroendosmosis is caused by the presence of fixed charges on the IPG strip in the electric field. During the equilibration step (with SDS), the IPG strips acquire negative net charges (their amino groups are neutral, and their carboxylic groups are negatively charged) *(46)*. As these charges cannot migrate (toward the anode) in the electric field (because they are fixed on the gel matrix), then a water flow toward the cathode (counter flow of H_3O^+ ions) takes place as compensation, leading to protein loss. The addition of urea and glycerol to the equilibration buffer reduces this effect during protein transfer from the IPG strip to the SDS-PAGE gel (*see* **Fig. 26.4** and **Note 30**).

by vacuum aspiration to bypass any physical contact with the gel or membrane (such as residues found on latex gloves) that can create staining artifacts, and/or damage the gel or membrane. At all stages of the procedure, continuous and gentle agitation of the dishes on a reciprocal ("ping-pong") shaker should be performed to ensure a uniform treatment of the gel or membrane and prevent its adhesion to the container. Rapid agitation and orbital shakers should be avoided, as these can lead to gel or membrane breakage. The container must be covered with a tight-fitting lid to prevent contamination with human epidermal proteins (keratins).

33. Ultrapure water should be changed several times until proteins appear as blue spots on a clear background.
34. The glutardialdehyde cross-links the proteins to the 2-DE gel by reacting with free amino groups of proteins, that is, the α-amino (N-terminal) and ε-amino (lysine side chain) groups.
35. For convenience, this step can be prolonged up to 3 days.
36. Develop the gel until protein spots are a homogeneous yellow-brown color on a clear background. Do not overdevelop the gel.
37. Protect the gel from light from this step to digital imaging. Use dark polypropylene or polyvinyl chloride (PVC) dishes

or cover the container with aluminum foil to protect the dye from bright light.

38. For convenience, the stained gel can be stored, before acquiring an image, in 2% (v/v) acetic acid at 4°C in the dark for a week with minor loss of signal intensity.

39. The very bright orange fluorescence generated by the SYPRO Ruby protein gel stain makes manual excision of protein spots easy on a UV transilluminator.

40. Polyvinylidene difluoride (PVDF) membranes can also be used as a substitute. However, because these are hydrophobic membranes, they must be prewetted with methanol before wetting with transfer buffer (whose composition is 10 mM 3-[cyclohexylamino]-1-propanesulfonic acid (CAPS), pH 11.0, 10% (v/v) methanol).

41. The gel and membrane cannot be separated once these touch. Hold the plastic gel holder cassette (containing the blot sandwich) firmly to ensure a tight contact between gel and membrane and avoid any distortion of the electroblotted protein spots.

42. It is convenient to cut a small piece from the 2-D gel, nitrocellulose membrane, and autoradiography film to define their orientation for further analysis. For example, cut the corners of the nitrocellulose membrane and autoradiography film in such a way that their cut edges are on the higher right-hand corner of the original 2-DE gel (corresponding with the highest M_r and pI values).

43. In a proteomic experiment, there can be variability both in the biological phase (e.g., in cell growth) and in the technical phase (e.g., in sample preparation, 2-DE, and/or immunodetection, among others) *(14)*. For this reason, the expected results of a single proteomic experiment are inaccurate and inadequate. Experimental replication should be performed in these two phases (**Fig. 26.12**). However, it is of note that biological replications are statistically more efficient than technical replications. In fact, no replication in the biological phase can multiply the number of false-positive results. Technical variability in the present immunoproteomic strategy can be found, for example, in the number of immunoreactive spots detected and/or in the variance of the measured signal.

44. If desired, this blocking step can be carried out overnight at 4°C.

45. Dot-blot or SDS-PAGE analyses should be performed previously to empirically establish the optimal dilution of the primary and secondary antibody. It is essential not to use

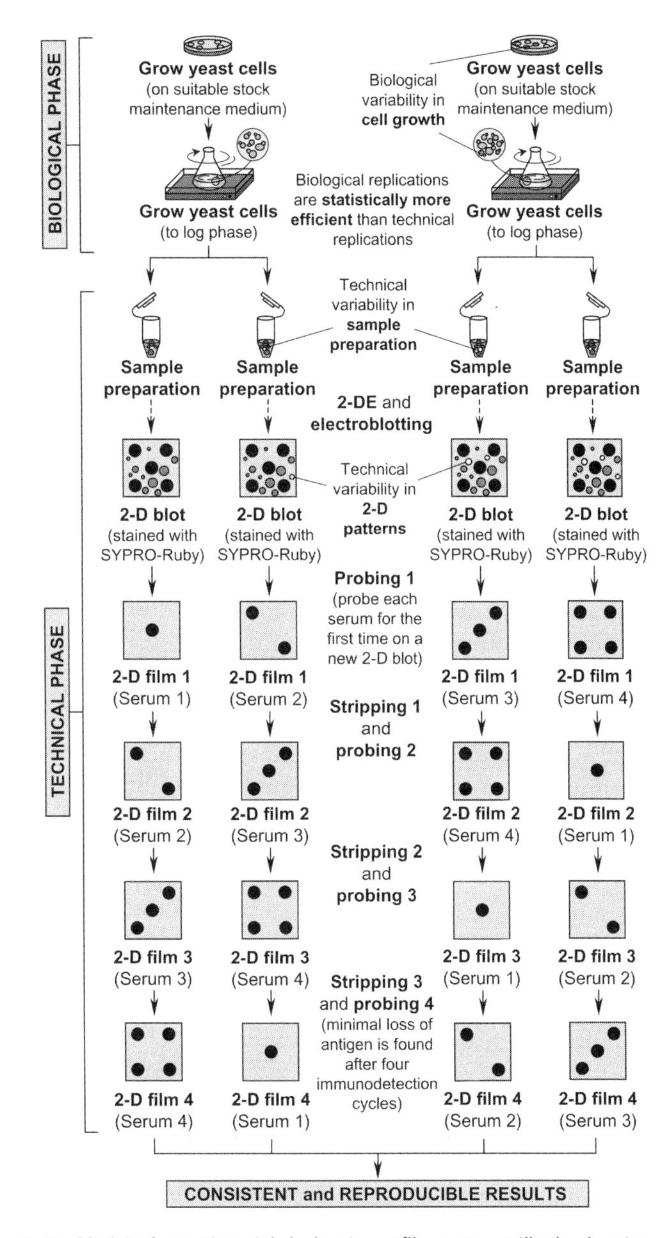

Fig. 26.12. Model of experimental design to profile serum antibody signatures using biological and technical replication. Replication has to be carried out in the biological and technical phases of an immunoproteomic experiment so that expected results are consistent and reproducible (*see* **Note 43**). Each serum sample should be probed for the first time on a 2-D blot that has not been stripped previously. Minimal loss of antigen is found after four immunodetection cycles.

sodium azide as a preservative for immunoblotting buffers because it inhibits HRP.

46. It is recommended to use heat-sealable plastic bags for all antibody incubation steps in order to reduce the volumes that are needed for the primary and secondary antibody

solutions and, consequently, save the clinical sample amount (which is often limited) and money, respectively. When using plastic or glass containers, the antibody solution volume should be increased to 100 mL.

47. If the serum sample used as primary antibody is, for instance, from mouse, then HRP-labeled anti-mouse (rather than anti-human) IgG (or appropriate Ig, according to the specific objectives of the experiment) antibody must be employed *(19, 23)*.
48. Detection reagents (stored at 2°C to 8°C) should be equilibrated to room temperature prior to using them for reproducible results. All the surface of the membrane must be covered with the solution. Use ~7.5 mL of each detection reagent for a 16 cm × 18 cm membrane.
49. Blue-light sensitive autoradiography films (like Hyper-film ECL) must be used because the maximum light emission produced by the ECL reaction is at a wavelength of 428 nm (**Fig. 26.5B**). Wear powder-free gloves when handling films and detection reagents. Optimal exposure parameters should be determined empirically (e.g., from 15 s to several minutes). To address this, begin first with short exposure times, develop the film, and, according to the results, continue with longer exposure times. It is also worth mentioning that the signal can be detected with an appropriate charged-coupled device (CCD) camera or scanning instrument (in a chemiluminescence detection method) in a dark closet rather than with film exposure (as described here).
50. After each immunodetection, membranes can be stored at 2°C to 8°C (in a refrigerator) in a container with tight-fitting lid containing TBS solution.
51. Membranes can be dehybridized using more stringent conditions. Place the 2-D blot in a container with tight-fitting lid, and incubate it in alternative stripping buffer (62.5 mM Tris-HCl, pH 6.7, 2% [w/v] SDS, and 100 mM 2-mercaptoethanol) at 70°C for 30 to 45 min.
52. The digital images should be acquired as a grayscale TIFF file with a sufficient resolution.
53. The SYPRO Ruby protein gel/blot stain has two excitation peaks (at ~280 and ~450 nm) that make it suitable both for UV transillumination-based systems with appropriate CCD cameras and for laser scanning instruments. The very bright orange fluorescence that is generated by SYPRO Ruby protein gel stain allows manual excision of protein spots on a UV transilluminator.
54. The digitized images from the 2-D blots should be rotated left-right with an appropriate program (like the Quantity One software; *see* **Note 7**) to have all images (2-D gels, blots

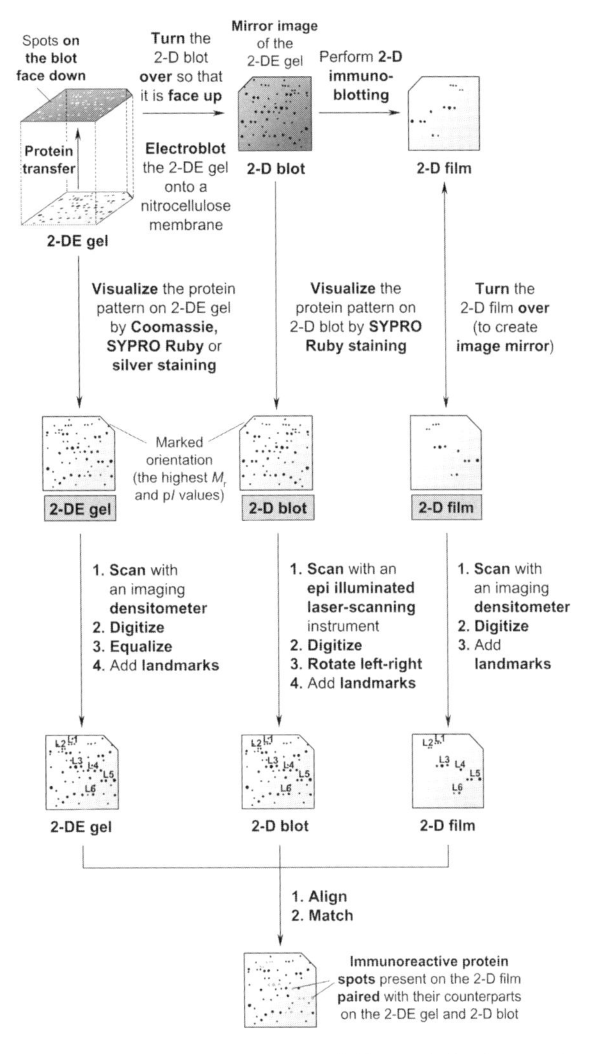

Fig. 26.13. Alignment of the digitized 2-D gel, blot and film images for matching process. *White spots* show spots that can only be visualized on the blot face down (as these are only visible on one face), and *black spots* indicate spots that can be visualized either on the two gel or film faces or on the blot face up (*see* **Note 54**). The marked orientations (cut edges) in all 2-D images correspond with the highest M_r and p*I* values (*see* **Note 42**). The size of 2-DE gel, which is larger than the 2-D blot or film as a result of the staining procedure, should be equalized to that of the 2-D blot image. The landmarks must be recognizable in all 2-D images (*see* **Note 55**).

and films) at the same orientation. This is because the 2-D blot has the spots only on one face, and the transferred proteins appear on the blot as the mirror image of the original 2-DE gel (**Fig. 26.13**). It can be bypassed by turning the 2-DE gel over to create mirror image when the transfer sandwich is prepared. It is of note that this problem does not occur with the 2-D film images as the generated spot signals are visible on their two faces.

55. Landmarks, which should be present in all 2-D gel, blot, and film images, can be added manually. Then, align all 2-D digitized images and proceed to automatic spot matching with ImageMaster 2D Platinum software or another suitable program (*see* **Note 7**) to obtain the immunoreactive spots present on the autoradiography film paired with their counterparts both on the appropriately stained, preparative 2-DE gel and on the SYPRO Ruby-stained blot (**Fig. 26.13**).
56. The discovered markers must be validated because these could be false-positives regardless of the data quality. There are three phases in order for a candidate marker to be introduced into routine clinical practice: the discovery phase (candidate marker), prototype development phase (validated marker), and product development phase (commercial product) *(21, 53–55)*.
57. There are two approaches for testing the hypothesis *Hj* (i.e., that the antibody reactivity does not differ according to the host-fungus interaction) and deciding if this can be rejected or not (**Fig. 26.6**) *(8, 14)*:
    (a) Univariate analyses (spot-by-spot approach), which are performed on the observed data on each individual protein spot (spot *j* only).
    (b) Multivariate analyses (global approach), which consider all the observations together.
58. While similar samples cluster close to each other, dissimilar samples cluster far from each other (**Fig. 26.6**). In theory, all the samples from the host-commensal interaction (healthy or nondisease samples) should cluster together, and all the samples from the host-pathogen interaction (disease samples) should cluster together. Unsupervised learning methods can also enable the identification of subgroups.

Acknowledgments

We thank the Merck, Sharp & Dohme (MSD) Special Chair in Genomics and Proteomics, Comunidad de Madrid (S-SAL-0246-2006 DEREMICROBIANA-CM), Comisión Interministerial de Ciencia y Tecnología (CICYT; BIO-2003-00030 and BIO-2006-01989), and Fundación Ramón Areces for financial support of our laboratory.

References

1. Pfaller, M. A. and Diekema, D. J. (2007) Epidemiology of invasive candidiasis: a persistent public health problem. *Clin. Microbiol. Rev.* **20,** 133–163.
2. Calderone, R. A. and Fonzi, W. A. (2001) Virulence factors of *Candida albicans*. *Trends Microbiol.* **9,** 327–335.

3. Hube, B. (2004) From commensal to pathogen: stage- and tissue-specific gene expression of *Candida albicans*. *Curr. Opin. Microbiol.* 7, 336–341.
4. Pitarch, A., Nombela, C., and Gil, C. (2006) *Candida albicans* biology and pathogenicity: insights from proteomics. *Methods Biochem. Anal.* **49,** 285–330.
5. Garber, G. (2001) An overview of fungal infections. *Drugs* **61,** 1–12.
6. Rex, J. H., Walsh, T. J., Sobel, J. D., Filler, S. G., Pappas, P. G., Dismukes, W. E., and Edwards, J. E. (2000) Practice guidelines for the treatment of candidiasis. Infectious Diseases Society of America. *Clin. Infect. Dis.* **30,** 662–678.
7. Pitarch, A., Abian, J., Carrascal, M., Sanchez, M., Nombela, C., and Gil, C. (2004) Proteomics-based identification of novel *Candida albicans* antigens for diagnosis of systemic candidiasis in patients with underlying hematological malignancies. *Proteomics.* **4,** 3084–3106.
8. Pitarch, A., Jimenez, A., Nombela, C., and Gil, C. (2006) Decoding serological response to *Candida* cell wall immunome into novel diagnostic, prognostic, and therapeutic candidates for systemic candidiasis by proteomic and bioinformatic analyses. *Mol. Cell Proteomics.* **5,** 79–96.
9. Martinez, J. P., Gil, M. L., Lopez-Ribot, J. L., and Chaffin, W. L. (1998) Serologic response to cell wall mannoproteins and proteins of *Candida albicans*. *Clin. Microbiol. Rev.* **11,** 121–141.
10. Ellepola, A. N. and Morrison, C.J. (2005) Laboratory diagnosis of invasive candidiasis. *J. Microbiol.* **43,** 65–84.
11. Krah, A. and Jungblut, P. R. (2004) Immunoproteomics. *Methods Mol. Med.* **94,** 19–32.
12. Caron, M., Choquet-Kastylevsky, G., and Joubert-Caron, R. (2007) Cancer immunomics using autoantibody signatures for biomarker discovery. *Mol. Cell Proteomics.* **6,** 1115–1122.
13. Seliger, B. and Kellner, R. (2002) Design of proteome-based studies in combination with serology for the identification of biomarkers and novel targets. *Proteomics.* **2,** 1641–1651.
14. Chich, J. F., David, O., Villers, F., Schaeffer, B., Lutomski, D., and Huet, S. (2007) Statistics for proteomics: experimental design and 2-DE differential analysis. *J. Chromatogr. B Analyt. Technol. Biomed. Life Sci.* **849,** 261–272.
15. Haoudi, A. and Bensmail, H. (2006) Bioinformatics and data mining in proteomics. *Expert. Rev. Proteomics.* **3,** 333–343.
16. Phan, J. H., Quo, C. F., and Wang, M. D. (2006) Functional genomics and proteomics in the clinical neurosciences: data mining and bioinformatics. *Prog. Brain Res.* **158,** 83–108.
17. Wilson, R. A., Curwen, R. S., Braschi, S., Hall, S. L., Coulson, P. S., and Ashton, P. D. (2004) From genomes to vaccines via the proteome. *Mem. Inst. Oswaldo Cruz* **99,** 45–50.
18. Sette, A., Fleri, W., Peters, B., Sathiamurthy, M., Bui, H. H., and Wilson, S. (2005) A roadmap for the immunomics of category A-C pathogens. *Immunity.* **22,** 155–161.
19. Pitarch, A., Diez-Orejas, R., Molero, G., Pardo, M., Sanchez, M., Gil, C., and Nombela, C. (2001) Analysis of the serologic response to systemic *Candida albicans* infection in a murine model. *Proteomics.* **1,** 550–559.
20. Pardo, M., Ward, M., Pitarch, A., Sanchez, M., Nombela, C., Blackstock, W., and Gil, C. (2000) Cross-species identification of novel *Candida albicans* immunogenic proteins by combination of two-dimensional polyacrylamide gel electrophoresis and mass spectrometry. *Electrophoresis* **21,** 2651–2659.
21. Pitarch, A., Nombela, C., and Gil, C (2007). Reliability of antibodies to *Candida* methionine synthase for diagnosis, prognosis and risk stratification in systemic candidiasis: a generic strategy for the prototype development phase of proteomic markers. *Proteomics, Clin. Appl.*, **1,** 1221–1242.
22. Fernandez-Arenas, E., Molero, G., Nombela, C., Diez-Orejas, R., and Gil, C. (2004) Contribution of the antibodies response induced by a low virulent *Candida albicans* strain in protection against systemic candidiasis. *Proteomics.* **4,** 1204–1215.
23. Fernandez-Arenas, E., Molero, G., Nombela, C., Diez-Orejas, R., and Gil, C. (2004) Low virulent strains of *Candida albicans*: unravelling the antigens for a future vaccine. *Proteomics.* **4,** 3007–3020.
24. Pitarch, A., Pardo, M., Jimenez, A., Pla, J., Gil, C., Sanchez, M., and Nombela, C. (1999) Two-dimensional gel electrophoresis as analytical tool for identifying *Candida albicans* immunogenic proteins. *Electrophoresis* **20,** 1001–1010.

25. Pitarch, A., Nombela, C., and Gil, C. (2006) Contributions of proteomics to diagnosis, treatment, and prevention of candidiasis. *Methods Biochem. Anal.* **49,** 331–361.
26. Pitarch, A., Molero, G., Monteoliva, L., Thomas, D. P., López-Ribot, J. L., Nombela, C., and Gil, C. (2007) Proteomics in *Candida* species, in *Candida: Comparative and Functional Genomics* (d'Enfert, C. and Hube, B., eds), Caister Academic Press, UK, pp. 169–194.
27. Sahin, U., Tureci, O., Schmitt, H., Cochlovius, B., Johannes, T., Schmits, R., Stenner, F., Luo, G., Schobert, I., and Pfreundschuh, M. (1995) Human neoplasms elicit multiple specific immune responses in the autologous host. *Proc. Natl. Acad. Sci. U.S.A.* **92,** 11810–11813.
28. Lee, S. Y., Obata, Y., Yoshida, M., Stockert, E., Williamson, B., Jungbluth, A. A., Chen, Y. T., Old, L. J., and Scanlan, M. J. (2003) Immunomic analysis of human sarcoma. *Proc. Natl. Acad. Sci. U.S.A.* **100,** 2651–2656.
29. Rollins, S. M., Peppercorn, A., Hang, L., Hillman, J. D., Calderwood, S. B., Handfield, M., and Ryan, E. T. (2005) *In vivo* induced antigen technology (IVIAT). *Cell Microbiol.* **7,** 1–9.
30. Hardouin, J., Lasserre, J. P., Canelle, L., Duchateau, M., Vlieghe, C., Choquet-Kastylevsky, G., Joubert-Caron, R., and Caron, M. (2007) Usefulness of autoantigens depletion to detect autoantibody signatures by multiple affinity protein profiling. *J. Sep. Sci.* **30,** 352–358.
31. Bradford, T. J., Wang, X., and Chinnaiyan, A. M. (2006) Cancer immunomics: using autoantibody signatures in the early detection of prostate cancer. *Urol. Oncol.* **24,** 237–242.
32. Hess, J. L., Blazer, L., Romer, T., Faber, L., Buller, R. M., and Boyle, M. D. (2005) Immunoproteomics. *J. Chromatogr. B Analyt. Technol. Biomed. Life Sci.* **815,** 65–75.
33. Chaffin, W. L., Lopez-Ribot, J. L., Casanova, M., Gozalbo, D., and Martinez, J. P. (1998) Cell wall and secreted proteins of *Candida albicans*: identification, function, and expression. *Microbiol. Mol. Biol. Rev.* **62,** 130–180.
34. Lopez-Ribot, J. L., Casanova, M., Murgui, A., and Martinez, J. P. (2004) Antibody response to *Candida albicans* cell wall antigens. *FEMS Immunol. Med. Microbiol.* **41,** 187–196.
35. Pitarch, A., Nombela, C., and Gil, C. (2008) Cell wall fractionation for yeast and fungal proteomics. *Methods Mol. Biol.* **425**, 217–239.
36. Pitarch, A., Nombela, C., and Gil, C. (2008) Collection of proteins secreted from yeast protoplasts in active cell wall regeneration. *Methods Mol. Biol.* **425**, 241–263.
37. Pitarch, A., Sanchez, M., Nombela, C., and Gil, C. (2003) Analysis of the *Candida albicans* proteome. I. Strategies and applications. *J. Chromatogr. B Analyt. Technol. Biomed. Life Sci.* **787,** 101–128.
38. Rupp, S. (2004) Proteomics on its way to study host-pathogen interaction in *Candida albicans. Curr. Opin. Microbiol.* **7,** 330–335.
39. Thomas, D. P., Pitarch, A., Monteoliva, L., Gil, C., and Lopez-Ribot, J. L. (2006) Proteomics to study *Candida albicans* biology and pathogenicity. *Infect. Disord. Drug Targets.* **6,** 335–341.
40. Gorg, A., Obermaier, C., Boguth, G., Harder, A., Scheibe, B., Wildgruber, R., and Weiss, W. (2000) The current state of two-dimensional electrophoresis with immobilized pH gradients. *Electrophoresis* **21,** 1037–1053.
41. Valdes, I., Pitarch, A., Gil, C., Bermudez, A., Llorente, M., Nombela, C., and Mendez, E. (2000) Novel procedure for the identification of proteins by mass fingerprinting combining two-dimensional electrophoresis with fluorescent SYPRO red staining. *J. Mass Spectrom.* **35,** 672–682.
42. Shevchenko, A., Wilm, M., Vorm, O., and Mann, M. (1996) Mass spectrometric sequencing of proteins silver-stained polyacrylamide gels. *Anal. Chem.* **68,** 850–858.
43. Okamura, H., Sigal, C. T., Alland, L., and Resh, M. D. (1995) Rapid high-resolution western blotting. *Methods Enzymol.* **254,** 535–550.
44. Pitarch, A., Sánchez, M., Nombela, C., and Gil, C. (2003). Analysis of the *Candida albicans* proteome. II. Protein information technology on the Net (update 2002). *J. Chromatogr. B Analyt. Technol. Biomed. Life Sci.* **787,** 129–148.
45. d'Enfert, C., Goyard, S., Rodriguez-Arnaveilhe, S., Frangeul, L., Jones, L., Tekaia, F., Bader, O., Albrecht, A., Castillo, L., Dominguez, A., Ernst, J. F., Fradin, C., Gaillardin, C., Garcia-Sanchez, S., de Groot, P., Hube, B., Klis, F. M., Krishnamurthy, S., Kunze, D., Lopez,

M. C., Mavor, A., Martin, N., Moszer, I., Onesime, D., Perez, M. J., Sentandreu, R., Valentin, E., and Brown, A. J. (2005) CandidaDB: a genome database for *Candida albicans* pathogenomics. *Nucleic Acids Res.* **33,** D353–D357.

46. Westermeier, R. and Naven, T. (eds.) (2002) *Proteomics in Practice: A Laboratory Manual of Proteome Analysis.* Wiley-VCH, Weinheim, Germany.
47. Choe, L. H. and Lee, K. H. (2000) A comparison of three commercially available isoelectric focusing units for proteome analysis: the multiphor, the IPGphor and the protean IEF cell. *Electrophoresis* **21,** 993–1000.
48. Palagi, P. M., Hernandez, P., Walther, D., and Appel, R. D. (2006) Proteome informatics I: Bioinformatics tools for processing experimental data. *Proteomics* **6,** 5435–5444.
49. Hochstrasser, D. F., Harrington, M. G., Hochstrasser, A. C., Miller, M. J., and Merril, C. R. (1988) Methods for increasing the resolution of two-dimensional protein electrophoresis. *Anal. Biochem.* **173,** 424–435.
50. Necas,O. (1971) Cell wall synthesis in yeast protoplasts. *Bacteriol. Rev.* **35,** 149–170.
51. Pitarch, A., Sanchez, M., Nombela, C., and Gil, C. (2002) Sequential fractionation and two-dimensional gel analysis unravels the complexity of the dimorphic fungus *Candida albicans* cell wall proteome. *Mol. Cell Proteomics* **1,** 967–982.
52. Heukeshoven, J. and Dernick, R. (1985) Simplified method for silver staining of proteins in polyacrylamide and the mechanism of silver staining. *Electrophoresis* **6,** 103–112.
53. Zolg, J. W. and Langen, H. (2004) How industry is approaching the search for new diagnostic markers and biomarkers. *Mol. Cell Proteomics* **3,** 345–354.
54. Pitarch, A., Jiménez, A., Nombela, C., and Gil, C. (2008) Serological proteome analysis to identify systemic candidiasis patients in the intensive care unit: Analytical, diagnostic and prognostic validation of anti-*Candida* enolase antibodies on quantitative clinical platforms. *Proteomics Clin. Appl.* **2**, 596–618.
55. Pitarch, A., Nombela, C., and Gil. (2008) The *Candida* immunome as a mine for clinical biomarker development for invasive candidiasis: From biomarker discovery to assay validation. In *Pathogenic Fungi: Insights in Molecular Biology.* (Eds., San-Blas, G. and Calderone, R.). Caister Academic Press, Norfolk, UK, pp. 103–142.

INDEX

A

D

G

H

Q

R

S

T

Printed in the United States of America